Symbiotic Nitrogen Fixation Technology

BOOKS IN SOILS AND THE ENVIRONMENT

Additional Volumes in Preparation

Symbiotic Nitrogen Fixation Technology

edited by

Gerald H. Elkan
Department of Microbiology
North Carolina State University
Raleigh, North Carolina

MARCEL DEKKER, INC. **New York and Basel**

Library of Congress Cataloging in Publication Data

Symbiotic nitrogen fixation technology.

(Books in soils and the environment)
Includes index.
1. Nitrogen--Fixation. 2. Rhizobium. I. Elkan, Gerald H. II. Series.
QR89.7.S94 1987 589.9'504133 87-8987
ISBN 0-8247-7751-4

MARCEL DEKKER, INC.
270 Madison Avenue, New York, New York 10016

Current printing (last digit):
10 9 8 7 6 5 4 3 2 1

PRINTED IN THE UNITED STATES OF AMERICA

Preface

In the 100 years since Frank described one symbiotic nitrogen fixation system, the legume-*Rhizobium* interaction, interest in this field has grown rapidly. The types of studies have been cyclical in nature, involving a cross section of disciplines. As a spin-off from the explosives industry after World War II, the availability of cheap nitrogenous fertilizers caused much of the nitrogen fixation research to become more theoretical in the developed world. The advent of the energy crisis in 1973 reawakened interest in the agricultural potential of nitrogen fixation. Thus, today the research has cross-disciplinary aspects ranging from molecular to agronomic studies.

The need for a source book on nitrogen fixation technology became evident to me when I was asked as sole resource person to design a relevant research program in this field in eastern Venezuela. As a microbiologist, in my usual environment, I normally depend on colleagues for research help involving other disciplines. Luckily, I carried my copy of *A Manual for the Practical Study of Root Nodule Bacteria* by J. M. Vincent, an elegantly useful source book, now out of print. The experience awakened in me the need for a single source/ reference book summarizing the major techniques used in the study of symbiotic biological nitrogen fixation. Although most of the research in this field involves the legume-*Rhizobium* symbiosis, we have included chapters covering other major symbioses as well. In fact, the chapter on associative nitrogen fixation systems does not involve a true symbiotic relationship but is included because of increasing interest in this system. Limiting the book to symbiotic nitrogen fixation technology was necessary for practical considerations, although many of the techniques discussed are applicable to free-living procaryotic nitrogen fixers as well.

Over 50 years ago, E. B. Fred wrote that the phenomenon of symbiotic nitrogen fixation had aroused the curiosity of bacteriologists, chemists, botanists, agriculturalists, and agronomists. Today, biological nitrogen fixation involves even more cross-disciplinary research. The purpose of this volume, then, is to serve as a comprehensive reference for symbiotic nitrogen fixation technology for investigators in those diverse scientific disciplines whose curiosity in the phenomenon has been aroused.

Gerald H. Elkan

Contents

Contributors

Frederick A. Bliss Professor of Horticulture, Department of Horticulture, University of Wisconsin–Madison, Madison, Wisconsin

B. Ben Bohlool Director, NifTAL Project, Department of Agronomy and Soil Science, University of Hawaii, Paia, Hawaii

Richard A. Date Principal Research Scientist, Division of Tropical Crops and Pastures, Commonwealth Scientific and Industrial Research Organisation, St. Lucia, Queensland, Australia

Allan R. J. Eaglesham Associate Plant Physiologist, Boyce Thompson Institute for Plant Research, Ithaca, New York

*Suzanne M. Ferrenbach** Research Assistant, Department of Biology, The University of Mississippi, University, Mississippi

Dennis D. Focht Professor of Soil and Environmental Sciences, Department of Soil and Environmental Sciences, University of California–Riverside, Riverside, California

Alan H. Gibson Senior Principal Research Scientist, Division of Plant Industry, Commonwealth Scientific and Industrial Research Organisation, Canberra, A.C.T., Australia

*Present affiliation: Senior Research Assistant, Division of Infectious Diseases, Department of Medicine, University of Tennessee Health Science Center, Memphis, Tennessee

Richard A. Glenister Research Specialist, Boyce Thompson Institute for Plant Research, Ithaca, New York

*Jake Halliday** NifTAL Project, Department of Agronomy and Soil Science, University of Hawaii, Paia, Hawaii

David Gareth Jones Professor of Plant Pathology and Soil Microbiology, Department of Agricultural Botany, University College of Wales, Aberystwyth, Dyfed, Wales

Harold H. Keyser Soil Microbiologist, Nitrogen Fixation and Soybean Genetics Laboratory, Agricultural Research Service, U.S. Department of Agriculture, Beltsville, Maryland

B. Dov Kishinevsky Head, Department of Legume Inoculation, The Volcani Center, Bet Dagan, Israel

L. David Kuykendall Microbiologist, Nitrogen Fixation and Soybean Genetics Laboratory, Agricultural Research Service, U.S. Department of Agriculture, Beltsville, Maryland

Thomas A. LaRue Biochemist, Boyce Thompson Institute for Plant Research, Ithaca, New York

Joe Eugene Lepo Assistant Professor of Microbiology, Department of Biology, The University of Mississippi, University, Mississippi

Thomas A. Lumpkin Assistant Professor/Agronomist, Department of Agronomy and Soils, Washington State University, Pullman, Washington

Mark Poth Microbiologist, Pacific Southwest Forest and Range Experiment Station, Forest Service, U.S. Department of Agriculture, Riverside, California

Juan Carlos Rosas† Research Associate, Department of Horticulture, University of Wisconsin—Madison, Madison, Wisconsin

R. Stewart Smith Applied Development Manager, Research and Development, The Nitragin Co., Inc., Milwaukee, Wisconsin

Mark D. Stowers‡ Senior Project Leader, Microbiology Division, NPI, Salt Lake City, Utah

---

Present affiliations:
*Manager, Industrial and International Business Development, Biological and Chemical Sciences, Batelle Columbus Laboratories, Columbus, Ohio
†Associate Professor, Department of Agronomy, Escuela Agricola Panamericana, Tegucigalpa, Honduras
‡Project Manager, Bio-Products Division, Eastman Kodak Company, Rochester, New York

*Robert G. Upchurch** Senior Molecular Biologist, Crop Science Laboratory, Allied-Signal Corporation, Syracuse, New York

Arthur G. Wollum II Professor of Soil Microbiology, Department of Soil Science, North Carolina State University, Raleigh, North Carolina

Johnny C. Wynne Professor of Crop Science, Department of Crop Science, North Carolina State University, Raleigh, North Carolina

David A. Zuberer Associate Professor of Soil Microbiology, Department of Soil and Crop Sciences, Texas Agricultural Experiment Station, Texas A&M University, College Station, Texas

---

**Present affiliation*: Assistant Professor, Department of Plant Pathology, North Carolina State University, Raleigh, North Carolina

Symbiotic Nitrogen Fixation Technology

1

Collection, Isolation, Cultivation, and Maintenance of Rhizobia

RICHARD A. DATE *Commonwealth Scientific and Industrial Research Organisation, St. Lucia, Queensland, Australia*

JAKE HALLIDAY* *NifTAL Project, University of Hawaii, Paia, Hawaii*

**Present affiliation*: Batelle Columbus Laboratories, Columbus, Ohio

I. INTRODUCTION

When a legume is introduced into a new environment, as part of a development project or as part of a plant introduction and species evaluation program, it may fail to encounter effective native strains of *Rhizobium* or to nodulate effectively with strains used in locally prepared inoculants. These problems may be avoided if strains of rhizobia that are genetically and geographically affiliated with the host plant are made available in the inoculant. There are two principal sources of new strains of rhizobia: (a) isolates made from nodules collected in regions from which the legume originated, and (b) existing collections of *Rhizobium*.

Soil microbiologists working in this area recognize two broad classes of bacteria forming nitrogen-fixing nodules on the roots of legumes: fast- and slow-growing types. The slow-growing group has been recommended for generic status (Jordan, 1982, 1984). Thus *Rhizobium* is used for all those isolates recognized as fast-growing and *Bradyrhizobium* for those of the slow-growing group. This reclassification is not universally accepted (Hernandez and Focht, 1984); however, the methodology for collection, isolation, cultivation, and maintenance is the same. For this reason *Rhizobium* will be used in this chapter to refer to both *Rhizobium* and *Bradyrhizobium*.

For those institutions involved in active plant introduction and species evaluation work, experience suggests that plant collectors should obtain samples of nodules from all legumes for which seed is collected, frequently, however, combined collection is not possible because collecting expeditions are made at plant maturity to facilitate seed collection, a time at which the legume frequently has a reduced complement of nodules. It is therefore imperative that plant collection site information be sufficiently well documented to permit the exact location to be revisited in a more suitable season for nodule sampling.

Some of the practical problems involved and guidelines for collecting nodules, preserving them during a collecting trip, and isolating and preserving the rhizobia obtained from the nodules are outlined in this chapter as a guide to the soil microbiologist or agronomist (plant introductionist) undertaking the isolation of rhizobia for the first time. The methodologies described define principles involved and can be altered to suit individual requirements and facilities.

II. COLLECTION

A. Presampling Considerations

Collecting nodules for new legumes is expensive and may involve overseas travel and the cooperation of a number of researchers. It is important when contemplating an expedition to remember that many collections have been assembled already. It is worthwhile studying the content of existing collections. Most collections have catalogs that are available on request, but some idea of the content may be obtained from the World Catalogue of Rhizobium Collections (McGowan and Skerman, 1982b) and the World Directory of Collections of Cultures of Microorganisms (McGowan and Skerman, 1982a). If potentially useful strains are not available, it is important to determine where to collect. If the legume has been collected previously, nodule collection should concentrate on the same area or at least in areas of similar soils and climate. The records of plant explorers who have visited the region previously are a valuable source of background information and can be used to identify local areas for collection of nodules. If no previous plant germplasm collections have been made, nodule collectors should refer to the various lists of plant flora for likely distribution of species. In these cases the collector is obliged to review available literature on soils, vegetation, and climate to localize the areas for collection, since edaphic factors have been shown to be closely associated with finding suitably adapted and effective strains of *Rhizobium* (Date et al., 1979).

A provisional itinerary should be drawn up well ahead of time and passed on to local contacts, who may be able to make arrangements to the proposed route and provide a means of transport.

B. Selection and Sampling

It is important to avoid collection from roadsides and other disturbed sites, since soil conditions, plant performance, and nodulation may not be typical of the locality. Collectors should seek out legumes that appear successful in native pastures and/or stable natural ecosystems.

It is seldom necessary to excavate whole plants to retrieve nodules. Careful removal of soil around the root crown to locate young adventitious roots, followed by removal of the roots with a knife, often yields best results. If this method proves unsuccessful, and it depends very much on the species involved, deeper excavation of the root system may be necessary. Use a digging implement to remove a sod approximately 25 cm in diameter containing roots to a depth of 10 to 15 cm. Exposure of the root system is greatly facilitated by immersion of the sod in water and allowing the soil to fall away. Alternatively, and especially for samples from species such as *Leucaena*, *Macroptilium*, and *Neonotonia* in which nodules become detached readily from their fine tertiary roots, the soil and roots can be washed in a sieve (1- to 2-mm openings) either by "dunking" or with running water to remove the soil. Exposed nodules can then be collected with forceps. Even without water it should be possible to examine the root system for nodules by sequentially fragmenting the soil by hand. Use a sharp knife or fine scissors to sever the root about 0.5 cm on either side of the site of nodule attachment.

The location of nodules on the root system is very species dependent. For tap-rooting plants such as *Trifolium subterraneum*, nodules can be found within 2 to 4 cm of the crown on tap and lateral roots, whereas nodules of stoloniferous legumes are usually concentrated superficially within 1 to 2 cm of the surface, attached to adventitious roots. In most *Stylosanthes* species nodules are more or less equally distributed along the length of the primary and lateral roots in the top 10 to 15 cm. In *Leucaena*, nodules seldom can be found under mature stands, and the best hope of success is with isolated seedlings and young plants growing in full sunlight.

Do not expect to locate nodules on the main root stock of perennial species. Nodules of most legumes have a finite life span, and a maturing root that may have borne nodules when younger no longer has the anatomy that would permit reinfection and nodulation. Similarly, do not expect to find nodules on a plant that has been pulled out of the ground by force. In most species the connection between the nodule and root is extremely tenuous.

Sample only fresh, firm nodules and avoid taking samples of those that are damaged or decaying. A sample of at least 20 to 25 nodules per plant is recommended. The reason for the large sample is that not all strains of rhizobia forming nodules on the host are effective in nitrogen fixation; thus it may be necessary to make isolations from a number of nodules to ensure success in obtaining effective isolates. In one recorded instance with *Stylosanthes* only 1 in 20 of the nodules sampled yielded an effective strain of *Rhizobium* (RAD, unpublished data). However, this was an extreme case in our experience and we would recommend that 5 to 10 nodules from each sample be processed, leaving the remainder as backup for further attempts if the initial isolations are unsuccessful (see Section III.C). This is especially important where nodules have been obtained during long-term trips and revisiting the site for further collection is not possible. All nodules from a single plant represent one unit of collected material and may be stored in the same container. Nodules from different plants should not be combined because they may represent different soil environments even if only several meters apart and in extreme cases (e.g., misidentification) could lead to a mixture of nodules from different species. Label the sample with a specimen number, and where appropriate the accession number of the plant colleciton, which relates it to the documentation of its origin.

C. Postsampling Program

Having lcoated a plant from which to collect nodules, subsequent procedures depend very much on the duration of the collecting trip.

Short-Term Trips

When it is possible to lodge the sample with the isolation laboratory the same or the next day, it is simplest to excavate the whole plant together with its nodulated root system in soil and pack it firmly into a polythene bag for transport to the laboratory.

Long-Term Trips

For trips of more than 2 days, the nodules collected must be prevented from decomposing and protected from invasion by soil microorganisms which interfere with subsequent isolation procedures. This is usually achieved by preserving the nodules in a vial containing a desiccant. Collection vessels (Figure 1) must have caps that provide an airtight seal. Desiccant (anhydrous calcium chloride or silica gel) occupies one-fourth of the volume of the container and is held in place by a cotton wool plug. The volume of nodules and/or roots should not exceed the volume of desiccant.

When collecting trips are longer than 2 to 3 weeks in duration it is advisable to dispatch samples to a laboratory for isolation of the rhizobia or to make isolations en route. The advantages of isolating en route are:

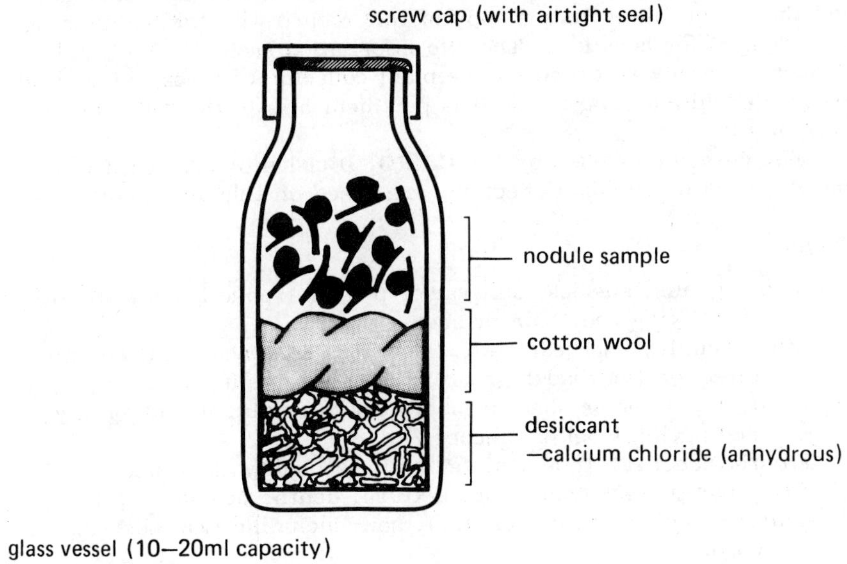

Figure 1 Airtight container for preserving, storing, and transporting nodule samples.

1. It eliminates the risk of loss in the mail and avoids the complications of quarantine.
2. Isolation from some nodule types (especially the very small nodules of *Stylosanthes* species) may be more successful from fresh than from desiccated nodules.

A plant collecting expedition with a narrow and clearly defined objective, such as a systematic collection of lines of *Stylosanthes capitata* from a remote region of Brasil, represents considerable investment, not only in direct costs for the participants but also of time in the planning and arranging, which may involve personnel from several collaborating organizations. In such a case the extra effort and sacrifice of the time required to ensure successful isolation of strains of *Rhizobium* en route becomes well justified. Ideally, such an expedition would have an experienced soil microbiologist as a member of the collecting team. He or she would make primary isolations en route, using a field-adapted procedure such as those outlined in Section III.B.

It is essential to provide adequate documentation of the origin of the nodules, of the host, and of the collection site. A convenient form for recording data pertaining to the strain, its origin, cultural characteristics, and effectiveness is that used by the World Data Centre on Microorganisms [WFCC Form SCC-4IR)]. A partially completed copy of such a form is provided in Figure 2. It is recommended that as many as possible of these descriptors be completed by the collector and that he or she include a copy of this record with the nodule sample when lodged for isolation. The site information (soil, climate, and species) normally recorded by the plant collector with respect to seed samples (Reid and Lazier, 1979) is pertinent also to the collection of *Rhizobium*.

The descriptors used by the CSIRO, Division of Tropical Crops and Pastures Rhizobium Collection, are based on this and include:

Origin

Host: genus, species, cultivar or plant introduction, or any other identifying accession number
Site: country, nearest town, kilometers from nearest town and direction (or road/highway), latitude, longitude
Collector: name or name of laboratory of origin, including any identifying label or number
Environment: altitude, rainfall and seasonality, land use
Soil: parent rock, soil type, texture, depth, drainage, pH
Source: nodules (fresh or dry), soil (including trap host), culture

Cultural Characteristics

Growth rate: fast, intermediate, slow, or very slow

Growth reaction: acid or alkali producing on yeast mannitol medium, special medium requirements
Colony characteristics: shape, color, texture, and size

Special Features and Usage

Symbiotic capacity: nodulation, including host used for authentication test; effectiveness, including host range and category (highly effective, effective, partially effective, ineffective)
Tolerances and specific requirements: includes, for example, requirements for low pH, tolerance of aluminum and manganese, high soil temperature or high levels of soil nitrate, survival on seed and in peat
References: significant publications in which information about the strain is reported

III. ISOLATION

Successful isolation depends very much on the quality of the nodule sample. Damaged nodules are not satisfactory for direct isolation and should be used (after mild surface sterilization) to inoculate a seedling host to obtain fresh nodules from which to isolate. Dried nodule samples should be allowed to fully reimbibe in water before surface sterilization, although in some instances we have successfully surface sterilized the dry nodules and then allowed them to imbibe in sterile water for 30 to 60 minutes before making the isolation.

Although the routine method of isolation is usually satisfactory, it is a wise precaution to begin with half the nodule sample. If isolation is unsuccessful, appropriate amendments can be made to the method (see Section III.C).

A. Laboratory Methods

Indirect via Single-Colony Growth

The general procedure is described by Vincent (1970). Each laboratory will use modifications to suit its own purpose and available materials. The general procedure (see the top half of Figure 3) is as follows:

1. Wash the nodules in tap water to remove gross soil contamination and drain or transfer to tissue paper to dry.
2. Immerse the nodules briefly (5 to 10 sec) in 95% ethanol, and then in sterilant and leave for 3 to 4 min. Sterilant can be sodium or calcium hypochlorite (7 ot 8%) solution; 3% hydrogen peroxide (H_2O_2); or 0.1% acidified mercuric chloride solu-

NOMENCLATURE DATA

Genus	Species, subspecies, infrasubspecies	Type
RHIZOBIUM	trifolii	B

=100 Author(s) of the name (*with date(s)*) (*e.g. Rhizobium phaseoli* Dangeard 1926)

=110 World Directory Collection Number	Collection Acronym	Collection Accession Number
57	CB	1990

HISTORY

=120 Received as (Genus, species, subspecies, infrasubspecies)	=125 Date received (Day / Month / Year)
Rhizobium trifolii	/ 8 / 55
=130 Received from (*Name and address of person or collection*) As	=140 Strain designation
J. Brockwell, Div. Plant Industry, CSIRO, Canberra	CC 248
=150 Who received it from (*Name and address of person or collection*) As	=160 Strain designation
H. V. Rees, Tasmanian Dept Agriculture, Launceston, Tas.	TA1
=170 Who received it from (*Name and address of person or collection*) As	=180 Strain designation

=185 Any previous history (include accession number)

=190 Also held in following permanent collections (*Include acronym and accession number, e.g.* ATCC 12345)
SU 659 (Uni of Sydney), NA14-1 (NSW Dept Agric)
ACR 221 (Rothamsted Collection of Rhizobium)

ORIGIN

=200 Isolated from

=201 Genus Trifolium =202 Species subterraneum =203 Cultivar

=210 Part of plant (*Tick applicable block*)

☑ Root nodule ☐ Other (specify) ....

=220 Original collection site =221 Australia (Country) 975 mm (May–Sept. rainfall)

=222 Tasmania (State/Province) =223 Bridport (Nearest town) =224 ____ (Distance and direction from town)

=225 20 m (Altitude) =226 41° 00′ ″ S (Latitude (indicate North or South)) =227 147° 24′ ″ E (Longitude (indicate East or West))

Collection site soil properties

=350 Soil family Sandy loam
=352 pH (water)
=354 pH (N KCl)
=356 pH (.01 M $CaCl_2$)
=358 Phosphorous (ppm)
=360 P-extraction solution used
=362 Salinity (m moles/cm)
=364 Base saturation (%)

=230 Isolated by (*Name and address*) As	= 235 Strain designation	=240 Date (Day / Month / Year)
H. V Rees, Tasmanian Dept. Agric.	TA1	/ / ?
=250 Identified by		=260 Date (Day / Month / Year)
		/ /

Figure 2 Example of a partially completed WFCC Form SCC-4(R) for recording data relevant to the origin and main characteristics of a strain of *Rhizobium*.

MAINTENANCE WFCC Form SCC-4(R)

=270 Method of preservation *(Please tick whichever is applicable)* and =280 Temperature of storage

Method	Tick	°C	Method	Tick	°C
Original source material		°C	Agar culture	✓	5 °C
Lyophilised culture	✓	20-25 °C	Liquid nitrogen		°C
In soil		°C	Under oil		°C
Desiccated on beads		°C	Other *(Specify)* ______		°C

=290 Preservation suspending medium
10% sucrose + 5% peptone

=300 Growth medium *(Name and reference, or give details on separate page)*
Yeast mannitol (Vincent, J.M. IBP Manual No 15 (1970))

=310 Temperature
26 °C

=320 Growth conditions *(e.g. aerobic, anaerobic, special gas mixture, light)*

=322 Incubation time *(Tick applicable block)*
Slow (> 5 days) ☐ Fast (< 5 days) ☑

=327 Period between subcultures

SPECIAL FEATURES AND USAGE *(e.g. for citric acid production, Type strain etc.)*

=370 Other legumes nodulated	=373 Effectiveness *(E, PE, I)*	=376 Suitability as inoculant *(Excellent, good, fair, poor)*
T. repens	E	Excellent (T. repens – T. pratense)
T. incarnatum	E	
T. fragiferum	E	
T. pratense	E	
T. ambiguum	I	
T. semipilosum	I	

Tolerance to stress: Indicate either (a) *range* or (b) *high, medium, low* tolerance

=380 As free-living organisms:
pH 4·8 – 7·5, Al^{+++} low (< 5 μM monomeric), Mn^{++} ______, NaCl ______, Temp. ______ °C

=390 As symbionts in host nodules:
pH 4·8 – 5·0 upper limit not known., Al^{+++} low, Mn^{++} ______, NaCl ______, Temp. ______ °C

=330 Other special features *(CHO metabolism, etc.)* Effective on wide range clovers. Nodulates T. subterranean (cv. Woogenellup) sparsely in plant-tubes.

=335 Reaction in standard YMA ☑ Acidic ☐ Basic

REFERENCES *(Journal, volume, pages, year)*

=340 Gibson, A.H. Aust. J. Biol. Sci. 19:499 (1966); Aust J. Agric. Res. 19:907 (1968); Aust. J. Agric Res. 19:891 (1968); Brockwell, J. J. Aust. Inst. Agric. Sci 34: 224 (1968); Dudman, W.F. Aust J. Agric. Res. 19:739 (1968); Chatel, D.L. Soil Biol Biochem 5:443 (1973).

If space is inadequate to reply to any question, please submit additional information on separate pages identified by appropriate reference number *(e.g. =310)*.

E = Effective PE = Partially effective I = Ineffective

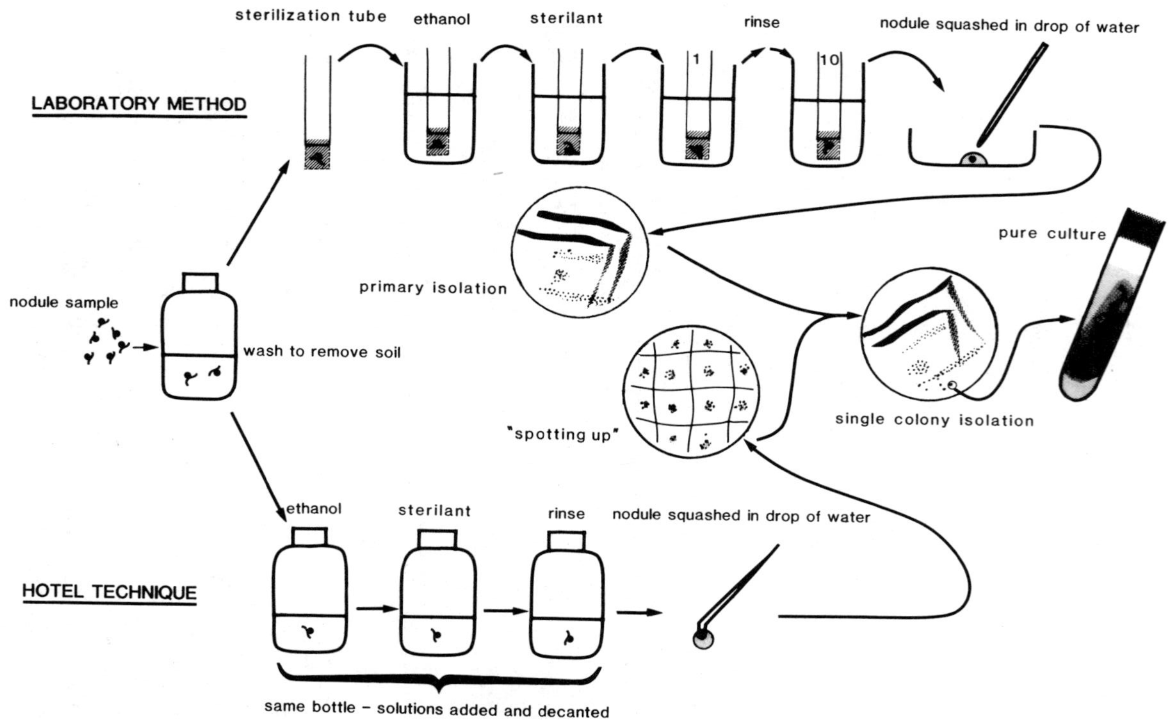

Figure 3 Sequence of steps for surface sterilization of nodules and isolation of rhizobia. The upper section describes a procedure for use in laboratories. Nodules are placed in a small section of glass tubing, one end of which is tied off with muslin. This unit is then immersed in the various solutions and the rhizobia isolated as described in the text. The lower section depicts the routine used by collectors using the "hotel" technique. The nodule washing and surface sterilization steps of this technique may be replaced by the *Rhizobium* collector's isolation kit.

tion, made up as 1 g of $HgCl_2$, 5 ml of concentrated HCl in 1 liter of water. Because mercury compounds are highly toxic and highly residual in the environment, they should be used only when other sterilants are unavailable or do not work. Correct disposal of washings and mercury is essential.

3. Remove sterilant and wash nodules at least five times in sterile water.
4. Crush each nodule to give a turbid suspension and transfer (with a flamed loop or glass rod) a drop to an agar plate. Flame the loop or glass rod after each nodule.
5. Streak the drop of suspension onto the agar surface in a way that progressively dilutes the suspension.
6. Incubate the plates in an inverted position at 26°C and check for growth typical of *Rhizobium* along the streak lines. Well-isolated single colonies should be picked off and restreaked into fresh plates to obtain pure cultures of the presumptive strain of *Rhizobium*. It is possible that more than one "typical" colony type may appear on a plate streaked from a single nodule, and each of these types should be taken to pure culture and held for characterization.

Basically, this is the procedure described by Fred et al. (1932). Some of the variations currently in use are described by Vincent (1970), Brockwell (1980), Somasegaran et al. (1982), and Broughton et al. (1975).

Where it is necessary to maintain the identity of a large number of individual nodules, surface sterilization can be achieved with apparatus similar to that described by Gault et al. (1973). The apparatus consists of wells cut in a Perspex tray to which an applicator is fitted. The combination permits surface sterilizing and washing solutions to flow over nodules in the wells. After surface sterilization nodules are crushed in the wells and streaked in the usual way.

Direct via Micromanipulation

Simple, inexpensive micromanipulator equipment (Skerman, 1968) has been used for isolation and identification of a number of microorganisms (e.g., Franzmann and Skerman, 1981). Single-colony and effectiveness variants of several strains of *Rhizobium* (H. V. A. Bushby, personal communication) as well as isolates from squashed nodule material have been obtained using this technique. Nodules were surface sterilized and spotted-up on a suitable medium in the usual way. Single cells were then isolated by micromanipulation from the suspension, but it is better to incubate overnight and isolate from a microcolony because of the difficulty of distinguishing between viable cells and nonviable bacteroids. See Section IV for a discussion on the viability of bacteroids.

B. Field Techniques

The "Hotel" Technique

The need for a simplified and portable system of isolating from nodules collected during long-term field collecting trips has been brought about by frequent failures to obtain viable isolates of *Rhizobium* from dried and reimbibed nodules. This is particularly important when attempting to isolate from nodules of plants such as *Arachis* and *Stylosanthes*. The following modifications of the standard isolation procedure and equipment have been used successfully on several occasions in recent years. The essential features are miniaturization, simplification, and improvization for east of portability.

The modifications provide for mass culture from individual nodules and for processing 8 to 20 nodules per sample collected.

Equipment

1. Small (3 to 5 cm in diameter) plastic petri dishes prepoured with medium [yeast extract-mannitol agar (YMA) or other as required] containing 0.002% cycloheximide (Actidione; Upjohn Co.).
2. Reserve of sterile medium in small lots (20 to 30 ml serves for 10 to 15 small plastic petri dishes) and supply of sterile petri dishes and disposable plastic pipettes. Additional plates can be prepared simply by melting and pouring the agar. Facilities may vary from pressure cookers to a used fruit can on an open fire at a campsite.
3. Fine-tipped (bent) forceps (two pairs), scalpels, and scissors.
4. Small (50-ml) beakers (three or four). Drinking glasses and glass ashtrays in hotel rooms make useful containers.
5. Alcohol burner, 100- to 200-ml bottle of alcohol (vodka, tequila, whisky, etc., are useful substitutes) for surface sterilization of nodules and flaming forceps; matches or lighter.
6. Screw-capped containers with preweighed calcium hypochlorite to give a 7.5% solution when water is added.
7. Screw-capped containers (e.g., 5-ml-capacity McCartney bottle or Vacutainer tubes 100 mm × 16 mm) for washing and surface sterilizing nodules.
8. Several bottles (100 ml) of sterile water. Bottled mineral water is a satisfactory substitute.
9. A portable electric (or battery) lamp is very useful, as hotel lights are frequently less than satisfactory for fine work. Isolation may necessarily be by day on camping expeditions.
10. Supply of plastic bags and petri dishes.
11. Adhesive tape for resealing bags and petri dishes.

Experience indicates that this equipment will occupy about 25% of the traveler's "suitcase space."

Nodule samples collected during the day should be kept moist (wet tissue paper, not free moisture) in plastic bags and in the coolest place possible while traveling (e.g., in the shade of a vehicle or tree when stopped—but not to be forgotten).

Procedure. See Figure 2.

1. Add water to the tube or screw-capped bottle containing the calcium hyperchlorite, shake, and allow to stand for 5 to 10 min. Decant supernatant to a clean screw-capped tube (or bottle), shake to wet the inside of the tube, and allow 5 min to sterilize the interior of the tube.
2. While the hypochlorite is standing, nodules are excised from the roots and washed by shaking vigorously in a small bottle. Transfer the nodules and wash water to a beaker (or ashtray), rinse the bottle clean, and return nodules individually, with forceps, free of soil.
3. Rinse the nodules in alcohol, decant, and add hypochlorite solution; agitate and decant after 4 or 5 min. Rinse once in sterile water or bottled mineral water. If neither is available, invert the bottle, leaving nodules adhering to vessel wall to allow liquid to drain and the chlorine to evaporate.
4. Transfer nodules to a sterile petri dish and squash one by one with the forceps, transferring a small drop of juice (nodule extract) to the agar by lightly touching the agar surface in the center of the area premarked on the petri dish. In this way 8 to 20 nodules from the one sample can be spotted-up on one petri dish. If desired, there is sufficient juice to spot-up the same nodule on more than one isolating medium. Forceps are flammed in alcohol between nodules.
5. The plates then "incubate" while traveling. It is convenient to retain the plastic bag in which the petri dishes were purchased for repacking them for traveling. We also recommend that each petri dish be "sealed" or "held together" with a strip of one of various available sealing films. A few (three to six) small holes should be made around the circumference with the points of flamed forceps to allow gas transfer. Inoculated plates can be held in this way for 15 to 25 days or longer, depending on growth rate of isolates and provided that they remain relatively free from contaminating fungi. On reaching the base (or home) laboratory, further processing of samples is made by streaking the growth from each spot onto separate petri dishes to obtain single, isolated colonies in the usual way.

Rhizobium *Collector's Isolation Kit*

The *Rhizobium* collector's isolation kit (Figure 4) is an alternative to the "hotel" technique and was designed (by J.H.) for the nonmicrobiologist undertaking simultaneous plant (seed) and nodule (*Rhizobium*) collection on extended field trips or at locations where there are no microbiological facilities. Routinely, the kit is prepared by microbiologist colleagues of the collector. In the field the collector places nodules in the sterilizing chamber, through which the various sterilizing and washing solutions are dispensed via a polypropylene syringe and appropriate positioning of the two-way valve. Isolation from the nodule is then completed by crushing nodules as described previously and as illustrated in Figure 3. With the exception of the alcohol and agar plates, the entire unit may be sterilized by autoclaving.

C. Problems

Given the variation in shape, structure, and age of nodule samples from which recovery of rhizobia is attempted, it is unlikely that resultant isolation plates will permit single colonies to be observed in every case. The problems can be due to:

1. *Contamination of the plate.* Contamination is caused particularly by fungi which may have survived in small recesses of the nodule surface or which may have been resident in the cortical tissues of the nodule and so escaped the sterilant. In such cases the problem can be reduced by increasing the sterilization time and/or using a different sterilant. Some workers have sectioned nodules, taking samples from the center for streaking.

Antifungal agents cycloheximide at 100 to 200 μg/ml (Vincent, 1977), pimaricin at 250 to 500 μg/ml (Bushby, 1983), and benomyl (benzimidazole) at 50 to 100 μg/ml (A. Diatloff, personal communication) have been used successfully to reduce fungal growth. Not all rhizobia are tolerant of pimaricin and benomyl and should be used only when their noninclusion makes isolation of rhizobia from nodules unsuccessful.

2. *Absence of growth on the plates.* It is not a common problem and indicates lack of viability of the rhizobia in the nodule or use of too high a concentration of sterilant or too much exposure to the sterilant. This is particularly true for isolations from reimbibed dry nodules. If excessive exposure to sterilant is the problem, variation of time of exposure and concentration of sterilant should avoid the problem. It is important to note that at least 20 to 25 days of incubation of the streaked plates should elapse before pronouncing no rhizobial growth. We have observed that rhizobia from nodules of *W. capitata* growing in acidic (pH 4.2 to 4.4) soils, require at least 12 days to become visible and up to 20 days for "normal" colony growth (of 1 mm)

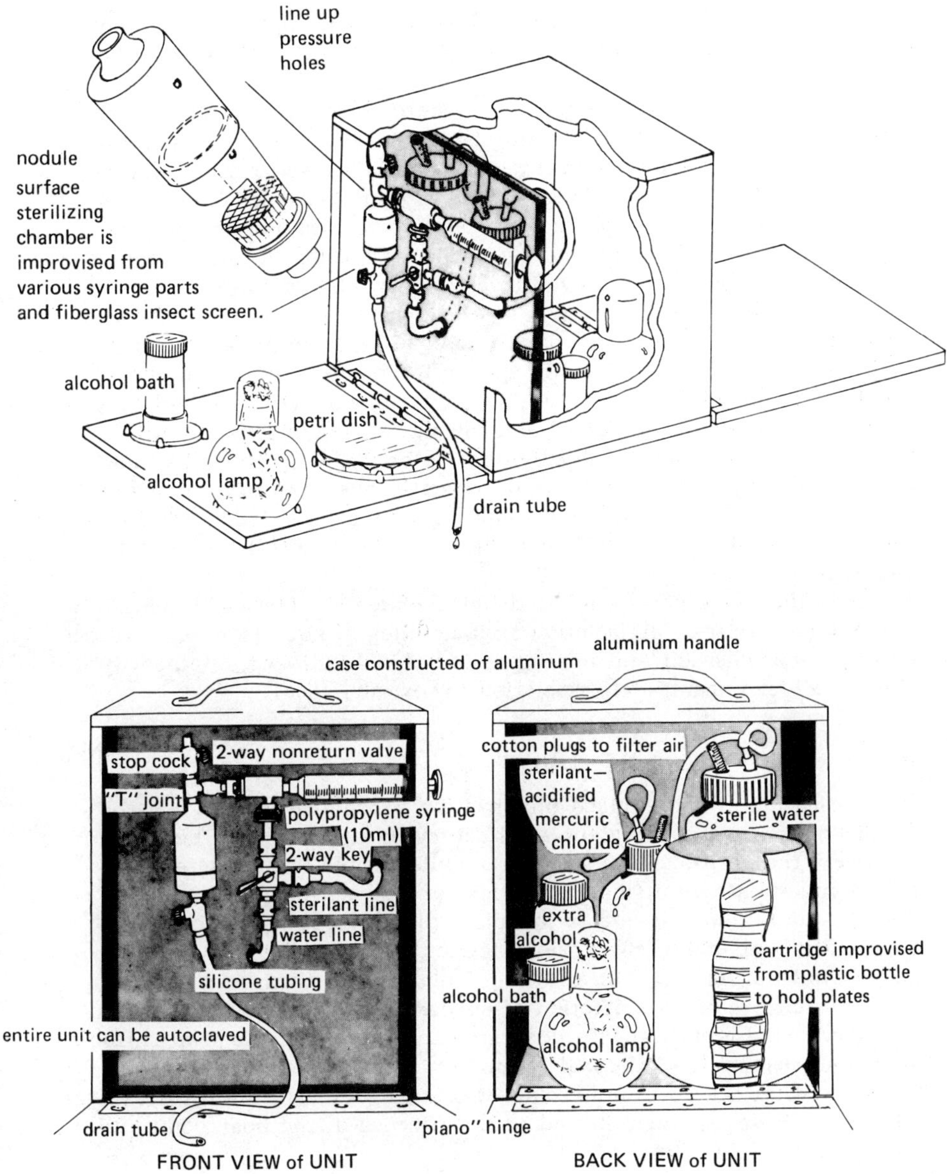

Figure 4 *Rhizobium* collector's isolation kit.

when isolations are made on acidified YM medium. Note that no visible colonies appeared even after 25 days on YMA at pH 6.8 (see also Section IV).

3. *Mixed cultures.* Very often more than one colony type is observed on the streaked plates and one or more of these are characteristically rhizobial colonies. Very often one is a fast-growing, gummy colony, whereas the other is a smaller, slower-growing type. Each appears to be of single-colony type, and both form nodules on a test plant. Some earlier claims for fast-growing rhizobia forming nodules on plants traditionally observed to have only slow-growing strains have been shown to be cultures contaminated by cells of the slow-growing type. The fast-growing rhizobia-like colonies represent a strain with cells morphologically similar to rhizobia but which are not capable of independently nodulating a host legume (Dakora and Vincent, 1984). There are, however, fast-growing strains of rhizobia which nodulate legumes traditionally thought to be nodulated only by slow-growing strains. For example, the recently observed occurrence of fast-growing strains for *Glycine soja* (Keyser et al., 1982, 1985) and earlier observations for *Cicer arietinum* (Vincent, 1977) and *Lotus pedunculatus*.

It is therefore paramount that new isolates be thoroughly checked if any doubt arises. Reisolation from nodules of such isolates is recommended or separation/and testing of individual cultures obtained by single-cell micromanipulator isolation (Skerman, 1968).

D. Recovery by Plant Infection

If modification of the surface-sterilization procedure fails to control contamination enough to allow satisfactory isolation, it may be possible to "trap" the rhizobia by using the nodule sample to inoculate the homologous host, an affiliated species, or a promiscuous test plant. Use the nodule homogenate to inoculate plants raised aseptically in tube culture or one of the various techniques described by Vincent (1970) and Norris and Date (1976). Isolation can then be attempted with a much greater expectation of success, using fresh nodules. As a last resort this indirect method can also be employed to trap rhizobia from soil samples from the collection site. It is to be borne in mind, however, that the resultant isolate will not necessarily be the strain whose effectiveness contributed to the success of the host plant and prompted its collection.

E. Note on the Viability of Bacteroids

Rhizobia occur in the nodule as both normal vegetative rod-shaped cells and in their nitrogen-fixing form as bacteroids. These may be morphologically indistinguishable from the normal vegetative cell in the nodules of such plants as *Glycine max* and *Macroptilium atropurpureum* but are distinctly misshapen in various enlarged forms in *Trifolium, Medicago,* and *Pisum* species. The bacteroid generally is thought to be nonviable (i.e., not capable of multiplication as a free cell in a growth medium), despite some views to the contrary for bacteroids from nodules of *Trifolium repens* and *T. subterraneum* (Gresshoff et al., 1977) and *G. max* (Gresshoff and Rolfe, 1978; Tsien et al., 1977). However, recent evidence (Cao et al., 1984) has demonstrated that the morphologically enlarged and misshapen bacteroids of *Astragalus sinicus*, *Medicago*, *Trifolium*, and *Vicia* are not viable, whereas in *G. max,* where vegetative cells and bacteroids show no apparent morphological differentiation, most cells are viable. Similar evidence has been obtained by Zhou et al. (1984, 1985) with *T. repens* and *G. max*.

Our own observations support the view of Zhou et al (1984) that mature morphologically differentiated bacteroids, in our case the spherical bacteroids of *Stylosanthes* nodules, are not capable of multiplication.

F. Authentication of Isolates

All isolates must be confirmed as *Rhizobium* before being included in a collection. Confirmation is dependent on demonstration of nodule-forming ability on a test host under bacteriologically controlled conditions, that is, satisfaction of Koch's postulates that the casual organism can be isolated in pure culture and that it can cause nodules to be formed on aseptically growing plants. Note that the formation of nodules is the only criterion required in this test. The effectiveness of the association is not important to the confirmation, but it is recorded qualitatively on a relative basis as no nodulation, parasitic, ineffective, partially effective, effective, or highly effective.

The test host should be that from which the original nodule was obtained. Alternatively, a "guinea pig" host can be used, where experience indicates that the species will cross-nodulate. For example, *Macroptilium atropurpureum*, *M. lathyroides*, and *Teramnus uncinatus* serve as test hosts for the true "cowpea-type" legumes, *Stylosanthes guianensis* cv. Endeavour or Cook and *S. hamata* cv. Verano for many species of *Stylosanthes* and *Trifolium burchellianum* var. *Johnstonii* for a range of east African clovers. *Glycine usseriensis* is a convenient species for strains from soybean (Brockwell et al., 1975). Isolates from *Leucaena*, *Lotononis*, *Cicer*, *Lotus*, and some *Stylosanthes* (e.g., *S. capitata*) should be tested on homologous hosts.

IV. CULTIVATION AND NOTES ON MEDIA

Most rhizobia grow on yeast extract-mannitol mineral salts media at pH 6.5 to 6.8; however, in some specialized cases an alternative carbon source is required and/or a pH environment of 4.5 to 5.0 is necessary. Colony and cell characteristics are variable and can change with change in composition of the medium and the physical growth environment. In this seciton we describe briefly colony and cell characteristics and highlight a few aspects of media composition that have led to successful isolation and growth of rhizobia that failed to grow on standard YM medium.

A. Colony Characteristics and Cell Morphology

Isolates of *Rhizobium* obtained from nodules can be described according to their growth characteristics in solid and liquid media. The size, shape, color, and texture of colonies and ability to alter medium pH are generally stable characteristics useful in defining strains or isolates. Typical information for these characteristics when grown on standard yeast-mannitol medium (see Vincent, 1970) is described below:

Shape

Usually, flat (——) to domed (⌒) and even conical (▲) on agar surface. When growing subsurface in the agar, colonies are typically lens shaped.

Color and Texture

Colonies may be white-opaque or milky- to watery-translucent. The opaque colony growth is usually firm with little gum, whereas the milky and watery colonies are often gummy and soft.

Colonies may be glistening or dull, evenly opaque or translucent, but may develop darker centers or riblike markings with age. Red and pink (e.g., from *Lotononis*) and pale yellowish to creamy colored (e.g., from some *Stylosanthes*) colonies are not unusual. If Congo red is included in the medium, colonies appear pinkish white and generally do not take up Congo red. This is not a stable characteristic and varies with the concentration of Congo red, age of culture, and whether plates have been exposed to light.

Growth Rate

Compared with other bacteria, even fast-growing rhizobia are slow growing. Mean generation times are 2 to 4 hr for fast-growing types and 6 to 8 hr for most slow-growing strains, but a wide variation in generation times exists. Generally, 2 to 4 (e.g., from most *Trifolium* and *Medicago* species), 3 to 5 (e.g., from *Leucaena*, *Psoralea*, *Sesbania*), 5 to 7 (e.g., from *Macroptilium atropurpureum*, *Desmodium*, *Galactia*), to 7 to 12 (e.g., some *Stylosanthes*, *Lupinus*) days are required to achieve maximum colony size on agar or growth in liquid medium. Actual time will vary according to temperature of incubation (optima 26 to 30°C), origin (culture or nodule), aeration (in liquid cultures), and composition of the medium.

Size

When well separated on agar plates, colony size may vary from 1 mm for many slow-growing strains (e.g., from *Stylosanthes*, *Zornia*, *Aeschynomene*, *Lupinus*) to 4 to 6 mm for faster-growing strains (e.g., from *Trifolium*, *Medicago*, *Leucaena*, *Psoralea*, *Sesbania*). In crowded plates colonies remain small and discrete but coalese to confluent growth when colonies touch.

pH Change in Meduim

Ability to produce acid or alkali end products is dependent on composition of medium and conditions of incubation. In yeast-mannitol medium with free exchange of air (gases) most intermediate- and slow-growing strains produce alkali, whereas the faster-growing strains produce acid. Qualitatively, pH change can be readily observed by inclusion of bromothymol blue in the medium. The quantitative change in pH of this medium varies from ±0.2 to 1.2 units for agar culture but may be more (±3 units) in liquid cultures.

Morphology

Cells of *Rhizobium* are gram-negative rods varying from 0.5 to 0.9 μm wide to 1.2 to 3.0 μm long. They are non sporeforming and older cells are filled with nonstraining, highly refractile granules of polymeric β-hydroxybutyrate, which gives an uneven banded appearance to the stained cytoplasmic materials.

B. Cultivation and Notes on Media

The routine medium for isolation from nodules is a yeast extract-mannitol agar (YMA). Inclusion of bromothymol blue pH indicator is recommended and offers the following benefits: the blue color resulting from the alkalizing effect of most slow-growing strains of rhizobia often appears before growth is visible on the plates; most of the common contaminants are acid producing; colonies are more easily characterized against the blue background of the indicator containing the medium; and since the standard medium has a green color (neutral), a mistake in its preparation is immediately apparent. The incorporation of Congo red, which also has fungicidal properties, in the medium has been used to distinguish between *Rhizobium* and contaminants, but interpretation requires caution since some rhizobia absorb Congo red.

Rhizobia of both the fast- and slow-growing types vary in their tolerance to pH and pH-related factors for growth and nodulation (e.g., Munns, 1978; Keyser and Munns, 1979a,b). With this in mind we have observed that when making isolations from nodules of a particular region in search of strains with tolerance to acidity, growth of the isolate was enhanced if the medium was modified to resemble more closely the soil conditions. In fact, some strains from very acid soils made little or no growth on standard medium but fared much better when the medium was acidified and maintained at that pH during growth of the culture (Date and Halliday, 1979). This has two important implications:

1. A strain of *Rhizobium* tolerant of, or adapted to, a particular soil stress should be maintained and cultured for inoculant production on a suitably modified medium on which its growth is optimal. This avoids the possibility that during conservation on nonstress media, nontolerant variants within the culture might become dominant.
2. The standard practice of searching for stress-tolerant *Rhizobium* in existing collections of strains that were originally isolated onto standard, nonstress YMA may not be the most valid or fruitful approach.

The availability and use of antifungal agents has improved the proportion of successful recoveries from reimbibed nodule material (see Section III.C), but no universally useful selective medium is available for general use with rhizobia because of the wide variation in susceptibilities and tolerance of different strains (Pattison and Skinner, 1974).

Fast-growing rhizobia utilize a wider range of carbon sources than do slow-growing strains. Both utilize glutamate as a source of nitrogen. The extent of utilization of carbon and nitrogen sources usually determines the extent of change in pH of the medium during growth. We have observed frequently that rhizobia grow on some media because they are able to change microenvironment pH in the medium or within a colony to a more favorable level. This is not as easily achieved in liquid media; thus some caution is necessary when selecting for growth at particular pH values that the medium is in fact sufficiently buffered to maintain the desired pH.

V. MAINTENANCE

Strains of rhizobia that are regularly subcultured on agar slopes are subject to genetic variation with respect to such characters as colony form, uptake of Congo red, antigenic constitution, and symbiotic capacity. It is, therefore, paramount that strains be conserved in a stable and nonmultiplying form that minimizes chances of variation, loss of viability, and contamination by other microorganisms.

Preservation has two aspects: the maintenance of an isolate for further study, and the conservation of germplasm. The method of preservation will vary according to which of these aspects is applicable. It is essential that a "mother culture" collection be maintained as an ultimate source. Individual workers or laboratories should maintain secondary collections for routine use and in doing so will protect the mother collection from indiscriminate use. A system of this type (Date, 1969; Thompson, 1980) is used in Australia by the Legume Inoculant Quality Control Laboratory for the issue of cultures to manufacturers of inoculants. Cultures supplied to manufacturers in any month or year are the same number (two or three) of transfers away from the original culture, thus minimizing variation and/or contamination, while retaining continuous evaluation of main characteristics.

A. Methods of Preservation

The methods used to preserve cultures vary from screw-cap tube or paraffin-oil-covered agar cultures, in the simplest form, to dried forms on porcelain beads, lyophilized (freeze-dried), or low-temperature storage of broth cultures suspended in glycerol at -20°C or at the liquid nitrogen temperature of -196°C. The choice of method will depend on experience, finances, and facilities available. Details of the various agar and dried culture techniques are described by Vincent (1970), Norris and Date (1976), and Dye (1980). The method for storage under liquid nitrogen (Bridges, 1966; Lapage et al., 1970) is similar to that for storage of semen.

B. Notes on Methods

For screw-capped and paraffin-oil-covered agar cultures it is critical that the screw cap be firmly secured and that the oil is at least 2 cm above the agar to prevent drying out of the agar. Rhizobia are very susceptible to drying conditions and lose viability rapidly. Cultures should be stored in a cool (15°C), dry area. If refrigerated (5°C) storage is available, this is preferable.

The viability of rhizobia under paraffin oil can be variable. Means and Erdman (1963) observed that 75% of cultures (from 462 strains) failed to grow after 7 to 8 years. However, at the other extreme we have observed that 95% (from 107 strains) have survived 20 years without loss of symbiotic capacity. Similarly, agar cultures in screw-capped test tubes remain viable for 2 to 5 years. Drying of the agar is the main problem. The proportion of fast-growing strains that survive these conditions is less than that for slow-growing strains.

If the porcelain bead technique is used, viability should be checked regularly. Our experience (RAD) has been that strains require rebeading every 3 to 4 years to ensure that viable cells can be recovered. Loss of viability was greater for fast-growing strains than for slow-growing strains.

Freeze-drying or lyophilization is the most widely used method for preserving rhizobia. The basic procedure was described by Annear (1962) and Lapage et al. (1970) and more recently fully described and discussed by Dye (1980).

There are several essential features to note for freeze-drying:

1. Cultures are first frozen and then dried from the frozen state (under vacuum), thus avoiding damage to the structural integrity of the cells and a change in their solute concentration.
2. A suspending medium that provides protection during the freezing, drying, and reconstitution must be used. A 10% sucrose + 5% peptone suspending medium is now used by most *Rhizobium* workers when freeze-drying cultures to maintain osmotic levels and provide a final water content of 1%.
3. Dried cells become sensitive to oxygen at relative humidities below 10% and should be sealed under vacuum or inert gas (Bushby, 1983). Some workers include glutamate in the suspending medium as an antioxidant.

Survival of freeze-dried cells can be as high as 80% or as little as or even less than 1% immediately after drying. Numbers of recoverable cells decrease with time, but the decline is less in the dark and at low temperatures than in the light at room temperatures (Dye, 1980). The suspending medium has the greatest influence, but it is variable with strain of *Rhizobium* (Dye, 1982).

The process known as L-drying (Annear, 1962; Lapage et al., 1970) is drying under vacuum without freezing. The technique requires that the culture be absorbed onto a cotton plug or strip of filter paper before applying the vacuum. The influences of suspending medium and susceptibility to oxygen are similar to those for freeze-drying. Because no freezing occurs, this method may give better survival and recovery of cells than for freeze-dried cells.

Most laboratories use either freeze-drying or L-drying to preserve their strains. In both cases a small quantity of cultures is absorbed onto a strip of filter paper (in the ampoule) on which the culture's identification name/number is typed (with nontoxic carbon ribbon).

VI. CULTURE LABELING, RECORDS, AND DOCUMENTATION

It is essential that all strains can be positively identified by label. This label, usually consisting of a combination of letter prefix followed by a specific number (and sometimes a subnumber), must be that of the laboratory of isolation. A culture sent to another laboratory should be labeled with its original label whether or not the sending laboratory is the laboratory of isolation. The use of this label refers specifically to publication of data related to the strain and the exchange of strains between laboratories. Acceptance of this principle ensures that a receiving laboratory does not inadvertently receive the same strain with different identifying labels. Within a collection of cultures, however, it is frequently convenient to allocate a local accession number to a culture received from elsewhere to facilitate ordered handling of records and cultures in the collection.

Where possible, distinct codes should be used. For example, isolates made by the Centro International de Agricultura Tropical (CIAT) are numbered consecutively and prefixed with the letters CIAT; those from the USAID, University of Hawaii NifTAL laboratory, with the prefix TAL; the nitrogen fixation laboratory of EMBRAPA at the Universidade Federal de Rio de Janeiro uses the prefix BR; the NSW Department of Agriculture, the prefix NA; and the Division of Tropical Crops and Pastures, CSIRO, Brisbane, the prefix CB.

Our experience suggests that consecutive numbering of accessions is less confusing than placing strains of the same host origin in groups of hundreds or thousands, and that culture collections be identified by a two-letter prefix that identifies the institution and if possible the state or country of origin. For example, NA indicates a strain isolated by the New South Wales (N) Department of Agriculture (A), CB being CSIRO (C) Brisbane (B), or RCR the Rothamsted Collection of Rhizobium. The Australian code was first prepared and used in 1956 (K. C. Marshall, personal communication).

As much information as possible should be obtained for each strain, particularly concerning its host and site (soil and climate) of origin. Each collection will have its own form of physical record of information related to a strain. The emphasis given to the various descriptors will vary according to the curator's (researcher's) interests. Those that we believe are important to record are embodied within the WFCC Form SCC-4(R) (Figure 2). Similar formats are used by the CSIRO Canberra Collection of Rhizobium (CC) (Brockwell, 1980) and the Rothamsted Collection of Rhizobium (RCR) (Dye, 1980). For large collections these data are stored and accessed via a computer data base. Some collections have a single microcomputer dedicated to input, storage, and retrieval of information, both for permanent records and for daily interrogation for information relevant to a particular strain. In this way the data base can be used as an initial step in the selection of strains with particular characteristics for evaluation in an experimental program.

VII. EVALUATION OF STRAINS

For those collectors desirous of extending their programs to the evaluation of strains, reference to Vincent (1977), Norris and Date (1976), Brockwell (1980), Brockwell et al. (1968, 1982), or Dye (1980) will provide essential details of criteria and methods for controlled environment, greenhouse, and field evaluation.

The choice of method is frequently dependent on the scale of operations. Controlled environment and greenhouse facilities where available are used to reduce the number of strains tested in the ultimate field evaluation.

The criteria used in evaluation may vary according to program refinement. The following, in order of importance by group, are the criteria that we believe should be evaluated.

1. Effectiveness in nitrogen fixation with appropriate host legumes
2. Competitive ability to form nodules on a host in the presence of other rhizobia; ability to colonize the soil and rhizosphere and to survive in the soil
3. Specialized characteristics; for example, the ability to nodulate a legume in acid soils, tolerance to seed-applied fungicide, promptness to nodulation, ability to fix nitrogen at high (>30°C) or low (<10°C) soil temperatures, and ability to form nodules at high levels of available soil nitrogen
4. Characteristics related to production of inoculants: survival on seed, in peat, or as culture; ability to grow in mass fermentation

Not all of these criteria need be applied at the same time. The experimenter should be guided by results of need-to-inoculate trials (Norris and Date, 1976; Nutman, 1976; Date, 1977) as to which of these criteria are needed in any one selection program. In fact, the experimenter must ensure that responses to inoculation can be obtained for a given legume in a particular region before collection and evaluating new strains of *Rhizobium*.

REFERENCES

Annear, D. I. (1962). *Aust. J. Exp. Biol. Med. Sci. 40*: 1.

Bridges, B. A. (1966). *Lab. Pract. 15*: 418.

Brockwell, J. (1980). In *Methods for Evaluating Biological Nitrogen Fixation*, F. J. Bergersen, ed. John Wiley & Sons, Inc., New York, p. 417.

Brockwell, J., Dudman, W. F., Gibson, A. H., Hely, F. W., and Robinson, A. C. (1968). *Transactions of the 9th International Congress on Soil Science*, Vol. 2, p. 103.

Brockwell, J., Diatloff, A., Grassia, A., and Robinson, A. C. (1975). *Soil Biol. Biochem.* 7: 305.

Brockwell, J., Diatloff, A., Roughley, R. J., and Date, R. A. (1982). In *Nitrogen Fixation in Legumes*, J. M. Vincent, ed. Academic Press, Inc., New York, p. 173.

Broughton, W. J., Chan, P. U., Padmanabhan, S., and Tan, K. P. (1975). *Malay. Agric. J.* 4: 141.

Bushby, H. V. A. (1983). In *Nitrogen Fixation*, Vol. 2, *Rhizobium*, W. J. Broughton, ed. Clarendon Press, Oxford, p. 35.

Cao, Y., Zhou, J., and Chen, H. (1984). *Sci. Sin.* 27: 593.

Dakora, F. D., and Vincent, J. M. (1984). *J. Appl. Bacteriol.* 56: 327.

Date, R. A. (1969). *J. Aust. Inst. Agric. Sci.* 35: 27.

Date, R. A. (1977). In *Exploiting the Legume-Rhizobium Symbiosis in Tropical Agriculture*, J. M. Vincent, A. S. Whitney, and J. Bose, eds. University of Hawaii, College of Tropical Agriculture Miscellaneous Bulletin 145.

Date, R. A., and Halliday, J. (1979). *Nature* 277: 62.

Date, R. A., Burt, R. L., and Williams, W. T. (1979). *Agro-Ecosystems* 5: 57.

Dye, M. (1980). In *Recent Advances in Biological Nitrogen Fixation*, N. S. Subba Rao, ed. Edward Arnold Publishers Ltd., London, p. 435.

Dye, M. (1982). *J. Appl. Bacteriol.* 52: 461

Franzmann, P. D., and Skerman, V. B. D. (1981). *Int. J. Syst. Bacteriol.* 31: 177.

Fred, E. B., Baldwin, I. R., and McCoy, E. (1932). *Root Nodule Bacteria and Leguminous Plants.* University of Wisconsin Studies 52, Science No. 5.

Gault, R. R. , Byrne, P. T., and Brockwell, J. (1973). *Lab. Prac.* 22: 292.

Gresshoff, P. M., and Rolfe, B. G. (1978). *Planta* 142: 329.

Gresshoff, P. M., Skotnicki, M. L., Eadie, J. F., and Rolfe, B. G. (1977). *Plant Sci. Lett.* 10: 299.

Jordan, D. C. (1982). *Itn. J. Syst. Bacteriol.* 32: 136.

Jordan, D. C. (1984). Family III. Rhizobiaceae Conn 1938.321. In *Bergey's Manual of Systematic Bacteriolody*, 9th ed., N. R. Krieg and J. G. Holt, eds. Williams & Wilkins, Baltimore, p. 234.

Keyser, H. H., and Munns, D. N. (1979a). *Soil Sci. Soc. Am. J.* 43: 500.

Keyser, H. H., and Munns, D. N. (1979b). *Soil Sci. Soc. Am. J.* 43: 519.

Keyser, H. H., Bohlool, B. B., Hut, T. S., and Weber, D. F. (1982). *Science* 215: 1631.

Keyser, H. H., Sadowsky, M. J., and Bohlool, B. B. (1985). Fast-growing soybean rhizobia. In *World Soybean Research Conference III, Proceedings*, R. Shibles, ed., p. 926.

Lapage, S. P., Shelton, J. E., Mitchell, T. G., and Mackenzie, A. R. (1970). In *Methods in Microbiology*, J. R. Norris and D. W. Ribbons, eds. Academic Press, Inc., New York, Vol. 3A, Chap. 2, pp. 136-228.

McGowan, V., and Skerman, V. B. D., eds. (1982a). World Directory of Collections of Cultures of Microorganisms, 2nd ed. World Data Center on Microorganisms, University of Queensland, St. Lucia, Queensland, Australia.

McGowan, V., and Skerman, V. B. D., eds. (1982b). World Catalogue of *Rhizobium* Collections, 2nd ed. World Data Center on Microorganisms, University of Queensland, St. Lucia, Queensland, Australia.

Means, U. M., and Erdman, L. W. (1963). *Soil Sci. Soc. Am. Proc. 27*: 305.

Munns, D. N. (1978). Effects of soil acidity. In *Mineral Nutrition of Legumes in Soils of the Tropics and Subtropics*, C. S. Andrew and E. J. Kamprath, eds., CSIRO, Melbourne, p. 247.

Norris, D. O., and Date, R. A. (1976). In *Tropical Pasture Research Principles and Methods*, N. H. Shaw and W. W. Bryan, eds., Bulletin Commonwealth Bureau of Pasture Field Crops 51, p. 134.

Nutman, P. S. (1976). In *Symbiotic Nitrogen Fixation in Plants*, P. S. Nutman, ed., Cambridge University Press, Cambridge, p. 211.

Pattison, A. C., and Skinner, F. A. (1974). *J. Appl. Bacteriol. 37*: 239.

Reid, R., and Lazier, J. R. (1979). In *Handbook for the Collection, Reservation and Characterization of Tropical Forage Germplasm.* G. O. Mott and Alejandro Jimenezc, eds. Centro Internacional de Agriculture Tropical, Cali, Colombia, p. 15.

Skerman, V. B. D. (1968). *J. Gen. Microbiol. 54*: 287.

Somasegaran, P., Hoben, H., and Halliday, J. (1982). *The Niftal Manual for Methods in Legume-Rhizobium Technology*, University of Hawaii, Honolulu.

Thompson, J. A. (1980). In *Methods for Evaluating Biological Nitrogen Fixation*, F. J. Bergersen, ed. John Wiley & Sons, Inc., New York, p. 489.

Tsien, H. C., Cain, P. S., and Schmidt, E. L. (1977). *Appl. Environ. Microbiol. 34*: 854.

Vincent, J. M. (1970). *A Manual for the Practical Study of the Root-Nodule Bacteria* (IBP Handbook 15). Blackwell Scientific Publications Ltd., Oxford.

Vincent, J. M. (1977). In *A Treatise on Dinitrogen Fixaiton*, Sec. 3, *Biology*, R. W. F. Hardy and A. H. Gibson, eds. John Wiley & Sons, Inc., New York, p. 277.

Zhou, J. C., Tchan, Y. T., and Vincent, J. M. (1984). In *The Seventh Australian Legume Nodulation Conference*, I. R. Kennedy and L. Copeland, eds. AIAS Occasional Publication 12, p. 175.

Zhou, J. C., Tchan, Y. T., and Vincent, J. M. (1985). In *World Soybean Research Conference III, Proceedings*, R. Shibles, ed., p. 918.

2

Collection, Isolation, Cultivation, and Maintenance of *Frankia*

MARK D. STOWERS* *NPI, Salt Lake City, Utah*

---

*Present affiliation: Eastman Kodak Company, Rochester, New York

I. INTRODUCTION

Biological nitrogen fixation is restricted to prokaryotes, some of which form a mutually beneficial symbiosis with higher plants. One of the most studied symbiotic nitrogen-fixing interactions is the *Rhizobium*-legume symbiosis. While the *Rhizobium*-legume symbiosis accounts for the fixation of tremendous amounts of atmospheric nitrogen, 100 to 200 kg N/ha (Gibson et al., 1982), other symbiotic associations contribute significantly to their particular ecosystem.

Of the nonleguminous nitrogen-fixing interactions, the actinorhizal symbiosis is the best studied, and several recent reviews are available (Akkermans and Roelofsen, 1980; Akkermans and van Dijk, 1981; Silvester, 1977; Torrey, 1982). Actinorhizal plants (woody dicotyledenous plants nodulated by the filamentous bacterium *Frankia* of the Actinomycetales) represent over 170 species among 22 genera in eight families (Table 1) (Akkermans and Roelofsen, 1980; Becking, 1981; Torrey; 1978). The term "actinorhiza" has been designated to root nodules formed by *Frankia* (Figure 1). The annual nitrogen fixation rates for

Table 1 Actinorhizal Plants

Family	Genera
Betulaceae	*Alnus*
Casuarinaceae	*Casuarina, Allocasuarina*
Coriariaceae	*Coriaria*
Datiscaceae	*Datisca*
Elaeagnaceae	*Elaeagnus, Hippophae, Shepherdia*
Myricaceae	*Comptonia, Myrica*
Rhamnaceae	*Ceanothus, Colletia, Discaria, Trevoa*
Rosaceae	*Cercocarpus, Cowania, Dryas, Purshia*

Source: Data compiled from Akkermans and van Dijk (1981), Akkermans and Houwers (1979), Akkermans and Roelofsen (1981), and Righetti and Munns (1982).

Figure 1 Actinorhiza of *Alnus rubra*.

actinorhizal plants can range from 2 to 362 kg N/ha (Silvester, 1977), comparable to the rates for leguminous plants (Gibson et al., 1982).

Actinorhizal plants are most valuable in forestry and land reclamation, although some species are important in ornamental horticulture. *Alnus* species have been proven valuable to forestry throughout the world as pulpwood, fuelwood, and in reforestation (Resch, 1979; Rottink et al., 1979; Smith, 1978; Zavitkovski et al., 1979). In the U.S. Pacific Northwest over 60% of the hardwoods produced are red alder (*Alnus rubra*) (Worthington et al., 1962; Smith, 1978). Red alder yields between 15 and 25 tons of biomass per hectare per year (Smith and DeBell, 1973; Gordon, 1975), making this tree species the highest-yielding temperate-zone tree (Gordon and Dawson, 1979).

Casuarina species represent another important actinorhizal tree. *Casuarina* produces a high-quality fuelwood, can be used as poles or beams in construction, and is excellent in slope stabilization and windbreaks (BOSTID, 1984). *Casuarina* is also used throughout the tropics as an ornamental. Other actinorhizal plants have uses in the human diet (e.g., *Hippophae*, *Shepherdia*, and *Elaeagnus*) (Uphof, 1968) and as forage for livestock (*Ceanothus* and *Purshia*) (Lyon and Mueggler, 1968; Ferguson and Basile, 1967).

With the recent isolation of the symbiotic nitrogen-fixing endophyte *Frankia* (Callaham et al., 1978), the improvement of actinorhizal plants in terms of increased yields, increased soil fertility, and improvement of stand establishment can be pursued using pure culture inoculation techniques. *Frankia* microbiology is, however, in its infancy. *Frankia* has not been collected and isolated from all actinorhizal plants, nor have all isolates been verified as *Frankia* via Koch's postulates. In light of the beneficial applications of *Frankia*, the need to enlarge *Frankia* germplasm collections becomes essential in optimizing this symbiotic association. In this chapter we describe methodology used by this laboratory and others for the collection, isolation, cultivation, and maintenance of *Frankia* as a means for the development of superior strains for use in forestry, land reclamation, and ornamental horticulture.

II. COLLECTION OF ACTINORHIZAS

A. Preparation

Extensive preparation is required before embarking on a collection trip. Foremost in the preparation is the examination of technical information of a particular region with regard to known actinorhizal species, other vegetation, soils, topography, and climate. Such information can facilitate the collection of interesting and important germplasm in a minimum of time, allowing for the prioritization of sample sites. Sources of technical information vary among locations, but one can consistently rely on local universities, herbaria, botanical gardens, and regional and national parks.

Aside from the technical preparation required for a collection trip, general information must be gathered. While the general information described below may be obvious, it is essential that each detail be given the utmost consideration. The most critical general facts to be gathered when planning a collection trip are (a) the existence and accessibility of highways and roads; (b) the availability of potable water, edible food, and oil and gasoline for the vehicle; (c) overnight accommodations; and (d) the need and availability of guides familiar with the region.

Determination of trip duration is also critical, particularly in the preparation of necessary equipment and the logistics of returning samples for isolation. It is important to make a provisional itinerary that includes proposed routes and means of transportaiton. It is advisable to inform a local contact of your itinerary in case of emergency.

After obtaining technical and general information about the region of interest and having established the trip duration, the appropriate equipment can be planned and secured. An equipment checklist amenable to most colleciton trips is provided in Table 2. Other important articles to include are (a) letters of invitation; (b) permission for collection, if required; (c) letters for local persons who will cooperate; and (d) plant importation permits if returning actinorhizas to the United States.

Table 2 Equipment for Field Collection Trips

A. General

Appropriate clothing depending on weather conditions of the region
Backpack to carry collected materials and equipment

B. Site identification

Field book and pencil
Collection data sheets
Detailed road maps
Topographical maps
Compass and altimeter
Camera, extra film, tripod

C. Plant identification

Plant taxonomic keys
Hand lens

D. Germplasm collection

Shovel
Small trowel
Knife
Forceps
Pruning shears
Small plastic Ziploc bags for soil and root samples
Collection vials
Paper towels
Water container
Small cooler and/or access to a cold storage unit
Permanent marker or labeled bags
Waxed boxes for shipping or transporting materials

E. Herbarium specimen collection

Large plastic bags
Plant press

B. Collection of Germplasm

Data Collection

The most critical part of field collection of actinorhizas is the maintenance of detailed, descriptive data. A detailed data sheet should include information about the plant collected, what was collected, site location (topography, rainfall, vegetation type, land use), site description (slope, surrounding vegetation), soil description, and nodulation data

(Figure 2). Other important data to gather are those relating to the presence of insects and disease. It is extremely important that specific site data be collected. Data should be detailed enough to permit the relocation of a specific site or a particular plant.

General Information

General rules: (a) collecting from roadsides or other disturbed areas should be avoided (plant performance and nodulation may be atypical for the region); (b) well-established plants should be sampled; and (c) collection trips should coincide with the season of vegetative growth and adequate moisture (in the tropics, 4 to 8 weeks after the rainy season is preferred; in the temperature zones, early spring is desirable).

Field Collecting

The most important factor in collecting actinorhizas in the field is the preservation of the nodule. For same-day collection trips (i.e., when isolations are performed the same day as collection) the collection of nodulated root systems is the simplest method. When collections are being made from small plants, the whole plant can be excavated, packed in soil, and placed in a plastic bag. Similarly, sections of nodulated roots can be packed with soil and returned to the laboratory.

Most collection trips exceed 1 day in duration, and in those cases special care should be taken with regard to the preservation of the actinorhiza. Actinorhizas can be collected and stored in small, screw-capped vials containing a desiccant, usually silica gel, which generally occupies approximately one-third of the volume of the container. A glass wool plug separates the nodules from the desiccant. The volume of actinorhizas should not exceed the volume of the desiccant.

The location of actinorhizas varies among species, environment, and plant age. It is easier to locate and collect actinorhizas from young plants. Only actinorhizas that are fresh (usually a whitish appearance) and firm should be sampled; avoid collecting damaged or decaying actinorhizas.

On older plants actinorhizas may not readily be observed. One should examine areas of new root growth sometimes several centimeters away from the crown. Actinorhizas can be found on fine roots near the crown in older plants of *Alnus* and *Elaeagnus*.

General rules for collecting the samples are as follows:

1. Rarely is it necessary to excavate the entire plant. Careful removal of soil around the crown area can yield sufficient nodule material. Most actinorhizas will be found in the top 2.5 cm.
2. Removal of the plant by force is discouraged due to the extremely low success rate of isolation from these plants.

Collection Site # 1
Sample # 1

GENERAL INFORMATION & LOCATION

Date Collected 2/7/85
Genus & Species ALNUS JAPONICA
Local Name

State
County
Locality CENTRAL JAVA, CIBODAS MOUNTAIN GARDENS INDONESIA
Location on Topo Map

Plant Material Collected Vegetative Seed None
Nodules Present Yes No
Nodules Collected Yes No
Soil Collected Yes No
Roots Collected Yes No
Herbarium Collected Yes No
Photographs Taken Yes No #'s on film:

NATURAL HABITAT & AREA VEGETATION

Season Rainfall
Rainfall Seasonality NOV → APR
Altitude 1450 meters
Topography 1)Swamp 2)Flood Plain 3)Level 4)Hilly 5)Mountainous
6)Other
Vegetation Type 1)Forest 2)Grassland 3)Other
Land Use 1)recently disturbed area 2) roadside 3)settlement
4)other RAIN FOREST UNDISTURBED
Management 1)cutting 2)burning 3)natural habitat preserved

DESCRIPTORS FOR SPECIFIC COLLECTION SITE

Position in landscape
Slope 1)flat or almost 0-1° 2)gently sloping 1°-3° 3)sloping 3°-7°
4)moderately sloping 7°-14° 5)steep 14°-29° 6)very steep >29°
Slope Facing SOUTH
Site Ground Cover 1)Bare 2)Very Thin 0-19% 3)Thin 20-39%
4)Moderate 40-59% 5)Heavy 60-79% 6)Very Heavy 80-100
Degree Plant Shaded 1)no shade 2)very slight 3)slight 4)moderate
5)heavy 6)completely
Shaded By 1)grasses 2)shrubs 3)trees 4)topography
Target Species Life Form 1)trees >30m 2)trees 10-30m 3)trees 5-10m
4)shrubs 2-8m 5)subshrubs 0-2m 6)herbs >1m 7)herbs 0-1m
Target Species Growth Habit 1)prostrate 2)decumbent 3)erect
4)vine 5)sarmentose 6)liana 7)caespitose 8)rhizomatous
9)stoloniferous 10)rootstock
Target Species Rooting Habit 1)deep tap root 2)shallow tap root
3)shallow spreading 4)deep spreading 5)fibrous
Target Species Stage of Growth 1)vegetative 2)flowering 3)fruiting
Target Species Longevity 1)annual 2)biennial 3)perennial deciduous
4)perennial evergreen
Associated Species

Elaeagnus sp

DESCRIPTORS OF SOIL CHARACTERISTICS

Soil Texture 1)sandy 2)loamy 3)clayey 4)organic 5)rocky
Soil Surface Color 1)red 2)yellow 3)brown 4)gray
Soil Drainage 1)ponded 2)poorly drained 3)imperfectly drained
4)moderately well drained 5)well drained
6)somewhat excessively drained 7)excessively drained

NODULATION

Location of Nodules 1)crown area 2)tap root 3)lateral roots
Nodule Growth Form 1)clumps 2)scattered 3)small clumps w/ scattering

Figure 2 Sample data sheet for collection of actinorhizas in the field.

3. Do not combine samples. Individual actinorhizas from different plants can contain unique isolates.

III. ISOLATION

Numerous attempts to isolate the endophyte from root nodules of actinorhizal plants have been made over the past century (Baker and Torrey, 1979). Only recently have researchers been able, successfully and reliably, to isolate the symbiotic nitrogen-fixing endophyte (Callaham et al., 1978; Quispel and Tak, 1978; Lalonde and Calvert, 1979; Baker and Torrey, 1979; Berry and Torrey, 1979).

Fundamental to the isolation of *Frankia* from actinorhizal root nodules is the need to eliminate all contaminating microorganisms from the surface of the nodule. Several surface sterilants have been employed, such as bromine solutions (Burggraaf et al., 1981; Berry and Torrey, 1979), chlorine bleach (Baker et al., 1979; Diem and Dommergues, 1983), mercuric chloride (Callaham et al., 1978), Pittchlor (Berry and Torrey, 1979), hydrogen peroxide (Berry and Torrey, 1979), osmium tetroxide (Lalonde et al., 1981), chloramine T (Diem et al., 1982), and glutaraldehyde (Baker and O'Keefe, 1984). Each of these surface sterilants has been used in this laboratory, with varying success depending on the isolation procedure.

Five methods for the isolation of *Frankia* from actinorhizal root nodules are currently in use. Use of a particular isolation procedure varies among laboratories. Some isolation procedures work better than others with certain nodules; these general features are noted below.

A. Sucrose Density Gradient Centrifugation

A detailed discussion of the sucrose density gradient centrifugation isolation method can be found elsewhere (Baker and Torrey, 1979; Baker et al., 1979; Carpenter and Robertson, 1983; Baker and O'Keefe, 1984). This technique relies on the separation of *Frankia* from host-cell particles based on differences in size, density, and solubility in the gradient solution. Using a discontinuous three-layer sucrose gradient [30, 45, and 60% (w/v)], *Frankia* can be separated from contaminating host-cell debris. *Frankia* band at or near the interface of the 45 and 60% solutions (Baker and Torrey, 1979; Baker et al., 1979). This technique has worked well with isolations from various alder species; *Casuarina* spp., *Purshia tridentata, Myrica cerifera, Cowania mexicana*, and *Elaeagnus umbellata* (Baker and O'Keefe, 1984).

A recent modification of the sucrose density gradient centrifugation isolation technique has yielded better results and can allow direct isolation of *Frankia* from the soil.

Modified Sucrose Gradient Centrifugation Isolation Procedure (Baker and O'Keefe, 1984)

1. Surface-sterilize small nodule portions with 2% (w/v) glutaraldehyde containing Tween-20 as a surfactant.
2. Rinse thoroughly with sterile distilled water.
3. Crush nodule pieces in a sterile mortar and pestle.
4. Incubate the nodule suspension in a final volume of 0.7% (v/v) liquefied phenol for 10 min.
5. Layer suspension onto sucrose gradient and centrifuge in a swinging bucket rotor (100,000 × *g* for 3 hr or 600 to 1000 × *g* for 12 hr).
6. After centrifugation, inoculate selective medium (described in Section VI.B) with sucrose fraction containing *Frankia.*

The modified sucrose gradient centrifugation isolation procedure can also be applied to direct isolations from the soil (Baker and O'Keefe, 1984). Soil samples are prepared by adding 5 g of soil to 50 ml of sterile distilled water. The soil suspension is incubated with 0.7% (v/v) liquefied phenol and layered onto a sucrose gradient as described above. This technique has allowed for the isolation of *Frankia* from *Cercocarpus* and other actinorhizal hosts that have few or inaccessible actinorhizas for collection (Baker and O'Keefe, 1984).

B. Selective Incubation (Quispel and Burggraff, 1981)

This method involves the incubation of disinfected nodules in tubes containing a rich nutrient agar for 6 weeks. Tubes containing most contaminants can be easily identified on the surface of the agar and discarded. Tubes containing *Frankia* will show growth from the nodules. The *Frankia* are removed from the agar and subcultured. This method of isolation is time consuming. If incubation proceeds for longer than 3 months, the infectivity of the *Frankia* is significantly reduced (Quispel and Burggraaf, 1981); however, this procedure has worked well in this laboratory using osmium tetroxide as the surface sterilant with isolations from various *Alnus* spp., *Ceanothus* spp., *Elaeagnus* spp., *Myrica* sp., and *Datisca glomerata.*

C. Filtration (Benson, 1982)

Filtration is one of the simplest and most efficient isolation procedures for *Frankia.* It requires a dual-filtration system: one of approximately 52 μm pore size and another of 20 μm pore size. Large nodule particles are retained on the 52-μm filter and *Frankia* vesicle clusters are retained on the 20-μm filter. A description of this techniques follows.

Filtration Technique for the Isolation of Frankia *(Benson, 1982)*

1. Wash root nodules with a stream of distilled water.
2. Remove nodule lobes 1 mm from the tip of the nodule.
3. Wash nodule lobes with distilled water.
4. Place 4 to 15 nodule lobes in a polypropylene microfuge tube (1.5 ml capacity).
5. Add 1 ml of 20% (v/v) laundry bleach [1.05% (W/v) sodium hypochlorate] and agitate for 5 min.
6. Decant bleach and wash four times with 1 ml of sterile distilled water.
7. Homogenize nodule lobes in 5 ml of *Frankia* medium (see below) with a tissue homogenizer (7.5 ml capacity).
8. Gravity-filter homogenate through nylon screens of 50- and 20-μm pore size.
9. Collect nodule debris on the 52-μm filter and wash with 30 ml of *Frankia* medium.
10. Collect *Frankia* on the 20-μm filter and wash twice with 30 ml of *Frankia* medium.
11. Remove *Frankia* from the 20-μm filter with a Pasteur pipette and place in sterile test tubes.
12. Dilute suspension and pour-plate the endophyte.
13. Incubate until visible colonies are observed.
14. Remove colonies asceptically and homogenize.
15. Inoculate 1 ml of *Frankia* medium containing 0.3% (w/v) filter-sterilized sodium pyruvate with a single homogenized colony.

Frankia medium (Benson, 1980) contains (g/liter): K_2HPO_4, 3.0; KH_2PO_4, 2.0; $MgSO_4 \cdot 7H_2O$, 0.2; NaCl, 0.3; ferric sodium-ethylenediaminetetracetic acid (EDTA), 0.16; 1 ml of trace elements as described by Baker and Torrey (1979); and 1 ml of vitamin mixture (Tjepkema et al., 1980). Calcium carbonate is added to liquid cultures at 0.05 mg/ml. Carbon sources are added at 3 g/liter.

D. Serial Dilution

Serial dilution has been incorporated in isolation procedures for various microorganisms, including *Frankia* (Berry and Torrey, 1979). *Frankia* generally require heavy inoculum doses (10 to 20%) to establish successful cultures, a practice in conflict with procedures of serial dilution. Consequently, serial dilution is not routinely used for the isolation of *Frankia*.

E. Microdissection

This technique has been used by various *Frankia* researchers as the most effective means of obtaining isolates. Most contaminating microorganisms are located on the outer surfaces of the nodule. Removal of the outer cell layers can greatly facilitate the isolation of *Frankia*. A general microdissection procedure has been described by Diem and Dommergues (1983) for the isolation of *Frankia* from *Casuarina*. This procedure has been successful in this laboratory for the isolation of *Frankia* from *Alnus* spp., *Elaeagnus* spp., *Shepherdia argentea*, *Casuarina* sp., *Myrica* sp., *Datisca glomerata*, *Cercocarpus montanus*, *Ceanothus* spp., and *Purshia tridentata*.

Microdissection Procedure (Diem and Dommergues, 1983; with modifications, Stowers, unpublished)

1. Clean nodules with distilled water.
2. Remove nodule lobes 1 mm from the tip.
3. Peel outer cell layers with scalpel using a dissecting microscope.
4. Soak nodule core in 1.5% (w/v) calcium hypochlorate for 10 min.
5. Rinse eight times in sterile distilled water.
6. Cut nodule core in half and place one piece on Qmod agar plates (Lalonde and Calvert, 1979) and the other on basal agar medium containing 5 m*M* pyruvate and 7.5 m*M* NH_4Cl, as carbon and nitrogen sources, respectively.*

IV. IDENTIFICATION AND CHARACTERIZATION

A. Morphology

Frankia are primarily identified on the basis of size and morphology. *Frankia* form thin filaments 0.8 to 1.2 μm in diameter which are usually not septate. *Frankia* also form specialized structures: sporangia, spores, and vesicles (Figure 3). Sporangia can be intercalary or terminal. Vesicles, the proposed site of nitrogen fixation (Tjepkema et al., 1980; Murry et al., 1984b), are generally spherical or club-shaped, 2.5 to 5 μm in diameter.

B. Growth Characteristics

Frankia are able to use few carbon sources for energy and growth (Benson and Hanna, 1983; Lechevalier and Ruan, 1984; Hafeez et al., 1984; Lechevalier et al., 1983). Most isolates are able to use propionate, acetate succinate, fumarate, and malate and Tween-80, but fail to use hexoses, pentoses, or disaccharides. Nitrogen utilization is restricted to ammonium, nitrate, and aspartate (Shipton and Burggraaf,

---

*Several nodule core pieces can be placed on a single plate.

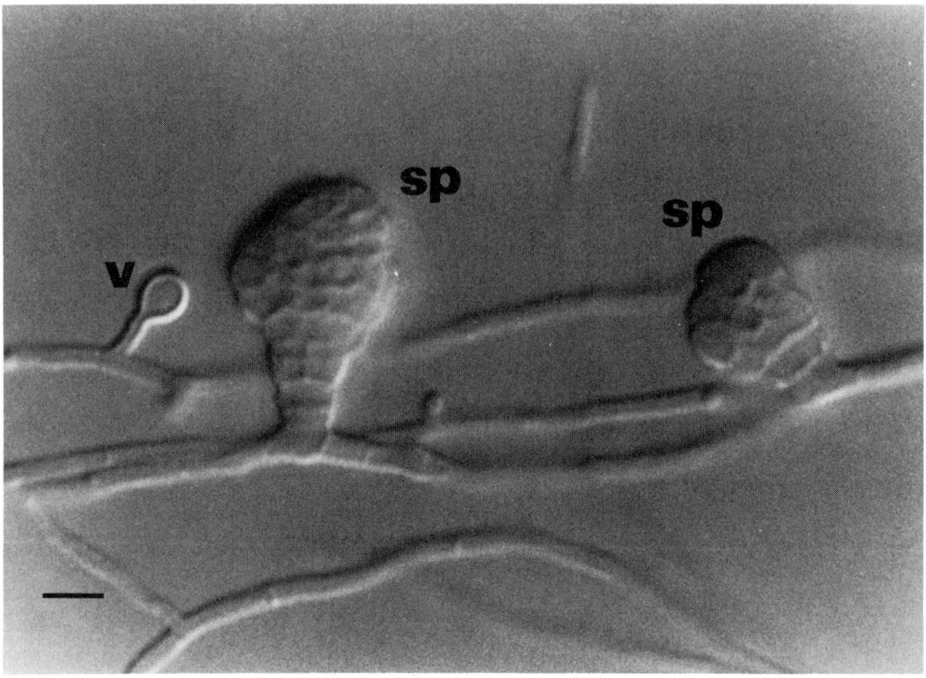

Figure 3 Photomicrograph of *Frankia* isolated from *Alnus viridis* spp. *sinuata* using Nomarski interference optics. sp, sporangia; v, vesicles. Bar = 5 μm.

1982), although urea can support growth of certain isolates. The optimum temperature is strain dependent, ranging from 25 to 36°C (Tisa et al., 1983). Most isolates can be routinely grown between 28 and 30°C. The pH optimum for *Frankia* lies between 6.5 and 7.0 (Burggraaf and Shipton, 1982). *Frankia* do not require a complete vitamin complement; however, some isolates show a stimulation of growth with the addition of biotin (1 mg/liter) (Shipton and Burggraaf, 1982). *Frankia* are routinely cultured aerobically and do not grow anaerobically (Callaham et al., 1978); however, some *Frankia* are considered strictly microaerophilic (Lechevalier et al., 1983).

C. Biochemical Properties

Structural features of *Frankia* have been used extensively as criteria for identification and classification. The presence of *meso*-diaminopimelic acid in the cell wall has been used to identify and delineate actinomycetes (Becker et al., 1964, 1965; Lechevalier and Fekete, 1971; Lechevalier et al., 1966). The presence of this compound has been detected in isolates from *Alnus* but not in the cell walls of endophytes from *Myrica* or *Casuarina* (Becking, 1977a,b; Van Dijk, 1978). Other structural features, such as cell surface components determined via immunodiffusion, have been used to group *Frankia* (Baker et al., 1981a,b).

Whole-cell sugars have also been used to characterize *Frankia*. All *Frankia* examined possessed rhamnose and 2-O-methylmannose (Mort et al., 1982). *Frankia* isolates from *Alnus* have unique whole-cell sugar patterns containing xylose and arabinose (Lechevalier and Lechevalier, 1979; Lechevalier et al., 1983; Baker et al., 1981a,b). Fucose has been detected in some *Frankia* isolated from *Ceanothus americanus* (Lechevalier and Ruan, 1984).

Sodium dodecyl sulfate polyacrylamide gel electrophoresis (SDS-PAGE) of total cell protein has been used extensively by Benson and his co-workers (1983, 1984) to distinguish isolates of *Frankia*. Several groups have been delineated among *Frankia* isolated from *Alnus incana* ssp. *rugosa* using SDS-PAGE. Further distinction can be made using two-dimensional polyacrylamide gel electrophoresis (2D-PAGE) (Benson et al., 1984).

All of the biochemical properties noted above require a certain laboratory sophistication. Simple biochemical tests have recently been developed for *Frankia*; however, these tests do not allow the distinction of particular isolates at the level of sensitivity required (Horriere, 1984).

Unlike *Rhizobium*, *Frankia* can be routinely cultured on aerobic nitrogen-free medium (Tjepkema et al., 1980; Gauthier et al., 1981b; Burggraaf and Shipton, 1983; Murry et al., 1984a,b). Growth of *Frankia* under nitrogen-limiting conditions leads to the formation of vesicles and nitrogen fixation (Fontaine et al., 1984). The ability of *Frankia* to derepress or induce nitrogen fixation activity in culture is a valuable biochemical property for the identification of these actinomycetes. A method for the depression or induction of nitrogenase in *Frankia* is listed below.

Induction of Vesicle Formation and Nitrogen Fixation Activity (Tjepkema et al., 1980; Murry et al., 1984a,b)

1. Collect filamentous mat of *Frankia* from a 3- to 4-week-old culture.
2. Homogenize mat.
3. Wash three times with a mineral salt solution from the defined medium (see below).
4. Measure the final packed cell volume.

5. Resuspend inocula in mineral salt solution and add known volume to test tubes containing 5 ml of defined medium (see below). *Note*: Generally, 0.8 ml packed cell volume (pcv) of *Frankia* can be used to inoculate 100 tubes (8 μl pcv per tube).
6. Measure nitrogen fixation activity via acetylene reduction (described in Chapter 12) and observe the number of vesicles microscopically. Maximum rates of acetylene reduction and maximum number of vesicles can be seen at 6 days with a 5 m*M* carbon source or 13 days with a 10 m*M* carbon source (Murry et al., 1984a).

Defined medium (Tjepkema et al., 1980) contains the following (g/liter): $CaCl_2PO_4$, 0.01; $MgSO_4 \cdot 7H_2O$, 0.1; $CaCl_2 \cdot 2H_2O$, 0.01; ferric sodium-EDTA, 0.01; 1 ml/liter of a trace element stock solution containing (g/liter): H_3BO_3, 2.86; $MnCl_2 \cdot 2H_2O$, 1.81; $ZnSO_4 \cdot 7H_2O$, 0.22; $CuSO_4 \cdot 5H_2O$, 0.08; $Na_2MoO_4 \cdot 2H_2O$, 0.025; 1 ml of a vitamin stock solution [containing (mg/100 ml); 10, thiamine HCl, 10; nicotinic acid, 05; pyridoxine HCl, 50]; and 1.2 g/liter of succinate.

V. AUTHENTICATION

A particular isolate's ability to form nodules on the roots of a host plant is the ultimate criterion of authenticity of *Frankia*. However, some isolates lack the ability to reinfect the host plant (Lalonde et al., 1975; Diem et al., 1982). These isolates show morphological similarity to other *Frankia* and are able to fix nitrogen in culture (Diem et al., 1982). In one case, isolates from *Casuarina equisetifolia* that failed to reinfect *C. equisetifolia* formed nodules on an unrelated actinorhizal plant, *Hippophae rhamnoides* (Gauthier et al., 1981a). The nature of this phenomenon is not clear, although it is apparent that nodule formation may not be a usable criterion for authenticity in all cases.

Because of the inconsistencies presented above, several workers have sought biochemical criteria for the authentication of *Frankia* and the delineation of *Frankia* into particular species (reviewed by Lechevalier, 1984). It has been concluded that an isolate can be determined to be *Frankia* based on five criteria (Lechevalier, 1984):

1. Morphology (production of sporangia and vesicles in culture)
2. Cell wall and membrane chemistry (cell wall type III and phospholipid type PI) (Lechevalier et al., 1983)
3. Serology
4. DNA homology
5. Infectivity and/or effectivity for a host plant

In this laboratory the criteria for authentication of *Frankia* have been simplified and are more selective, based on the desire to obtain infective and effective isolates for a strain seleciton program and ultimate inocula production. The criteria are as follows:

1. Morphology (production of sporangia and/or vesicles in culture)
2. Ability to fix nitrogen in culture
3. Ability to infect and form an effective symbiosis with a host plant

VI. CULTIVATION

A. Measurement of Growth

Various laboratory media have been used for the cultivation of *Frankia*. Culture media can be complex or defined. The evaluation of the usefulness of a particular culture medium has been hampered by the lack of a simple and convenient means for the determination of growth.

Presently, growth of *Frankia* can be measured using total protein, total ATP yield, or dry weight. Burggraaf and Shipton (1982) reported a highly significant correlation ($P < 0.01$) between dry weight and total protein, dry weight and total ATP, and total ATP and protein. In this laboratory total protein is used routinely to measure growth. This procedure, described below, uses the protein-dye-binding assay described by Bradford (1976).

Protein Determination for Cell Yield (Murry et al., 1984a; with modifications by Steele and Stowers, unpublished)

1. Harvest cells by centrifugation (2000 × *g* for 10 min).
2. Wash twice with 50 m*M* potassium phosphate buffer, pH 7.0.
3. Resuspend final pellet in 0.2 *N* NaOH in 50 m*M* potassium phosphate buffer.
4. Sonicate 3 min in 1-min intervals.
5. Place the cell suspension in a boiling-water bath for 10 min.
6. Cool to room temperature with ice.
7. Determine protein using the procedure of Bradford (1976) with bovine serum albumin as the standard.

B. Growth

Carbon Sources

Frankia are quite varied in their ability to use carbon sources for growth and energy (Table 3). Propionate and acetate are the most universally used carbon sources. Low levels of these organic acids (5.0 to 10.0 m*M*) have been determined to be optimal for growth, with levels greater than 10.0 m*M* showing growth inhibition (Akkermans et al., 1983). Other organic acids, such as succinate, fumarate, malate,

Table 3 Summary of Carbon Utilization by Various *Frankia* Isolates

Carbon source	Number of isolates tested	Number of isolates showing a growth response[a]
Starch	5	5(100)
Galacturonic acid	3	3(100)
Propionate	35	33(94)
Acetate	25	23(92)
Trehalose	7	6(86)
Fumarate	7	5(71)
Rhamnose	3	2(67)
Succinate	34	19(56)
Pyruvate	16	8(50)
Mannitol	4	2(50)
Fructose	4	2(50)
Glucose	37	18(49)
L-Malate	9	4(44)
Galactose	7	3(43)
Xylose	7	3(43)
Maltose	11	4(36)
Sorbitol	6	2(33)
Sucrose	23	5(22)
Glycerol	10	2(20)
L-Arabinose	22	2(9)
3-O-Methylglucose	4	0(0)

[a]Numbers in parentheses indicate the percent isolates showing a growth response.

Source: Data compiled from Akkermans et al. (1983), Burggraaf and Shipton (1983), Benson and Hanna (1983), Lechevalier et al. (1983), Shipton and Burggraaf (1983), Tisa et al. (1983), Hafeez et al. (1984), and Lechevalier and Ruan (1984).

and pyruvate, are also used by most isolates but not consistently with all *Frankia*. The utilization of sugars appears to be restricted to a few isolates, although a high percentage of a sample of isolates showed the ability to use trehalose, galacturonic acid, rhamnose, mannitol, fructose, and glucose. More information is needed with more isolates before precise carbon utilization patterns can be used to characterize *Frankia*. As a general rule, it is recommended that propionate or acetate be used as the sole carbon source for the cultivation of *Frankia*.

Nitrogen Sources

Nitrogen utilization is quite consistent among *Frankia* (Table 4). Almost all *Frankia* prefer ammonium chloride as the sole nitrogen source (Shipton and Burggraaf, 1982; Lechevalier et al., 1983; Shipton and Burggraaf, 1983). Some isolates show the ability to use amino acids such as aspartate and glutamate and inorganic sources such as ammonium nitrate and potassium nitrate.

Frankia are also capable of growing in the absence of combined nitrogen (Tjepkema et al., 1980; Gauthier et al., 1981b; Murry et al., 1984; Fontaine et al., 1984). *Frankia* form specialized structures called vesicles, which are thought to be the site of nitrogen fixation under ambient oxygen concentrations (Tjepkema et al., 9180; Murry et al., 1984a,b). The formation of vesicles and the reduction of nitrogen by *Frankia* in vitro is repressed in the presence of combined nitrogen (Gauthier et al., 1981b; Fontaine et al., 1984).

Vitamin Requirements

Frankia tested to date do not require vitamins for growth. Various vitamin combinations containing nicotinic acid, thiamine, pyridoxine, and vitamin B12, and mixtures containing nicotinic acid, thiamine, pyridoxine, pantothenate, riboflavin, folic acid, and biotin, have been used, resulting in no stimulatory responses (Shipton and Burggraaf, 1982). Biotin has been shown to stimulate the growth of some *Frankia* when propionate is used as the sole carbon source (Shipton and Burggraaf, 1982). Biotin stimulation is expected since *Frankia* metabolize

Table 4 Summary of Nitrogen Utilization by Various *Frankia* Isolates

Nitrogen source	Number of isolates tested	Number of isolates showing a growth response[a]
Ammonium chloride	10	9(90)
Urea	6	4(67)
Aspartate	8	5(63)
Glutamate	7	4(57)
Potassium nitrate	8	4(50)
Ammonium nitrate	6	3(50)
Uracil	6	3(50)
Thymine	6	0(0)

[a]Numbers in parentheses indicate the percent isolates showing a growth response.

Source: Data compiled from Shipton and Burggraaf (1982), Lechevalier et al. (1983), and Shipton and Burggraaf (1983).

propionate via propionyl-CoA carboxylase in the methylmalonyl pathway (Stowers et al., 1986).

Media Composition

Frankia have been cultivated in both simple defined and complex media. In this laboratory a defined medium (BAP) described by Murry and her co-workers (1984a) is routinely used. The medium contains (in m*M*) KH_2PO_4, 7; $K_2HPO_4 \cdot 7H_2O$, 0.1; $CaCl_2$, 0.07; ferric sodium-EDTA, 10 mg/liter; biotin, 450 μg/liter; and trace elements. Doubling times for isolates from *Alnus* spp. range from 8 to 24 hr, depending on the particular isolate.

Complex media have also been used extensively. Qmod medium (Lalonde and Calvert, 1979) has been used by many researchers. Qmod consists of the following (per liter): K_2HPO_4, 300 mg; NaH_2PO_4, 200 mg; $MgSO_4 \cdot 7H_2O$, 200 mg; KCl, 200 mg; yeast extract, 500 mg; Bacto-peptone, 5 g; glucose, 10 g; ferric citrate (1% solution of citric acid and ferric citrate), 1 ml; minor salts (g/liter: H_3BO_3, 1.5; $MnSO_4 \cdot 7H_2O$, 0.8; $ZnSO_4 \cdot 7H_2O$, 0.6; $CuSO_4 \cdot 7H_2O$, 0.1; $(NH_4)_6MO_7O_{24} \cdot 4H_2O$, 0.2; $CoSO_4 \cdot 7H_2O$, 0.01), 1 ml; $CaCO_3$, 100 mg; and lipid supplement [500 mg/liter lecithin, 22% (w/v) phosphatidyl choline in 50 ml of absolute ethanol and 50 ml of distilled water], 0.5 to 50 mg. In this laboratory Qmod does not support the growth of all isolates and is generally regarded as inferior to defined propionate medium (Murry et al., 1984a). Other complex media, such as tryptone broth (g/liter: Bacto-tryptone, 5; glucose, 10; $MgSO_4 \cdot 7H_2O$, 0.2; $CaCO_3$, 0.1; K_2HPO_4, 0.3; $NaH_2PO_4 \cdot 2H_2O$; and KCl, 0.2) has been used successfully for the culture of *Frankia* (Shipton and Burggraaf, 1982).

Selective Media

Since *Frankia* possess the ability to grow in the absence of combined nitrogen, several researchers have developed a selective medium for the isolation and culture of *Frankia*. Baker and O'Keefe (1984) have developed a selective medium for the isolation of *Frankia* from root nodules and directly from the soil. The medium consists of the following (g/liter): KH_2PO_4, 1.0; $MgSO_4 \cdot 7H_2O$, 0.1; $CaCl_2 \cdot 2H_2O$, 0.01; sodium propionate, 1.2; Hoagland's microelement stock (Hoagland and Arnon, 1950), 1 ml/liter; and ferric sodium-EDTA stock, 1.8 ml/liter; adjust pH to 6.8. This medium has been used to selectively cultivate *Frankia* from *Alnus*, *Casuarina*, *Purshia*, *Myrica*, *Cowania*, *Cercocarpus*, and *Ceanothus* (Baker and O'Keefe, 1984).

Aeration

Very few studies have investigated the effect of aeration on the cultivation of *Frankia*. It has been generally assumed that *Frankia* were microaerophilic (Becking, 1981). *Frankia* failed to grow anaerobically

(Callaham et al., 1978). In this laboratory and others *Frankia* isolates from *Alnus* spp. are routinely cultured in highly aerobic cultures (agitated and sparged) (Stowers et al., 1986; Fontaine et al., 1984). On the other hand, *Frankia* isolates from *Casuarina* appear to be obligately microaerophilic (N. Boonherd, personal communication; Diem and Dommergues, 1983). Much more information is needed to establish the nature of oxygen toxicity in *Frankia.*

Temperature Optima

Frankia are routinely cultured between 25 and 30°C (Becking, 1981). Temperature optima for *Frankia* have been investigated with few isolates (Burggraaf and Shipton, 1982; Tisa et al., 1983). Of the isolates tested, few isolates show the ability to grow at temperatures below 10°C and above 42°C. A commonly used isolate, CpIl, showed an unusual temperature optimum of 33 to 36°C (Burggraaf and Shipton, 1982; Tisa et al., 1983). At temperatures above 33°C, CpIl does not form sporangia (Tisa et al., 1983).

pH Optima

The pH optima for *Frankia* is within the range 6.5 to 7.0 (Burggraaf and Shipton, 1982). Increases in biomass have been observed with some isolates between pH 4.6 and 8.0 (Burggraaf and Shipton, 1982).

Inoculum Density

The inoculum density is a critical factor in the cultivation of *Frankia.* Low-level inoculum densities (<10% v/v) generally do not produce growing cultures. Inoculum densities above 10% are generally required. The metabolic basis of this phenomenon is not clear, but due to the nature of *Frankia* growth (i.e., budding from filament tips), a sufficient number of growing tips are required for successful cultivation. More effort should be directed toward the elucidation of the nature of growing propagules of *Frankia.*

VII. MAINTENANCE

Very little information is available with regard to long-term storage of *Frankia*, although several methods are available for the storage of *Frankia* (Becking, 1981; Carpenter and Robertson, 1983; Horriere et al., 1983).

A. Storage in Broth

In this laboratory *Frankia* cultures have been maintained in a defined propionate medium for 3 years. Cultures remain viable for 1 to 2 years when stored at 28°C. Similar results have been described by Carpenter and Robertson (1983). Other defined media have been used to store *Frankia* (see Becking, 1981). *Leptospira* broth (Ellinghausen and McCullough, 1965a,b) has been recommended (Becking, 1981). Other maintenance media include semisolid yeast malt medium (Shirling and Gottlieb, 1966) and diluted Bennett's agar (Horriere et al., 1983). The maintenance medium of choice is a yeast extract medium (0.5%; w/v; yeast extract Difco) in distilled water, pH 6.4.

B. Storage on Agar

It is generally thought that many *Frankia* are strict microaerophiles. To date only one report describes the maintenance of *Frankia* on agar surfaces (Horriere et al., 1983). Isolate AirIl, isolated from *A. incana* ssp. *rugosa*, has been maintained on agar slants. In this laboratory a survey of the ability of *Frankia* to grow on the surface of agar plates was examined. Of 12 isolates from *A. rubra*, all exhibited growth on Qmod or BAP agar (Stowers and Steele, unpublished).

C. Lyophilization (Freeze-Drying)

Many microorganisms can be stored as lyophilized cultures for long periods (>10 years). There are no reports available on the storage of *Frankia* by lyophilization, although the American Type Culture Collection (Bethesda, Maryland) supplies CpIl in a lyophilized form. More information is needed with regard to the suitability of lyophilization for the long-term storage of *Frankia*.

D. Cryogenic Storage

Carpenter and Robertson (1983) described a method for the storage of *Frankia* in liquid nitrogen. *Frankia* are grown in agar pour plates. Agar plugs containing individual colonies of *Frankia* are removed, covered with small amounts of sterile broth, and placed in small cyrotubes (2 ml capacity). The inoculated tubes are then quick-frozen in liquid nitrogen. Alternatively, *Frankia* can be prepared by adding 1 ml of *Frankia* cells to 20% (v/v) glycerol in broth, then transferred to cryotubes (Carpenter and Robertson, 1983).

E. Storage in Soil

The storage of *Frankia* in dried, sterilized soil is routinely performed in several laboratories. This preservation method has not been employed in this laboratory, in part due to the ease and reliability of storage cultures in broth culture. Carpenter and Robertson (1983) caution researchers with regard to storage in soil based on a recent report by Pridham and Lyons (1980), who did not recommend long-term storage of actinomycetes in soil.

F. Other Methods

Storage of *Frankia* in frozen, liquid culture has been used with several isolates in this laboratory with a great deal of success. At this time, long-term viability has not been examined.

Porcelain bead storage has also been investigated in this laboratory. *Frankia* are dried onto porcelain beads and placed in vials containing silica gel. Single beads can be removed asceptically and used to inoculate broth cultures. This procedure has been used extensively in the maintenance of rhizobia. *Rhizobium* retains viability for at least 5 years before transfer to new beads is required.

ACKNOWLEDGMENTS

The author wishes to thank the National Science Foundation (Grants PCM 8213855 and PCM 8313855); the U.S. Agency for International Development, Science Advisors Office, 4.396; and NPI for supporting this work. Additional appreciation is extended to D. B. Steele and S. K. Upham for critical reading of the manuscript and to B. Gandy for typing the manuscript.

REFERENCES

Akkermans, A. D. L., and Houwers, A. (1979). In *Symbiotic Nitrogen Fixation in the Management of Temperate Forests*, J. C. Gordon, C. T. Wheeler, and D. A. Perry, eds. Oregon State University Press, Corvallis, Oreg., p. 23.

Akkermans, A. D. L., and Roelofsen, W. (1980). In *Nitrogen Fixation,* W. D. P. Stewart and J. R. Gallon, eds. Academic Press, Inc., New York, p. 279.

Akkermans, A. D. L., and van Dijk, C. (1981). In *Nitrogen Fixation,* Vol. 1, *Ecology*, W. J. Broughton, ed. Clarendon Press, Oxford, p. 57.

Akkermans, A. D. L., Roelofsen, W., Blom, J., Huss-Danell, K., and Hartink, R. (1983). *Can. J. Bot. 61*: 2793.

Baker, D., and O'Keefe, D. (1984). *Plant Soil 78*: 23.

Baker, D., and Torrey, J. G. (1979). In *Symbiotic Nitrogen Fixation in the Management of Temperate Forests*, J. C. Gordon, C. T. Wheeler, and D. A. Perry, eds. Oregon State University Press, Corvallis, Oreg., p. 38.

Baker, D., Torrey, J. G., and Kidd, G. H. (1979). *Nature 281*: 76.
Baker, D., Lechevalier, M., and Dillon, J. (1981a). In *Current Perspectives in Nitrogen Fixation*, A. Bigson and W. Newton, eds. Australian Academy of Science, Canberra, p. 479.
Baker, D., Pengelly, W., and Torrey, J. G. (1981b). *Int. J. Syst. Bacteriol. 31*: 148.
Becker, B., Lechevalier, M., Gordon, R., and Lechevalier, H. (1964). *Appl. Microbiol. 12*: 421.
Becker, B., Lechevalier, M., and Lechevalier, H. (1965). *Appl. Microbiol. 13*: 236.
Becking, J. H. (1977a). In *A Treatise on Dinitrogen Fixation*, Sec. 3, *Biology*, R. W. F. Hardy and W. F. Silver, eds. John Wiley & Sons, Inc., New York, p. 185.
Becking, J. H. (1977b). In *Recent Developments in Nitrogen Fixation*, W. Newton, J. R. Postgate, and C. Rodriguez-Barrueco, eds. Academic Press, Inc., New York, p. 551.
Becking, J. H. (1981). In *The Prokaryotes*, M. Starr, H. Stolp, H. G. Truper, A. Balows, and H. G. Schlegel, eds. Springer-Verlag, New York, p. 1991.
Benson, D. (1982). *Appl. Environ. Microbiol. 44*: 461.
Benson, D. R., and Hanna, D. (1983). *Can. J. Bot. 61*: 2919.
Benson, D. R., Buchholz, S. E., and Hanna, D. G. (1984). *Appl. Environ. Microbiol. 47*: 489.
Berry, A., and Torrey, J. G. (1979). In *Symbiotic Nitrogen Fixation in the Management of Temperate Forests*, J. C. Gordon, C. T. Wheeler, and D. A. Perry, eds. Oregon State University Press, Corvallis, Oreg., p. 69.
BOSTID (1984). *Casuarinas: Nitrogen-Fixing Trees for Adverse Sites*. National Academy Press, Washington, D.C.
Bradford, M. M. (1976). *Anal. Biochem. 72*: 248.
Burggraaf, A., and Shipton, W. .A (1982). *Plant Soil 69*: 135.
Burggraaf, A., and Shipton, W. .A (1983). *Can. J. Bot. 61*: 2774.
Burggraaf, A., Quispel, A., Tak, T., and Valstar, J. (1981). *Plant Soil 61*: 157.
Callaham, D., Del Tredici, P., and Torrey, J. G. (1978). *Science 199*: 899.
Carpenter, C. V., and Robertson, L. R. (1983). In *Biological Nitrogen Fixation in Forest Ecosystems: Foundation and Applications*, J. C. Gordon and C. T. Wheeler, eds., Martinus Nijhoff/Dr. W. Junk Publishers, Dordrecht, The Netherlands, p. 89.
Diem, H. G., and Dommergues, Y. (1983). *Can. J. Bot. 61*: 2822.
Diem, H. G., Gauthier, D., and Dommergues, Y. R. (1983). *Can. J. Bot. 61*: 2822.
Ellinghausen, H. C., and McCullough, W. G. (1965a). *Amer. J. Vet. Res. 26*: 39.

Ellinghausen, H. C., and McCullough, W. G. (1965b). *Amer. J. Vet. Res. 26*: 45.

Ferguson, R. B., and Basile, J. V. (1967). *J. Range Manage. 20*: 380.

Fontaine, M. S., Lancelle, S., and Torrey, J. G. (1984). *J. Bacteriol. 160*: 921.

Gibson, A. H., Dreyfus, B., and Dommergues, Y. (1982). In *Microbiology of Tropical Soils*, Y. R. Dommergues and H. G. Diem, eds., Martinus Nijhoff/Dr. W. Junk Publishers, Dordrecht, The Netherlands, p. 37.

Gordon, J. G. (1975). *Iowa State J. Res. 49*: 267.

Gordon, J. G., and Dawson, J. O. (1979). *Bot. Gaz. 140* (Suppl.): 588.

Hafeez, F., Akkermans, A. D. L., and Chaudnary, A. H. (1984). *Plant Soil 78*: 45.

Hoagland, D. R., Arnon, D. I. (1950). *CA. Agric. Exp. Stat. 347*: 1.

Horriere, F. (1984). *Plant Soil 78*: 7.

Horriere, F., Lechevalier, M., and Lechevalier, H. (1983). *Can. J. Bot. 61*: 2843.

Lalonde, M., and Calvert, H. E. (1979). In *Symbiotic Nitrogen Fixation in the Management of Temperate Forests*, J. C. Gordon, C. T. Wheeler, and D. A. Perry, eds. Oregon State University Press, Corvallis, Oreg., p. 95.

Lalonde, M., Knowles, R., and Fortin, J. A. (1975). *Can. J. Microbiol. 21*: 1901.

Lalonde, M., Calvert, H. E., and Pine, S. (1981). In *Current Perspectives in Nitrogen Fixation*, A. Gibson and W. Newton, eds. Australian Academy of Science, Canberra, p. 296.

Lechevalier, M. (1984). *Plant Soil 78*: 1.

Lechevalier, M., and Fekete, E. (1971). In *Workshop Subcommittee on Actinomycetes*. American Society for Microbiology, Washington, D.C.

Lechevalier, M., and Lechevalier, H. (1979). In *Symbiotic Nitrogen Fixation in the Management of Temperate Forests*, J. C. Gordon, C. T. Wheeler, and D. A. Perry, eds. Oregon State University Press, Corvallia, Oreg., p. 111.

Lechevalier, M., and Ruan, J. S. (1984). *Plant Soil 78*: 15.

Lechevalier, H., Lechevalier, M., and Becker, B. (1966). *Int. J. Syst. Bacteriol. 16*: 151.

Lechevalier, M., Baker, D., and Horriere, F. (1983). *Can. J. Bot. 61*: 2826.

Lyon, J. L., and Mueggler, W. F. (1968). *J. Wildl. Manage. 33*: 538.

Mort, A., Normand, P., and Lalonde, M. (1982). In *Proc. Can. Soc. Microbiol.* (Abstract).

Murry, M., Fontaine, M. S., and Torrey, J. C. (1984a). *Plant Soil 78*: 61.

Murry, M., Fontaine, M. S., and Tjepkema, J. (1984b). *Arch. Microbiol. 139*: 162.

Pridham, T. G., and Lyons, A. J. (1980). In *Actinomycete Taxonomy*, A. Dietz and D. W. Thayer, eds. Society for Industrial Microbiology Press, Arlington, Va., p. 152.

Quispel, A., and Burggraaf, A. (1981). In *Current Perspectives in Nitrogen Fixation*, A. Gibson and W. Newton, eds. Australian Academy of Science, Canberra, p. 229.

Quispel, A., and Tak, T. (1978). *New Phytol. 81*: 587.

Resch, H. (1979). In *Symbiotic Nitrogen Fixation in the Management of Temperate Forests*, J. C. Gordon, C. T. Wheeler, and D. A. Perry, eds. Oregon State University Press, Corvallis, Oreg., p. 444.

Righetti, T. L., and Munns, D. (1982). *Plant Soil 65*: 383.

Rottink, B., Strand, R. F., and Bentley, W. R. (1979). In *Symbiotic Nitrogen Fixation in the Management of Temperate Forests*, J. C. Gordon, C. T. Wheeler, and D. A. Perry, eds. Oregon State University Press, Corvallis, Oreg., p. 14.

Shipton, W. A., and Burggraaf, A. (1982). *Plant Soil 69*: 149.

Shipton, W. A., and Burggraaf, A. (1983). *Can. J. Bot. 61*: 2783.

Shirling, E. B., and Gottlieb, D. (1966). *Int. J. Syst. Bacteriol. 16*: 313.

Silvester, W. B. (1977). In *A Treatise on Dinitrogen Fixation*, R. W. F. Hardy and A. Gibson, eds. John Wiley & Sons, Inc., New York, p. 141.

Smith, N. J. (1978). In *Utilization and Management of Alder*, D. G. Briggs, D. S. deBell, and W. A. Aktinson, eds. Pacific Northwest Forest and Range Experiment Station, Forest Service, U.S. Department of Agriculture, Portland, Oreg., p. 139.

Smith, J. H. G., and DeBell, D. S. (1973). *For. Chron. 49*: 31.

Stowers, M. D., Kulkarni, R. K., and Steele, D. B. (1986). *Arch. Microbiol. 143*: 319.

Tisa, L., McBride, M., and Ensign, J. C. (1983). *Can. J. Bot. 61*: 2768.

Tjepkema, J. D., Ormerod, W., and Torrey, J. C. (1980). *Nature 287*: 633.

Torrey, J. G. (1978). *Bioscience 28*: 586.

Torrey, J. G. (1982). In *Proceedings of the International Workshop on Biological Nitrogen Fixation Technology for Tropical Agriculture*, P. H. Graham and S. Harris, eds. Centro Internasional de Agricultura Tropical, Cali, Columbia, p. 427.

Uphof, J. C. (1968). In *Dictionary of Economic Plants*. Cramer, London.

van Dijk, C. (1978). *New Phytol. 81*: 601.

Worthington, N. P., Ruth, R. H., and Matson, E. E. (1962). U.S. Department of Agriculture Miscellaneous Publication 881.

Zavitkovski, J., Hansen, E. A., and McNeel, H. A. (1979). In *Symbiotic Nitrogen Fixation in the Management of Temperate Forests*, J. C. Gordon, C. T. Wheeler, and D. A. Perry, eds. Oregon State University Press, Corvallis, Oreg., p. 388.

3

Collection, Maintenance, and Cultivation of Azolla

THOMAS A. LUMPKIN *Washington State University, Pullman, Washington*

I. INTRODUCTION

The aquatic fern azolla has a high relative growth rate (RGR) and a symbiotic relationship with a nitrogen-fixing endophytic cyanobacteria called *Anabaena azollae*. This symbiotic relationship gives azolla a competitive advantage over other floating hydrophytes in environments such as rice fields which are relatively low in available nitrogen. The symbiosis is of interest to agriculturalists since incorporation and decomposition of an azolla crop in a rice field soil can result in an increase of available nitrogen to a companion or succeeding rice crop.

Azolla can accumulate more than 10 kg N/ha per day (25 tons/ha of azolla × 6% dry weight × 4% N × RGR 0.17 ton/ton per day), and thus has the potential of supplying the entire nitrogen requirement for a high-yielding rice crop within a few weeks. Under proper management and environmental conditions, this rate can be sustained because azolla reproduces by fragmentation and is not restrained by soil as are terrestrial plants. Azolla can also be used to suppress paddy weeds and is an excellent food for fish, pigs, and ducks (Lumpkin and Plucknett, 1982).

In the 1960s and 1970s, azolla was regularly cultivated on a large scale only in southeastern China (Beijing Agricultural University, 1979) and northern Vietnam (Vo and Tran, 1970). In these areas, azolla was used primarily as a green manure for rice and was viewed as a crop of rather low economic value. Only short periods, before or with the spring rice crop were allocated for its cultivation. Azolla would make use of short cool periods in the cropping sequence because some varieties had both (a) the ability to sustain a relatively high growth rate under cool temperatures (10 to 20°C) and (b) a potential for doubling their biomass in as little as 2 days (0.35 g/g per day) under more optimal conditions (Peters et al., 1980; Watanabe and Berja, 1983; Lumpkin and Bartholomew, 1986).

Tolerance of cool weather was the key to the success of azolla in parts of Vietnam and China. A high growth rate in cool seasons made azolla an economic competitor against alternative crops grown for food and a good biological competitor against other hydrophytes and algae. In addition, cultivation during cool seasons allowed azolla to avoid devastation by its thermophyllic insect pests.

The cool season (winter-spring) cultivation of *Azolla pinnata* var. *imbricata* in Vietnam's Red River Delta and irrigation system probably reached an area exceeding 500,000 ha during the late 1960s and has continued to be an important crop through the 1970s and 1980s because of a lack of chemical nitrogen fertilizers in Vietnam. In China, cool season cultivation of *A. pinnata* var. *imbricata* and *Azolla filiculoides* peaked in the late 1970s at over 1,000,000 ha and has since declined rapidly in the face of improved supplies of nitrogen fertilizer, low economic return, and the elimination of commune and research support systems through decentralization.

Outside China and Vietnam, azolla has met with scant success, primarily because of a lack of knowledge and problems with insects and diseases. However, the general level of interest and understanding about the agricultural uses of azolla has been enhanced by descriptive reviews and handbooks designed for scientists and researchers in developing countries (Peters, 1977; FAO, 1978; IRRI, 1979; Watanabe et al., 1981; Lumpkin and Plucknett, 1980, 1982; National Azolla Action Program, 1982; Van Hove et al., 1983; Lumpkin, 1984; Watanabe, 1984). These publications cover topics such as management practices, environmental requirements, identification of species, sexual and vegetative reproduction, alternative uses, and insect and disease pests.

For azolla to become an economically viable crop in rice-producing countries, numerous efforts and breakthroughs in azolla research must be made, especially in collection of germplasm with desirable traits, procedures to reduce or eliminate maintenance requirements, and a reduction in the labor and water requirements for cultivation. The intent of this chapter is to provide a short review of the current level of understanding about azolla and to comment on aspects constraining wider use.

II. COLLECTION

A. Identification

The name *Azolla* is derived from the Greek *azo*, to dry, and *ollyo*, to kill (Jaeger, 1978) and is a genus of heterosporous leptosporangiate ferns from aquatic and semiaquatic habitats. These habitats include ponds, ditches, canals, and paddy fields from temperate to tropical regions. Azolla is also a frequent pioneer in aquatic areas disturbed by man or animals. Countless places where agricultural runoff or urban effluent accumulate are seasonally covered by a thick mat of azolla, usually in conjunction or succession with other free-floating aquatics, such as *Lemna, Pistia, Salvinia, Spirodela, Ricciocarpus*, and *Riccia.*

Taxonomy

The genus *Azolla* was established by Lamarck in 1783 and is grouped with the genus *Salvinia* in the order Salvinales, but is separated into the monotypic family Azollaceae (Reed, 1954). Part of the reason for grouping *Azolla* with *Salvinia* is that both genera produce two distinct types of spores (heterospory), in contrast to most other ferns, which produce only one type of spore. In addition, the spores of *Azolla* and *Salvinia* are borne on special stalks (columnella) and are contained in special capsules called sporocarps.

The seven extant species of *Azolla* are shown in Figure 1. The conspectus of classification (Table 1) is consistent with recent reports (Moore, 1969; Fowler and Stennett-Willson, 1978; Lumpkin and Plucknett, 1980), except that *A. rubra* R. Brown has been retained in preference to the *A. filiculoides* var. *rubra* of Strasburger (1873) in

Azolla caroliniana

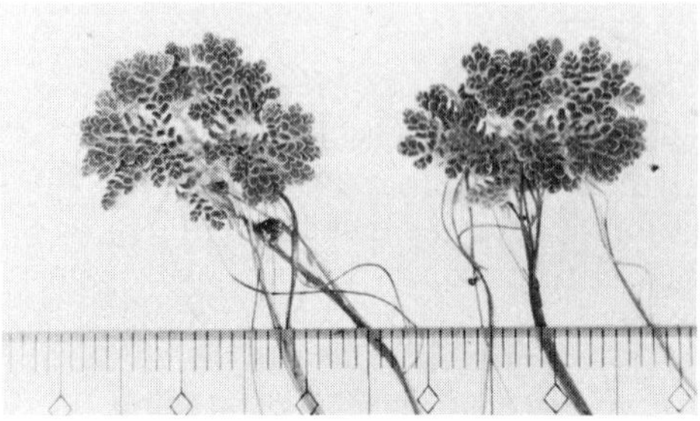

Azolla nilotica

Azolla filiculoides

Azolla pinnata var. *pinnata*

Azolla mexicana

Azolla pinnata var. *imbricata*

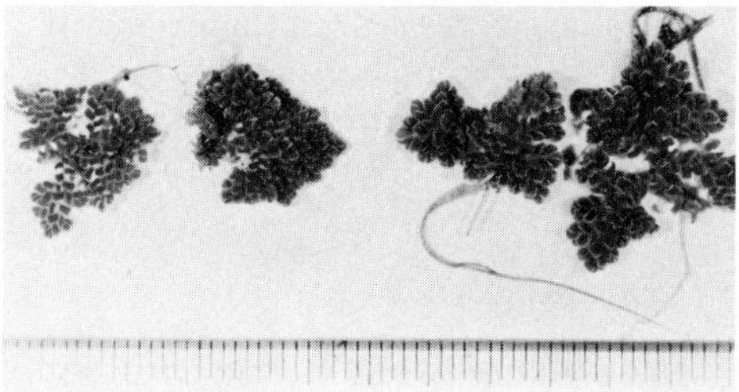

Azolla microphylla

Azolla rubra

Table 1 Conspectus of Classification

Division	Pteridophyta	
Class	Filicopsida	
Order	Salviniales	
Family	Azollaceae	
Genus	*Azolla*	
Sections	*Azolla*	*Rhizosperma*
Species	*A. caroliniana*	*A. nilotica*
	A. filiculoides	*A. pinnata*
	A. mexicana	
	A. microphylla	
	A. rubra	
Varieties		*A. pinnata* var. *imbricata*
		A. pinnata var. *pinnata*

light of the reports by Fowler and Stennett-Willson (1978), and Lumpkin and Plucknett (1982), and SEM studies of megaspores (Perkins et al., 1985). Unfortunately, several of the type specimens for species in the section *Azolla* do not represent the current description for those species (K. Fowler, personal communication). Several of the type specimens may be of the same species.

Morphology

The diploid spore-producing generation of azolla is called the sporophyte and consists of a horizontal to vertical main rhizome. Leaves on the rhizome are bilobed with alternate pinnation. Rhizomes also bear individual roots or root bundles at branch points. The main rhizome, when mature, may range in size form 0.5 to 7 cm in diameter with individual roots 1 to 5 cm long, except *A. nilotica* (Demalsy, 1953; Lumpkin, 1981), which can produce a trailing rhizome up to 40 cm or more long with root bundles up to 15 cm or more.

The main rhizome is usually achlorophyllous and has alternating branches with several orders of lateral branches. Species that can mature into a nearly vertical morphology when crowded, such as *A. filiculoides* and *A. nilotica*, have a more highly developed vascular

Figure 1 The seven species of the genus *Azolla*, including two varieties of *A. pinnata*. The plants as shown are nearly fully developed in size and are approximately life size, except *A. nilotica*, which should be 3.3 times larger for comparison. Note the millimeter scale below each plant for reference.

system, appearing like a amphiphloic siphonostele in mature plants. Well-developed plants of *A. nilotica* have lateral branches which often lack their own roots and are thus dependent on the root system of the main rhizome. Rhizomes lacks stomata, but occasionally have chloroplasts near the shoot tips. In section *Rhizosperma*, the rhizome has trichomes either on the ventral surface or on both surfaces, while plants in section *Azolla* lack trichomes on their rhizomes.

Roots are initiated at branch points along the rhizome and bear root hairs up to 1 cm long, which emerge from under a root cap. Root initials eventually give rise to a single-layered sheath over a two-layered cap, followed by a epidermal layer and two layers of cortical cells containing chloroplasts (Kawamatu, 1963). A rich microflora is usually found in association with the roots and have been partially described by Roger and Reynaud (1979).

The leaf consists of two lobes: a thick aerial dorsal lobe and a thin ventral lobe occasionally of a slightly larger size (Figure 2). The dorsal lobe is chlorophyllous, except in the transparent margin, and contains a colony of *Anabaena* (Figures 3 and 4) within a basal cavity connected to the atmosphere by a pore on the adaxial side. The surface of the dorsal lobe has an epidermis covered with vertical rows of single-celled stomata and trichomes of one or more cells. Dorsal lobe tissue consists of a single layer of elongate palisade parenchyma with prominent intercellular spaces, one or more layers of sparsely branched spongy mesophyll, and a single concentric vascular bundle.

The thin ventral lobe is nearly achlorophyllous with few stomata and trichomes and several chambers. A single vascular bundle consists of several tracheary and sieve elements, but lacks parenchyma cells. The ventral lobe probably provides bouyancy as a result of its convex surface touching the water. It may also function in absorption, since azolla plants are known to survive without roots.

All dorsal lobes produce a cavity for housing the *Anabaena* symbiont. As the leaf primordium differentiates at the growing point, a slight depression is formed near the base on the adaxial side of the dorsal lobe. A portion of the *Anabaena* colony generating at the shoot apex above the dorsal lobe primordia is scooped into the enlarging depression by a glove-shaped transfer hair (Calvert and Peters, 1981). This colony becomes entrapped in the cavity by a ring of meristematic epidermal cells originating around the circumference of the depression. The cells grow inward to cover the depression, leaving a pore over the center where they meet. The inside of the cavity is lined by a porous envelope which develops only if the symbiont is present (Peters et al., 1978).

Life Cycle

Although azolla normally reproduces vegetatively by fragmentation via an abscission layer that forms at the base of each branch, some species

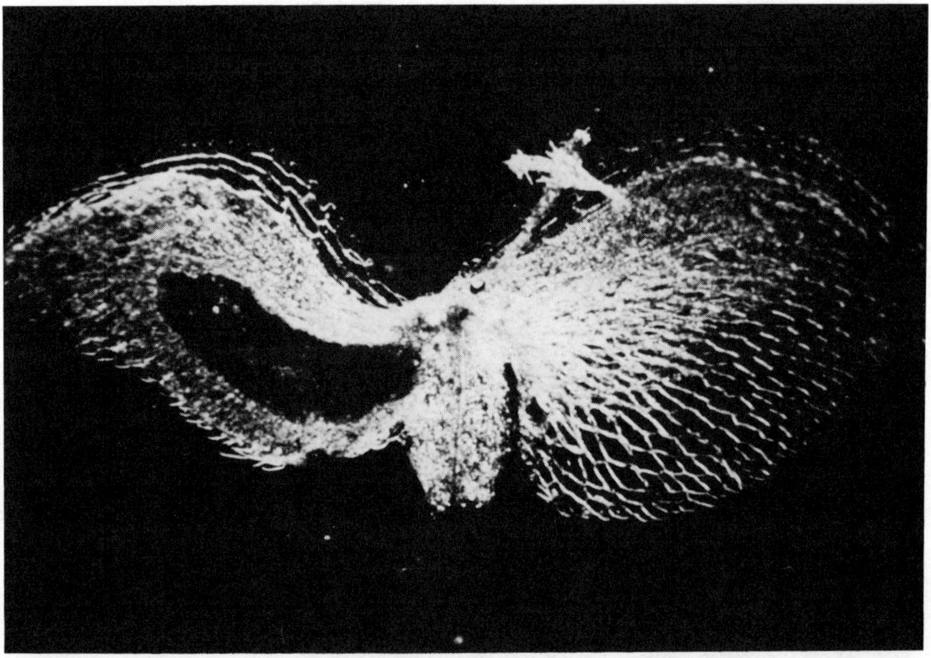

Figure 2 Complete leaf of *A. nilotica*, with the two lobes flattened for illustration. The smaller dorsal lobe contains the dark-colored cavity where the *Anabaena azollae* symbiont resides. The larger ventral lobe is normally boat shaped and often touches the surface of the water when the plants are not crowded.

occasionally express a gametophytic cycle (Lumpkin and Plucknett, 1982). The gametophytic cycle is usually absent in most situations for most species. Initiation of the cycle seems to be stimulated by a combination of environmental factors (Ashton, 1977), and is most often associated with the beginning or end of a period of environmental stress. Environmental factors also may affect the ratio of microsporocarps to megasporocarps.

Varieties of certain species (e.g., *A. filiculoides* and *A. nilotica*) become fertile only after attaining a mature morphology. Mature morphology is a precondition but not necessarily concurrent with fertility in these species. Like fertility, initiation of the mature morphology is also environmentally dependent. The failure to initiate mature morphology has a positive agronomic value because multilayered immature fronds have a higher nitrogen content and lower lignin content, and thus decompose more easily than do mature fronds.

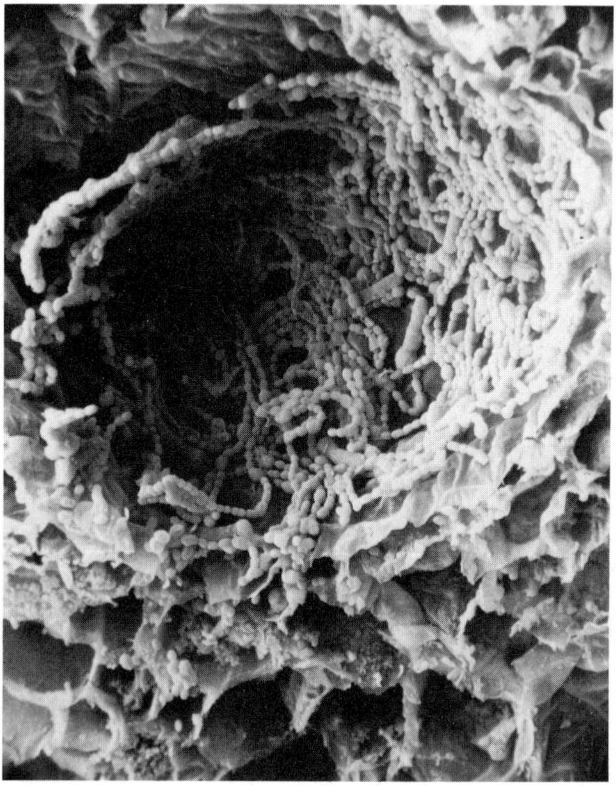

Figure 3 SEM photo (×300) showing the upper end of the dorsal lobe cavity of an *A. pinnata* leaf. The filamentous cyanobacteria and the larger fingerlike transfer hairs are clearly visible in the cavity.

When the gametophytic cycle is initiated, sporocarps are formed in pairs (tetrads on *A. nilotica*) after a few divisions by a fertile ventral lobe initial of the first leaf of a branch (Bonnet, 1957). The dorsal lobe of the fertile leaf retains its normal shape and function. The sporangial initials develop in a leptosporangiate fashion (Pfeiffer, 1907) and eventually give rise to two megasporangia, each containing 32 megaspores initials produced by meiotic division. If a megaspore is to develop within the megasporangium, all but one of the 32 megaspores initials abort. If all of the megaspore initials abort, microsporangia initials arise from basal outgrowths on the stalk of the megasporangium. The resulting pair (or tetrad) of sporocarps may be all of one sex or any mix (Figure 5).

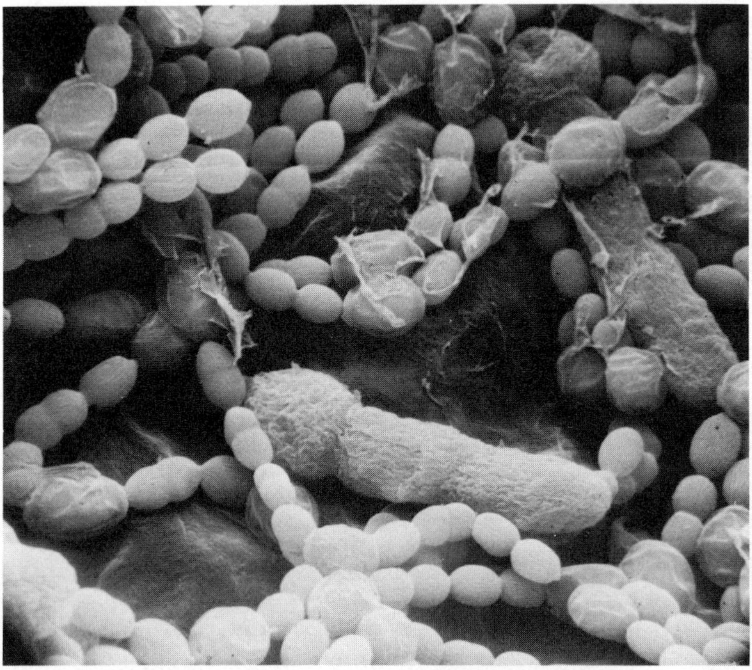

Figure 4 SEM photo (×1500) inside the leaf cavity of *A. pinnata*. *Anabaena azollae* filaments and two azolla transfer hairs are visible. The *Anabaena* is composed of numerous small vegetative cells, some in the process of division, and larger heterocyst cells, the site of nitrogen fixation.

Sporocarps mature on the plant in a week or more depending on enviornmental conditions. Megasporocarps are 0.4 to 0.6 mm in diameter and give the appearance of a dark cone (indusium) resting atop a sphere. The indusium overlies the alagal colony and the characteristic floats: three floats in section *Azolla* and nine floats in section *Rhizosperma*. After maturation, the megasporocarp becomes dehiscent and the megasporangial wall disintegrates, exposing the sporoderm (perispore), which is usually covered with rubberlike or resinlike hairs (filosum) (Brederoo et al., 1976). These hairs seem to function as entanglements for appendages (glochidia) attached to packets (massulae) of microspores, except for *A. nilotica*, which does not produce glochidia.

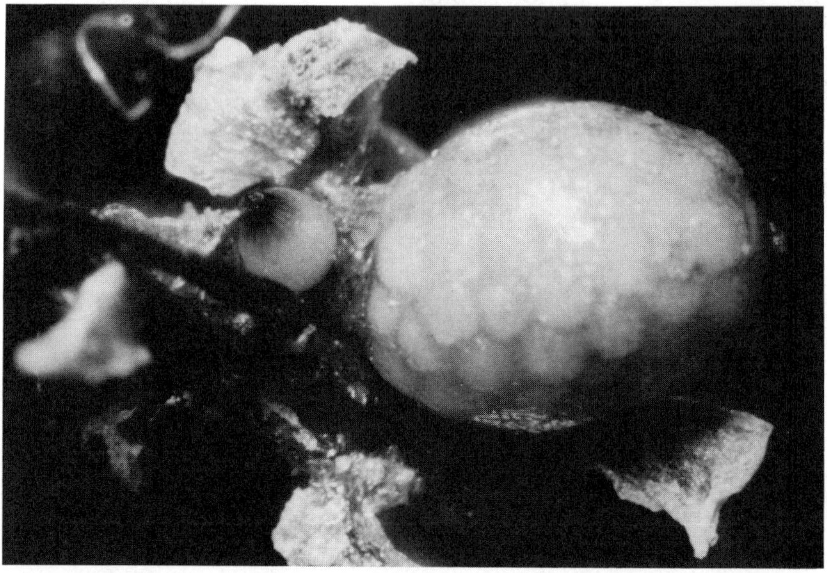

Figure 5 Pair of sporocarps (×30) on a rhizome of *Azolla filiculoides*. The smaller female megasporocarp is on the left and is capped by a dark red indusium. The large microsporocarp on the right contains numerous microsporangia which are visible under the epidermis, and these are further divided into massulae which contain the microspores.

Mature microsporocarps are globular in shape and are easily visible, measuring 2 mm tall by 1.5 to 1.75 mm in diameter. On a plant, a microsporocarp may completely conceal a megasporocarp when occurring as a mixed pair. A microsporocarp may contain up to 130 stalked microsporangia, which appear like a cluster of balloons tied to strings. Each microsporangium contains 32 or 64 microspores divided into aggregates of 3 to 10 alveolar massulae. Massulae may be bald (*A. nilotica*), partially covered with simple branched or unbranched glochidia (*A. pinnata*), or totally covered with arrowlike glochidia (section *Azolla*). After dehiscence, the wall of the microsporocarp disintegrates, releasing the microsporangia, which in turn disintegrate, releasing the massulae.

Massulae are swept about by underwater currents or along the surface by the wind. Massulae equipped with glochidia may become entangled or ensnared in hairs covering the sporoderm of respective megaspores (section *Azolla* and *A. pinnata*). Massulae of *A. nilotica* may become hooked on prostrate spines which cover the sporoderm of its megaspores. Microspores germinate and release flagellated anterozoids which escape through the gelatinous massula to fertilize an archegonium produced by the megaspore gametophyte. The tipping of the indusium is the first visible sign of fertilization and is caused by the emerging alga-free cotyledon (Figure 6). This can occur within 5 days or more after initial contact of mature gametophytes, depending on the water temperature. A portion of the generative *Anabaena* colony surviving under the indusium becomes entrapped in the shoot apex and succeeding true leaves. As the root appears and the first or second true leaf emerges through the cotyledon, the seedling floats to the surface, beginning sporophytic reproduction (Cheng et al., 1980; He and Lin, 1981; He et al., 1982; Lumpkin, 1985).

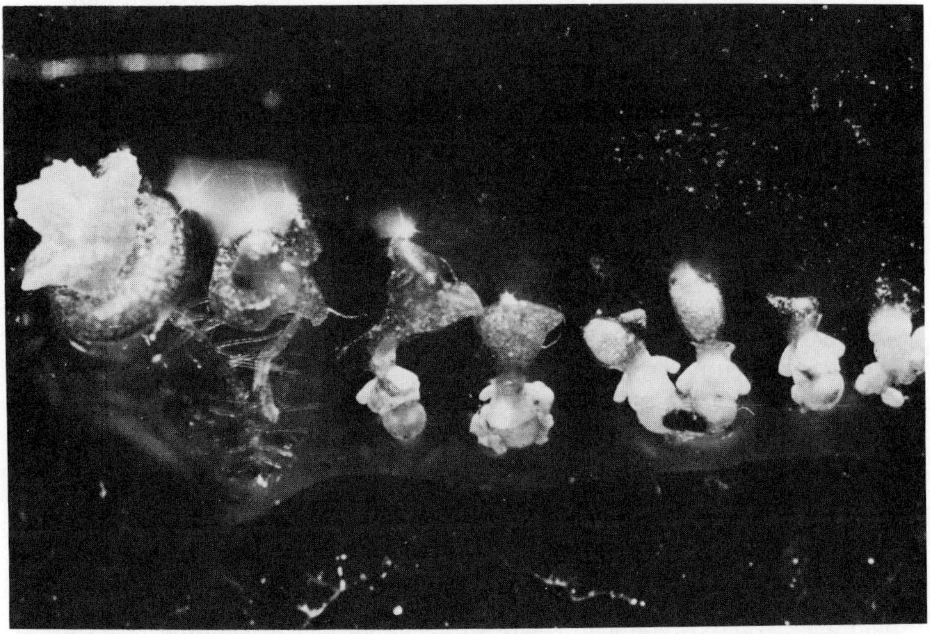

Figure 6 From right to left, stages in the development of azolla sporelings. Massulae containing microspores are attached to the base of the megaspore on the right. The first visible sign of fertilization and germination is the tipping of the indusium caused by emergence of the cotyledon through the columella, second from right. Later, to the left, the growing point emerges through the cotyledon and the root and root hairs develop.

B. Distribution

The distribution of *Azolla* species has been reviewed by Svenson (1944), Sculthorpe (1967), Moore (1969), Ott and Petrick-Ott (1973), and Lumpkin and Plucknett (1980). Sweet and Hills (1971) discussed the distribution of *A. pinnata* var. *pinnata* and *A. pinnata* var. *imbricata*.

The native distribution of *Azolla* species has been confirmed by the author through collection or by observation of herbarium specimens (Figure 7). The species *A. caroliniana* is distributed in the eastern half of the United States, through Central American, and in South America east of the Andes, but is often confused with immature fronds of *A. filiculoides*. *A. filiculoides* is found in the Rocky Mountain and Pacific coast states of the western United States and Canada, through Central America and most of South America, and may also be native to Japan. *A. mexicana* is found from the west coast of the United States east to the Mississippi River and into Mexico and Central America. Its occurrence has been reported but not confirmed in the northern half of South America. Very few confirmed reports of *A. microphylla* have been made due to problems with identification. Only the population found in the Galapagos Islands (Morton and Wiggins, 1971) has been confirmed, but accessions collected in Paraguay have similar morphology. *A. nilotica* occurs only in Africa as far south as Mozambique, north through the upper reaches of the Nile River drainage basin to Kosti in the Sudan, and from Kenya south along the east coast and west through the drainage basin of the Congo River to the southern west coast. *A. pinnata* is found in East and South Asia through equatorial Asia to northern Australia, and equatorial and southern Africa including Madagascar. *A. rubra* (*A. japonica*) is found only in higher latitudes of the western Pacific, exclusively in Japan, Korea, Australia, and New Zealand.

Azolla species have been dispersed throughout the world by a variety of mechanisms, of which man has become the most important. *A. filiculoides* and *A. caroliniana* were introduced into western Europe in the nineteenth century as ornamentals and have since spread eastward as far as the eastern USSR. *A. filiculoides* has also been introduced into South Africa (Ashton and Walmsley, 1984) and China, and *A. pinnata* into New Zealand. Accessions of all species have recently been introduced to research stations on all continents and are undergoing field evaluations which may result in establishment of adapted species.

C. Preparation and Shipment of Germplasm

Traditional methods and safeguards for preparation and shipment of seeds, plants, or cutting are not directly applicable to azolla since it is an aquatic plant and rarely produces spores. The only requirement for the import of azolla into the United States under a U.S. Department

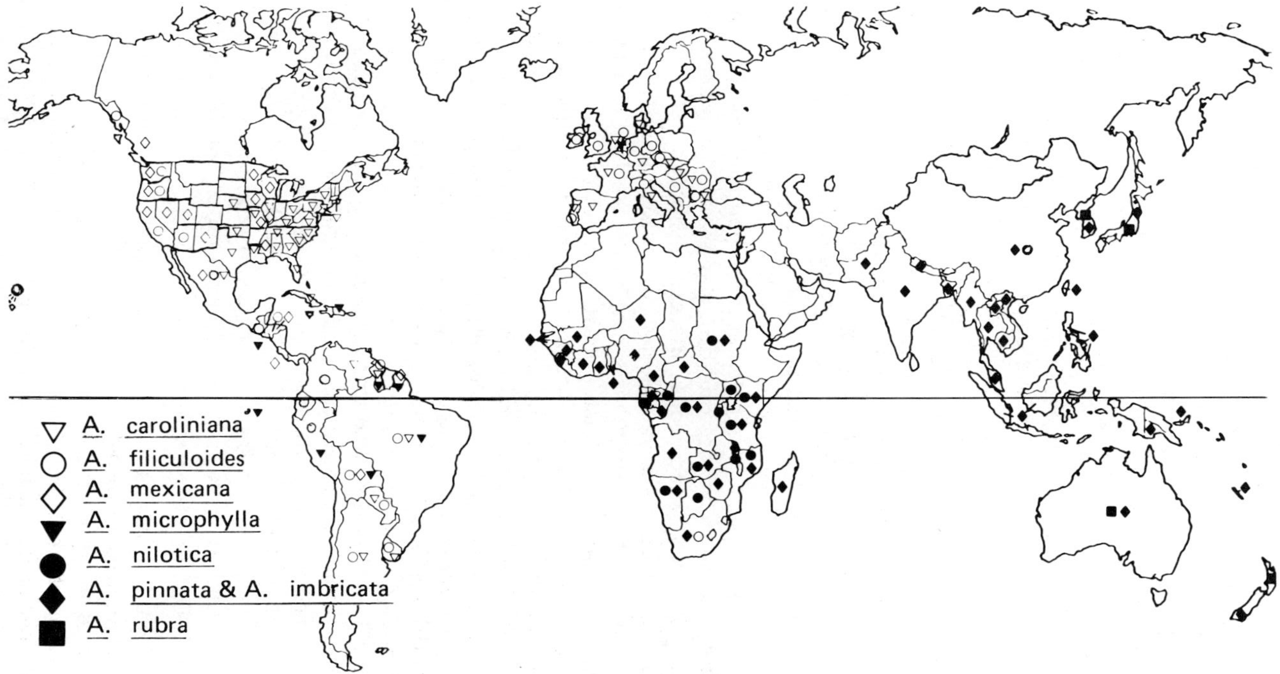

Figure 7 Global distribution of *Azolla* species. Identification and distribution of species in Central and South America is somewhat tentative since they are largely based on nonfertile herbarium specimens. All species in Europe are introduced.

of Agriculture (USDA) permit is that the plants be free of insects and soil. Methods and safeguards discussed by azolla researchers for shipment are fairly primitive, since the intensive study of azolla in the West is a fairly recent phenomenon.

In the transport of azolla, there is always the possibility that other organisms will be brought along as uninvited guests. Certainly many organisms have been carried along with samples of azolla which have been transported and introduced to new areas of the world, especially rice-growing areas. However, because of a lack of investigation, there has never been a report of any damage being done by the introduction of organisms carried with azolla, but the likelihood and possibility of damage are real, especially for pests of rice.

Ideally, only aseptic specimens of azolla should be transported between different environments, but this is probably unrealistic considering the difficulty of producing aseptic plants and the lax nature of most plant inspection and quarantine measures. Responsible preparation before shipment of azolla should include the application of insecticides and fungicides at both the point of departure and arrival, and culture in isolation at the point of arrival. Insecticides such as carbofuran and lindane, and fungicides such as pentachloronitrobenzene (PCNB) and benomyl have been used (Lumpkin and Plucknett, 1982).

In addition to weeds, weed seeds, insects, nematodes, and microbes, the transport of snails and leeches with azolla germplasm is a real concern. Some of these organisms may be destructive or may host-dangerous pathogens such as schistosomes. The culture of azolla on nutrient solution in quarantine is an important step that should be taken whenever azolla is collected outside the laboratory, to reduce the possibility of an unwanted introduction. Observation, pesticide application, and re-isolation of individual plants during a quarantine period can help to prevent the introduction of any unwanted organisms.

Small quantities of azolla can be prepared for shipment in various ways, but all methods must take into consideration the environments to which the azolla will be exposed during shipment. Regardless of the duration of shipment, azolla must not be allowed to desiccate and must not be exposed to temperatures below 0°C or above 35°C. Azolla should not be transported in water since turbulence resulting from transport will cause excessive fragmentation and often death.

Some of the simplest ways to package and maintain small quantities of azolla for shipment include the following:

1. *Bulked azolla.* Azolla can be placed directly in a plastic bag or other container without water or other amendments. This method seems to be the most effective. For periods of transport longer than 1 day, an ice chest or portable refrigerator should be used to maintain a temperature of 5 to 10°C. Under

this cool condition, it may be possible to keep the plants alive up to a month or more, but with reduced viability over time.

2. *Azolla in gel.* Surface-sterilized azolla can be placed on a liquid medium which is then stabilized with a gelling agent. This method was tried for preservation of azolla germplasm (Bai et al., 1981) but has proved troublesome. However, transport on gel can be effective, especially when ice or refrigeration is not available.
3. *Spores.* If fertile azolla is found, whole plants bearing sporocarps can be air dried and the dried material can be shipped. After arrival, the dried material can be placed in pure water in indirect sunlight to induce germination of the spores, fertilization, and development of seedlings. A few days after the dried material is placed in the water, the container should be lightly agitated to cause the male and female spores to make contact. Within 2 to 3 weeks small seedlings should be present and can be moved to nutrient solution.

III. MAINTENANCE

At present the sexual cycle is insufficiently understood to make possible the use of spores as planting material or for germplasm maintenance. Since only vegetative material can be used for these purposes, a certain amount of azolla must be cultured throughout the year. This requirement significantly increases the cost of using azolla as a green manure, and the cost of maintaining a germplasm collection.

A. Germplasm Collections

Germplasm collections with accessions representing all known species are currently maintained at the International Rice Research Institute in the Philippines and with their collaborators in England, at Washington State University, the University of California at Davis, the Catholic University of Louvain-la-Neuve in Belgium, and at the Azolla Research Center in Fuzhou (Figure 8) and other places in China.

All of these institutions use culture on nitrient solution as their primary method for germplasm maintenance. Some institutes maintain surface-sterilized azolla on nutrient solution under aseptic conditions in flasks. Nickell (1961) produced *Anabaena*-free azolla by using antibiotics and maintained them under aseptic conditions on agar, but this method is currently unacceptable for germplasm collections since the *Anabaena* are an essential part of the collections. As mentioned above, Bai et al. (1981) tested this method for preservation of surface-

Figure 8 Azolla germplasm collection at the National Azolla Research Institute, Fuzhou, China. Accessions are maintained on nutrient solution culture as shown, and aseptically on nutrient solution in ventilated flasks.

sterilized germplasm, and although it was successful for short- and medium-term preservation, it proved more troublesome than culture on nutrient solution for long-term preservation.

Van Hove (personal communication) at the Catholic University of Louvain-la-Neuve in Belgium also uses nutrient solution maintenance, but has developed a partially automated continuous-circulation system for controlling solution level and pH, filtering recycled solution, and reducing biomass. Other systems have been suggested but have never been studied, including cryopreservation and culture under low light/ low temperature.

B. Nurseries for Cropping Systems

Perhaps the greatest limitation to the wider use of azolla as a green manure for rice is the need to maintain a certain quantity of azolla in nurseries during the off-seasons. This maintenance requirement might be avoided if azolla could be propagated from spores (Figure 9). Azolla is capable of producing spores, but no methods are known for inducing spore formation, large-scale harvesting, or for utilizing them

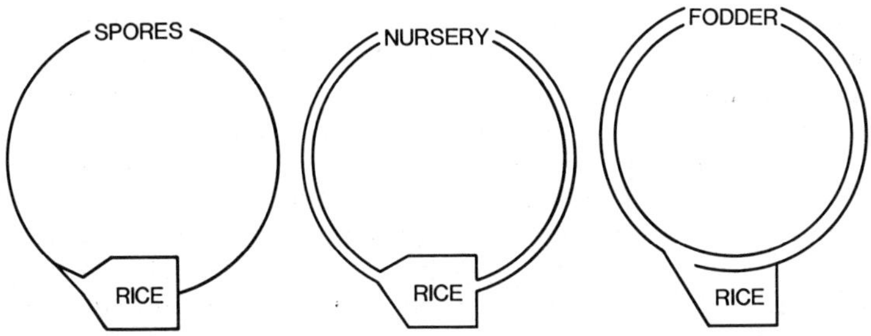

Figure 9 Three basic systems for maintaining azolla planting material during the off-season. The use of spores is currently not practical, but the other two systems can be combined in application. If the use of spores becomes practical, 2 to 4 weeks must be allowed for germination and the development of full-sized plants.

in cropping systems, although Chinese scientists have been studying these problems (Ke et al., 1981; Wang et al., 1981; Lumpkin, 1985). Methods for inducing spore formation must be developed because few of the accessions that are promising for cultivation in the tropics produce spores or produce them at the time when needed.

To maintain azolla during the off-season, special nurseries must usually be constructed. In China the construction of nurseries is integrated into a year-round process which divides the year into several phases designated as follows: overwintering, spring multiplication, mass cultivation in large fields, oversummering, and autumn multiplication. The Azolla must be protected from death or injury caused by extremes of cold or heat during overwintering and oversummering, and must also be multiplied in sufficient quantities in time for wide-scale monocropping in the field or for direct inoculation as an intercrop with the rice. Careful planning and management are required to satisfy these objectives. Maintenance nurseries must meet the specific needs of particular environments in regard to azolla, climate, available construction materials, and farming systems.

Overwintering Nurseries

At high latitudes or elevation it is often necessary to protect Azolla from cold or frost damage to ensure a sufficient supply for multiplication in the spring. In general, cold-tolerant varieties of azolla die when temperatures drop below -5°C for more than a few hours. Cold-tolerant varieties generally belong to the species *A. caroliniana, A. filiculoides, A. pinnata,* and *A. rubra.*

Where winters are severe, hot spring water or warm water from factories is useful for the overwintering of azolla (Ye and Wu, 1964).

Where warm water is not available, azolla can be grown in glass or plastic houses. Where winters are mild, azolla can be overwintered in the field or in small field structures. In areas where water does not freeze, ponds, lakes, canals, and large flooded rice fields are sometimes used for overwintering since the temperature in deep bodies of water fluctuates less than shallow water (Yuan, 1964).

The quantity of azolla kept for overwintering varies according to the area for cultivation, the season and duration of azolla field cultivation, the cropping system, and the particular method for overwintering. One hectare of spring-transplanted rice usually requires the overwintering of 400 to 600 kg of azolla, since little time is available for nursery multiplication. A crop of summer-transplanted rice requires only 7.5 to 15 kg of azolla for overwintering since the azolla can be multiplied during the spring for a longer period before large-scale field cultivation.

In most situations where overwintering is necessary some type of protective structure must be constructed. Azolla is moved into the protective structure when night temperatures begin to drop below 10°C. In these structures, azolla is maintained at a density of 0.75 to 1.0 kg/m^2 and water in the beds is kept at 6 to 7 cm deep. The beds are fertilized with superphosphate, ammonium sulfate, and straw ash.

In China, arch-shaped plastic greenhouses, lean-to plastic- or glass-roofed greenhouses, or lean-to plaster-roofed greenhouses were often used during winter. In some cold regions, the azolla is taken out of the water and stored under a thick mat of rice straw (Anwei Academy, 1977). Azolla was stored in a loosely piled stack 50 cm high. The pile was covered with 5 to 10 cm of straw ash and then routinely sprinkled with water to keep the azolla from drying out. A removable frame and thick insulating mat of straw were placed over the stack. During the first 10 days of storage the temperature inside the azolla stack was closely monitored to ensure that composting was not taking place. The temperature was maintained at around 5 to 8°C or lower, but not less than 0°C. Azolla was usually stored in these stacks during the coldest 50 to 60 days of winter. After this period 50 to 80% of the azolla was still alive. In late winter or early spring the azolla was moved into wet nursery beds for propagation. This was done before warm weather could cause the stored azolla to rot. Pits (Li, 1977) and plastic bags (Wu, 1978) were also used for out-of-winter storage.

Oversummering Nurseries

Oversummering techniques are used in locations where field temperatures are too high for azolla cultivation, or where azolla must be preserved through a hot season for later use (Lu et al., 1965; Feng, 1977). For example, when the azolla is cultivated as a green manure for the autumn rice crop, it is maintained in oversummering sites until it can be cultivated in the field.

Characteristics of a good oversummering site include good ventilation, direct sunlight, and cool flowing water. Ponds, canals, and the edges of placid rivers with low fertility and few fish can be used as maintenance sites. Floating and submerged weeds and algae must be removed. The water surface must be subdivided into small areas for management of azolla and to prevent winds from blowing azolla away or into a pile along the bank. Living fences can be used to subdivide the azolla and are prepared by tying large floating aquatic plants such as *Alternanthera, Eichhornia*, of *Pistia* together with a plastic rope. These floating fences are anchored by stone weights to the bottom or tied to stakes on the shore. Extra-long anchor ropes are used so that the floating fences can adjust to changes in the water level.

If azolla is oversummered in paddies, the water should be drained or cold water should be added when the paddy water temperature begins to exceed 35°C. The occurrence of insects must be closely monitored during summer because of their rapid reproductive rates at high temperatures. Azolla can also be oversummered between shading bands of rice or in the shade of tall herbaceous plants such as *Sesbania* that are planted along the edges of rivers, ditches, and ponds. *Alternanthera philoxeroides, Eichhornia crassipes*, and *Pistia stratiotes* are sometimes intercropped with azolla to provide shade.

IV. CULTIVATION

The cultivation of azolla is generally thought of as the production of biomass for use as a green manure, and in this section we discuss only the production of biomass. However, in essence, the purpose for cultivating azolla is to fix atmospheric nitrogen and make it available to rice and or other crops. Farmers usually feel that they are accomplishing this purpose by producing biomass, but the quantity of biomass does not indicate how much nitrogen has been fixed and how much of the azolla nitrogen has become available to the rice crop. In the cultivation of azolla, total nitrogen accumulation, portion derived from fixation, and decomposition rate must be considered in analyzing the value of biomass. If the azolla scavenges most of its nitrogen from its aquatic environment and decomposes very slowly, it may be of little value to a farmer. Nitrogen fixation by azolla and its decomposition in soil can best be studied through the use of ^{15}N (IAEA, 1985).

The decomposition rate of azolla in soil and the mineralization of its nitrogen have been studied somewhat in China. The decomposition of species such as *A. filiculoides* and *A. nilotica* which have more lignin in their mature stages is much slower than with species that can develop only overlapping horizontal layers, such as *A. caroliniana*, *A. mexicana*, and *A. pinnata*. Shi et al. (1980) labeled a species in the former group (*A. filiculoides*) with ^{15}N and used it to fertilize rice. After the incorporation of azolla, the first rice crop recovered only about 15 to 20% of the azolla nitrogen, and a second rice crop recovered only about 1 to 5% of the initial application.

A. Nursery and Field Multiplication

The goal of multiplication is to produce azolla as early and as fast as possible, so that the largest possible area of paddy fields can be inoculated with sufficient azolla. To meet this goal, intensive management and close coordination are essential.

In south China, spring-type nursery multiplication takes place during the winter, and from mid-March or mid-April to mid-June along the Yangtze River (Figure 10). When the average temperature rises to 14 to 20°C, azolla can double every 3 to 5 days. The azolla mat is continually subdivided to prevent crowding and consequent slowing of its growth rate. Whole nurseries are subdivided to provide inoculum for adjoining fields. Any extra azolla is used as a fodder or is composted for upland crops. Every available area is used for the extension of the nursery, and can include canals, ditches, ponds, and other bodies of water.

A few weeks before azolla inoculum is needed on a large scale to plant in rice production fields, it is often necessary to use large fields as multiplication nurseries. Rice fields must be used for this last stage of multiplication and are usually chosen because of close proximity to assured water supplies and suitability for growing azolla. As with maintenance nurseries, multiplication nurseries and fields must be partitioned to prevent wind and wave damage. Floating barriers, temporary low bunds, or rows of rice stubble can be used.

In an article on multiplication methods, the Subei Agricultural College (1971) has generalized the approaches taken for the multiplication of azolla: "first smaller land area, then larger land area; first shallow water, then deeper water; first nutrient-rich water, then nutrient poor water; first still water, then running water; first in water without fish, later in water with fish; and first nitrogen plus phosphorus fertilizer, later phosphorus only."

B. Field Cultivation and Incorporation

There are three different systems used for the cultivation of azolla as a green manure for rice: azolla is grown (a) as a monocrop and in-

Figure 10 Field-level multiplication nursery in Guangdong Province, China. The rice field is subdivided with temporary bunds to ease application of inputs and harvest of azolla and to reduce piling by the wind. The photo was taken from the edge of a canal, through which excess azolla was transported by barge to other fields.

corporated into the paddy mud as a basal green manure before the rice is transplanted; (b) as an intercrop with rice (Figure 11) and either incorporated as a topdressing manure or allowed to die naturally without incorporation; or (c) as both a monocrop and an intercrop.

The azolla cropping system used in any particular location may vary somewhat each year depending on the weather, the rice production system being used, tradition, and other crops in the cropping system. When azolla is monocropped in China and Vietnam, it is usually grown in the fields for 20 to 30 days before the rice and incorporated into the paddy mud once or twice during that period by plowing or harrowing. As an intercrop it grows beneath the rice in the interrow areas until shaded out by the developing rice canopy; usually

Figure 11 Azolla growing as an intercrop with rice at a research station in Thailand. Azolla was removed from around the central hill of rice seedlings. Azolla at this biomass density should be partially incorporated into the soil to reduce its inhibition of rice tiller emergence.

within 20 to 40 days. After transplanting the rice, the azolla intercrop is partially incorporated once or twice during hand-weeding operations, or, in a few locations, with a modified rice weeder. The last azolla mat that develops before the rice canopy completely closes is either allowed to die naturally due to fungal rot and light starvation, or is incorporated by hand before the maximum tillering stage of the rice.

Environmental Factors Affecting Cultivation

Environment is the primary influence on the use and productivity of azolla in rice production. For example, when irrigation water is available, relative growth rate, total biomass accumulation, and nitrogen concentration are higher in cool dry environments. This higher productivity of azolla usually promotes a greater increase in rice yields when compared to the wet monsoon season. Rice is usually more responsive to any fertilizer during dry weather, but this response is increased by higher azolla productivity. The reliability of azolla productivity during dry weather is also higher because of fewer insect and disease problems. Improved dry season rice yields resulting from azolla use were 5 to 69% greater than comparative zero nitrogen control plots in reports from Brazil, India, Pakistan, and Thailand (Mandal and Bharati, 1983; Barthakur and Talukdar, 1983; Kannaiyan et al., 1983; Subudhi and Singh, 1983; Haq and Rosh, 1984; Fiore, 1984). Wet-season rice yields have nearly the same range of yield improvement (-1 to 56%), but tend to have a lower average in reports from Burma, India, Sri Lanka, and Thailand (Swatdee and Seetanum, 1979; Kulasooriya et al., 1980; Kaushik and Venkataraman, 1981; Partohardjono et al., 1983; Subudhi and Singh, 1983; Win, 1984). Individual environmental factors have been studied for their influence on azolla and include water, temperature, humidity, light, and plant nutrition.

Water: The primary constraint to the use of azolla is its requirement for an aquatic habitat. An individual plant can survive for only a few moments on a dry surface under a tropical sun. It can survive for only a few days or a week on paddy soil that dries during intermissions between rains. Some varieties can survive indefinitely on moist shaded mud, but these will not multiply to any useful extent.

In places where rice is grown in paddies without water control (i.e., where rainfed bunded rice is grown) azolla would probably be an unreliable option as a green manure since a continuous supply of water from rain cannot be assured. With water control, azolla can usually be grown in nurseries and fields whenever desired during the cropping season, and the timeliness of inoculum needed for production fields can be assured.

The fact that azolla is produced vegetatively from whole plants compounds the problem with water. Some azolla must be maintained all year round and multiplied in nurseries prior to large-scale field cultivation. With the present system, a certain amount of water must be available all year for maintenance and multiplication. Water availability for these nurseries is especially critical if azolla is to be cultivated for rice at the start of a monsoon season. For cultivation as an intercrop, an area of azolla equal to 20% of the rice area has to be produced before the rice is transplanted.

Water control is also necessary for avoiding injury due to high or low temperatures. If air temperatures are too low or too high, water can be made deeper to provide a larger mass that will moderate changes in water temperature. If water temperatures are too high, cooler water can be introduced into the field to provide cooling, or water levels can be dropped to allow the azolla to come into contact with the relatively cooler soil. When temperatures are moderate, a fairly shallow depth of water, 2.5 to 5 cm, is most suitable for azolla growth.

Once an azolla crop is grown as a green manure, it must be incorporated into the soil in order to recover a significant portion of its nitrogen. Since the rhizome and leaves of azolla are hydrophobic, water must be completely drained from the field or azolla will be washed aside in the effort to incorporate. Farmers are reluctant to drain rainfed and even irrigated rice fields in order to incorporate an azolla crop. Incorporation of an azolla monocrop is possible with existing implements, but water is not usually available for monoculture. Incorporation of an intercrop is presently impractical because no efficient methods or implements are known.

Temperature and humidity: Although the need for water control greatly reduces the area under which azolla can be cultivated, the indirect effects of a humid tropical climate further reduce the possibility of successful cultivation. Many researchers suggest that high water or air temperatures are the most important factor preventing azolla use. Growth rates are depressed at temperatures above 35°C and even the most heat-tolerant varieties are killed at prolonged temperatures above 45°C. The temperature of shallow water standing over dark soil may exceed 40°C in hot weather. However, it is also possible to avoid high temperatures by draining, shading, or flushing. Thus high temperature will not necessarily prevent cultivation, although it certainly makes it more difficult.

Azolla species and varieties have different tolerances to temperature extremes. Some can survive a very wide temperature range (i.e., -5 to 35°C), whereas others can survive only within a very narrow temperature range. *A. filiculoides* and *A. rubra* (*japonica*) are quite tolerant of low temperatures, while *A. mexicana, A. microphylla, A. caroliniana, A. nilotica*, and certain varieties of *A. pinnata* possess greater tolerance to higher summer temperatures (Peters et al., 1980; Watanabe and Berja, 1983; Lumpkin and Bartholomew, 1986).

The worst threat from the climate comes from the stimulating effects of high temperature and humidity on the population dynamics of insects and fungi that attack azolla (Figures 12 and 13). Insects and fungi can be controlled by a comprehensive program of azolla management and frequent pesticide applications, but the environmental and economic consequences of frequent pesticide applications would make azolla cultivation unrealistic by any method of cost accounting.

Light: Most azolla species do not need full sunlight for optimum growth (Talley and Rains, 1980), except that full sunlight is helpful at high latitudes during the somewhat lower light intensities of the spring, autumn, and winter. For oversummering nurseries 25 to 35% of full sunlight is sufficient.

Figure 12 Larva of *Nymphula* spp. build cocoons of azolla, from which they move through a mat of azolla feeding on the buds. This individual abandoned its cocoon after being placed on a leaf of *Marsilea*. Besides species in the order Lepidoptera, azolla is known to be attacked by species in the Chironomidae, Coleoptera, Orthoptera, Homoptera, and Arrenurae.

Azolla grown in combination with rice will suffer and may die from insufficient light during the tillering stage of the rice plants. *A. caroliniana* seems to be somewhat tolerant of lower light intensities and fungal attack, and a small amount may survive under a full stand of rice through crop maturation. This may be an important characteristic for its use in dual rice/azolla production systems, or when it is desired to grow azolla after a single crop of summer rice.

Plant nutrition: Plant nutrition is very important for successful cultivation of azolla (Becking, 1979; Chen and Li, 1980; Yatazawa et al., 1980; Tang et al., 1984). Since azolla is an aquatic plant, it must obtain most required nutrients from the water. Water that flows from residential areas or from fertilizer-enriched agricultural lands will often contain sufficient nutrients to ensure good azolla growth. However, most sources of water do not provide sufficient nutrients or nutrients in the right combination to meet the nutritional needs of azolla.

Figure 13 Snails of the genera *Lymnaeidus*, *Planorbidus*, and *Limnaea* have been reported to attack azolla. Chemical methods for their control are not environmentally acceptable, although Bayluscid is effective. Azolla is currently being tested in China as a food for a large edible snail, species unknown.

The most important nutrient is phosphorus (Ma and Xu, 1962; Subudhi and Watanabe, 1981; Watanabe and Ramirez, 1984). Low phosphorus in the water often causes smaller, pink-to-red, less vigorous plants. Under severe conditions, the plants become highly compact and may develop a deep red color and very long roots. To overcome this problem, phosphorus fertilizers must be used, or the water level must be dropped to allow azolla roots to come into contact with phosphorus-containing soil. A water depth of 2 to 3 cm is sufficiently shallow to allow root/soil contact. Field cultivation of azolla usually requires the application of 0.5 to 1.0 kg P/ha per day. However, this does not necessarily mean that an increase in the total annual use of phosphorus is required. Phosphorus initially intended for direct application to the rice can be applied to azolla, which will enable it to grow and fix nitrogen that will be used by the rice. Once the azolla mat is incorporated into the soil and allowed to decompose, the phosphorus becomes available for the rice crop.

One kilogram of phosphorus applied to azolla can result in the fixation of about 5 to 10 kg of nitrogen by azolla. However, if 100% of the applied phosphorus could be absorbed by the azolla and the phosphorus content in the azolla was kept at a minimum for maximum nitrogen (about 0.25% P for 4% N on a dry weight basis), 1 kg of phosphorus would be sufficient for 16 kg of nitrogen (1:16 ratio). For some countries that have local phosphate deposits, supplying phosphorus to azolla to obtain nitrogen for the rice crop may be an economical proposition. For example, Vietnam is deficient in petroleum products for the production of nitrogen fertilizer, but has sufficient phosphate deposits to fuel its miniature "azolla nitrogen factories."

Azolla should be fertilized with a soluble phosphate source such as phosphoric acid or triple superphosphate. The dripping of a phosphorus solution into irrigation water or the use of small rafts of superphosphate would minimize losses of phosphorus through fixation by soil. Also, a 1 to 2% solution can be sprayed on azolla in areas where fixation and leaching are severe. The most phosphorus-efficient way to utilize phosphorus fertilizers is to multiply azolla in nurseries ponds that are lined with plastic. Through luxury consumption, azolla can be loaded with over five times its minimum phosphorus requirement. This stored phosphorus can be utilized by the plants for growth during cultivation in the paddy fields.

In certain depleted soils, azolla responds to the application of other nutrients, such as potassium and lime. In some deficient soils, the addition of a small amount of molybdenum and/or iron (Rains and Talley, 1979) has proved useful in increasing the rate of nitrogen fixation of azolla.

Management Factors Affecting Azolla

A number of special management steps are required for the successful cultivation of azolla for rice. Some of these are related to the requirements of the azolla. Others are related more to the needs of the rice crop.

Inoculation: To begin cultivation, the fields must be inoculated with azolla. Inoculation usually consists of broadcasting azolla plants from the nursery onto the water surface of a well-prepared field. The planting rate can vary according to the time of year, the amount of field space available, the time available until the planting of the rice crop, and the soil and water fertility conditions. For example, during cool weather in areas with a long fallow season, fields can be inoculated with a low level of azolla (>25 g/m^2) when there is sufficient time for development of a thick mat of azolla before the rice crop is planted. However, when azolla needs to be fitted into a very short fallow period between crops, much higher levels of inoculum (300 to 800 g/m^2) will be required to produce sufficient green manure.

To be used as an inoculum, azolla can either be removed from the nurseries (Figure 14) and carried to the field by hand, animal, or machine, or with proper planning, can be moved from the nursery through the irrigation system by laminar flow to the field, without being removed from the water. Researchers at the University of California at Davis, where low inoculum densities were used during the spring, experimented with the use of airplanes to broadcast the azolla inoculum. For upland transportation, azolla inoculum can be efficiently collected by first concentrating it by wind, water flow, or pushing, and then harvesting with a net or by developing a floating mechanized harvester. The same harvester could be designed to spread the azolla in a fashion similar to a mechanized manure spreader (see Talley and Rains in Lumpkin and Plucknett, 1982).

For guidance in selecting an inoculum application rate, the following points should be considered:

1. If azolla and land are scarce but labor is plentiful, inoculate at the rate of 0.5 to 0.8 kg/m^2 and subdivide or partially incorporate every 2 to 4 days. This system produces the highest yield in the shortest time, since azolla is kept in a linear growth phase. However, this

Figure 14 The use of plastic mosquito netting is one of the most effective ways to collect azolla from nurseries under a labor-intensive system, as shown here in Thailand. The water level must be raised so that a portion of the net can slip under the roots of azolla and avoid collecting mud.

high level of inoculation means that a higher labor requirement will be necessary for more frequent subdivision or incorporation.

2. If azolla is plentiful but labor and land area are scarce, inoculate at 0.3 to 0.5 kg/m^2 and subdivide or partially incorporate every 5 to 10 days. This is the best system for producing azolla for a large area. This moderate inoculation rate ensures better utilization of sunlight, field area, and water, and should bring about the best economic return in developing countries.

3. If azolla and land area are plentiful but labor is scarce, inoculate at the rate of 0.1 to 0.3 kg/m^2 and subdivide or harvest every 10 to 14 days. This system is used when azolla is grown on a large scale for soil incorporation as a basal manure. This low level of inoculation will save labor but may result in more insect damage since the prolonged growth of azolla allows for the population of insects to increase.

Monoculture: When azolla is monocropped in large fields, provisions should be made to prevent the wind from blowing the floating azolla inoculum from one side of the field to the other. Strong winds can result in piles of azolla being formed along the paddy bunds, and these will rot if not quickly dispersed. Azolla grows best if the plants are kept evenly spread over the water surface. To do this the Chinese often used single or double rows of rice about 1 to 2 m or so apart to provide bands of water surface for the azolla. Also, azolla can be grown in rice stubble, which serves to anchor the fronds. In Vietnam, windbreaks of sesbania or dense rows of rice were planted to protect azolla from wind damage. Where water is flowing from one field to another, it is necessary to place barriers made of reeds, twigs, netting, or other screening materials that will prevent the azolla from being washed out of the field.

Azolla grown as a monocrop or as an intercrop with rice should be incorporated into the soil before its maximum growth density (i.e., while it is still in the linear growth phase). Monocropped azolla can be incorporated into the soil with regular tillage equipment before the rice is planted or transplanted (Figure 15). This feature makes the use of monocropped azolla very feasible for all types of rice cultivation, whether transplanted, broadcast, or drill sown, and in temperate regions as well as the tropics. The system usually requires draining of the field to ensure contact of the azolla with the soil, and to prevent wave action caused by the machinery from moving the azolla around and causing incomplete and uneven incorporation patterns. Should it be desired to grow a second or third mat of azolla as a monocrop or intercrop with rice, enough azolla can be left in the field in puddles and low spots to ensure sufficient inoculum to initiate the next mat.

Before the incorporation of an azolla monocrop, the field should be completely drained for several days before the combined azolla incorporation-puddling operation is carried out. This will ensure that

Figure 15 A paddy harrow can be used to incorporate an azolla monocrop into the soil as a basal green manure. As shown here on a commune in Guangdong province, China, a small amount of water is left in the field so that some of the azolla will escape the incorporation process to serve as inoculum for an azolla intercrop with the rice crop.

all or most of the azolla is incorporated. However, if it is desired to grow the azolla as an intercrop in addition to the monocrop, the monocropped azolla mat should be incorporated immediately after draining the paddy while scattered puddles of water remain. In either case, monocropped azolla only or with an intercrop, the rice seedlings should not be transplanted until 4 or 5 days after the final monocrop incorporation. A thick mat of freshly incorporated azolla will aggravate the "transplanting shock" suffered by transplanted rice seedlings and result in a higher incidence of empty hills. The heavier the azolla mat, the longer the fallow period should be. During this period, it is best to keep the paddy drained to speed decomposition, unless an intercrop of azolla is also desired. In the latter case, several shallow irrigations will keep enough azolla alive to grow as an intercrop with the rice.

Interculture: When environmental or cultural limitations do not allow for the growing of monocropped azolla, it may be possible to grow azolla concurrently with rice. Azolla can be grown as an intercrop for use as a green manure, or when grown in early planted rice fields, to produce additional inoculum for late planted rice fields (Agriculture Technical Station, 1964). In fields where azolla is inoculated into transplanted rice to grow as an intercrop, an even distribution of azolla can be difficult. For this situation the Chinese flooded the fields to the point of nearly covering the rice seedlings before broadcasting and dispersing the azolla inoculum.

In China, intercropped azolla was usually incorporated by hand into the soil. This method was used because an effective hand tool for the various types of paddy soils and azolla mat thicknesses had not been developed. Hand incorporation is quite effective, but it is very labor intensive and backbreaking. Attempts at incorporating azolla into a flooded paddy are less effective than incorporating into drained soil, but the former method conserves precious water (Agriculture Bureau, 1964).

A small hand-operated (Figure 16) or gas-powered rotary rice weeder can be used to incorporate a thin mat of azolla into soils of suitable texture. If these weeders are used in both directions in a field (e.g., west to east and north to south), they can do a very complete job of incorporation. However, for this two-way incorporation technique to be used effectively, the rice must be planted equidistant, both in the rows and between the rows, to allow passage of the weeding machine.

The intercrop system brings new factors into the cultivation of both rice and azolla, including that of competition. Despite the fact that azolla floats on the water, whereas rice grows in the soil and in the vertical plane, competition does arise. There are noticeable and measurable effects of competition between an azolla intercrop and rice. Some effects are negative but transitory, while others are more long term (Experiment Station, 1978). First, azolla competes with rice for nutrients, including nitrogen. Azolla is a scavenger for combined nitrogen in the water, and thus reduces available nitrogen for rice. This period of competition lasts for about 10 to 20 days after azolla is planted, usually until after the first azolla incorporation, and was referred to in China as "early inhibition" of rice by azolla. However, this early competition usually disappears after nitrogen from the first incorporation of azolla becomes available to the rice. After an additional incorporation, the end result is usually increased rice yield from more panicles per hill and increased grain weight. The Chinese called this "late promotion" of rice by intercropped azolla; they therefore spoke of "early inhibition and late promotion" as the effects of intercropped azolla on rice.

The decomposition rate of azolla becomes quite important when azolla is intercropped with rice. The rate of decomposition is also affected by the azolla variety, temperature, the quantity of biomass,

Figure 16 The rotary rice weeder can be used to incorporate thin mats of intercropped azolla into the soil, as shown here being done by an

the C:N ratio of the azolla, and the soil environment. Excessive manuring with azolla can "overload" the soil and result in rice seedling fatalities early in the season, and lodging and a higher incidence of disease and insect in the later stages of rice development.

An azolla intercrop affects the water temperature of the rice field. A thick azolla mat does not allow sunlight to penetrate to the water, so the water remains cooler under azolla during the day than when azolla is absent. At night the azolla mat tends to prevent heat loss, whereas an open water surface looses more heat. The net result is that changes in water temperature are much less with an azolla mat than without. In temperate regions, an azolla mat keeps the water cold during the spring and thereby causes a setback in growth of young rice seedlings. This is the case in California where the irrigation water for rice is quite cool. Under such conditions an open water surface that would heat up during the day could be an advantage to the thermophyllic rice plants and may override any advantage of using azolla as an intercrop.

A heavy azolla mat growing under rice can also interfere with tiller emergence and development. Indeed, new tillers can be physically inhibited by a heavy azolla mat caused by excessively high inoculation levels and/or a failure to incorporate part of the azolla mat. However, this is usually not a problem when azolla is incorporated one or more times, since incorporation reduces mat density. Before the widespread use of herbicides, some Japanese rice farmers used to complain about the harm done to tillering by the naturally occurring azolla in their rice fields (Tuzimura et al., 1957). In China, where azolla was grown as an intercrop, peasants would plant up to 10 rice seedlings per hill to overcome the inhibitory effects of azolla.

V. OTHER USES

Several studies have investigated the weed-suppressing effect of an azolla mat (Satapathy and Singh, 1985). Janiya and Moody (1981) observed that *Monochoria vaginalis*, *Echinochloa glabrescens*, *Cyperus difformis*, and *Paspalum* sp. were suppressed by azolla, but *Scirpus maritinus* and *Echinochloa crus-galli* were unaffected. Srinivasan (1981) found that azolla could suppress *Marsilea quadrifolia.*

---

organic farmer in the Philippines. Unfortunately, the usefulness of the rotary weeder is limited because (a) the rice must be planted in rows, which is rarely the case in most developing countries; and (b) the fields must be drained or allowed to dry out so that the azolla does not escape incorporation. A new version has been recently developed by IRRI.

Figure 17 Azolla has long been recognized as a food for domesticated ducks in China and Vietnam, and for wild waterfowl in the United States. In Vietnam, birds are fed 0.1 to 0.3 kg of fresh azolla per day; however, for efficient production, azolla should not contribute more than 25% of the food intake.

Azolla has also been shown to have a beneficial effect on maize yields (Ferrera-Cerrato and Miranda, 1982) and on the population of an actinomycetes called "5406" commonly used for agriculture in China (Xiang and Li, 1981). Additional studies on the use of azolla as feed have been conducted with chickens (Castillo et al., 1981), ruminants (Dolberg et al., 1981; Scharpenseel et al., 1982; Waha, 1983), tilapia (Carraro, 1983), and grass carp (Cassani, 1981). Results are generally positive but show that the feeding of azolla must be limited in the diet since its constituents are not well balanced and have undigestible fractions (Figures 17 and 18).

VI. CONCLUSIONS

The nitrogen-fixing azolla is an excellent green manure, suppressor of certain paddy weeds, and fodder for pigs, ducks, and fish. How-

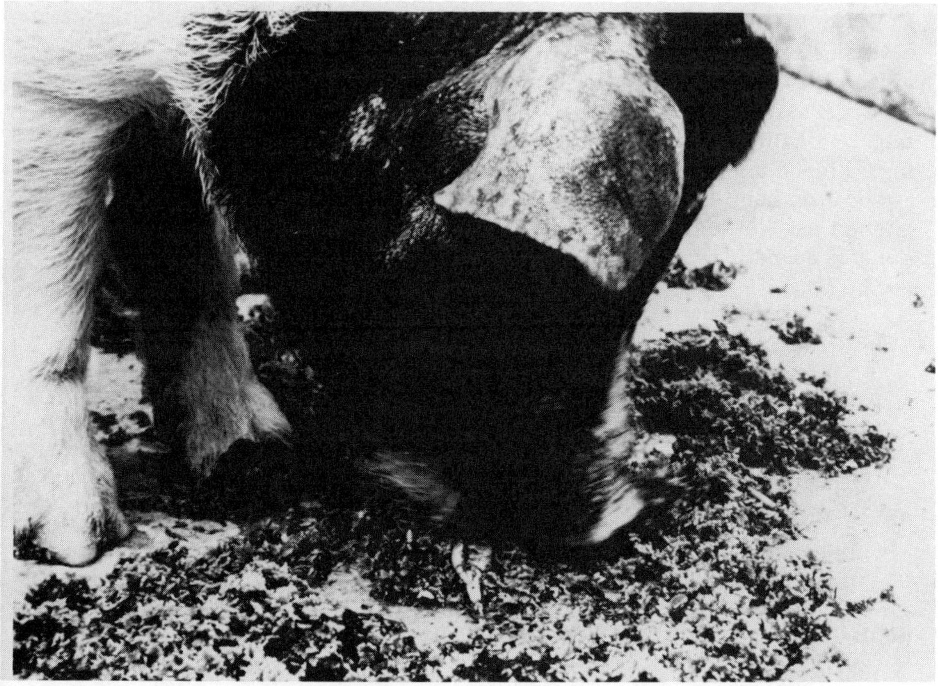

Figure 18 In China, azolla has been grown for centuries on canals and ponds as a fodder for pigs and is considered to be equivalent to sweet potato tops in feed value. The taste of azolla is not preferred by pigs, so it must be fed before more palatable foods. The addition of a small amount of salt to the azolla can improve its attractiveness.

ever, despite its previous cultivation and use in China and northern Vietnam, it has not been successful in the humid tropics, including southern Vietnam, because of problems with insects and disease, availability of water and phosphorus, and an intensive labor requirement. Its use in China has declined rapidly along with the use of other green manures because they were not economical in light of recent economic reforms.

Azolla cultivation is often promoted as a means of reducing the importation of nitrogen fertilizers or of the fossil fuel needed for their manufacture. However, any savings on nitrogen fertilizers can be quickly negated by imports to support additional pumping of irrigation water and the application of pesticides and phosphorus if these inputs become necessary for azolla's cultivation.

If cultivation is ever to be economically successful on a large scale, a strong research program and a number of scientific breakthroughs are necessary. As a part of a strong program, azolla germplasm must be aggressively collected, maintained, and characterized, and important breakthroughs will have to be made in the use of spores as planting material and the development of efficient implements and fertilizer application methods and formulations. These objectives can be accomplished most effectively by a well-focused and well-financed internationally coordinated research program.

REFERENCES

Agriculture Bureau, Wuxing County (1964). *Turang Feiliao 3*: 74 (in Chinese).

Agriculture Technical Station, Changding, Yuyao County (1964). *Turang Feiliao 3*: 65 (in Chinese).

Anwei Academy of Agricultural and Forestry Science (1977). *"Three Water One Duckweed" Cultivation Techniques (Water Lettuce, Water Hyacinth, Water Peanut, Red Azolla)*. Agriculture Publishing House, Beijing (in Chinese).

Ashton, P. J. (1977). Factors affecting the growth and development of *Azolla filiculoides*. *Natl. Weeds Conf. South Africa, 2nd*, pp. 249-268.

Ashton, P. J., and Walmsley, R. D. (1984). *Bot. J. Linn. Soc. 89*(3): 239.

Bai, K. Z., Berja, N. S., and Watanabe, I. (1981). *Int. Rice Res. Newsl. 6*(3): 21.

Barthakur, H. P., and Talukdar, H. (1983). *Int. Rice Res. Newsl. 8*(1): 20.

Becking, J. H. (1979). Environmental requirements of azolla for use in tropical rice production, in *Nitrogen and Rice*. International Rice Research Institute, Philippines, Manila, pp. 345-374.

Beijing Agricultural University (1979). *Fertilizer Handbook*. Agriculture Publishing House, Shanxi Province (in Chinese).

Bonnet, A. L. M. (1957). *Rev. Cytol. Biol. Veg. 18*: 1.

Brederoo, P., Meeuwissen-van der Vlugt, E. G. J., and Daems, W. T. (1976). *Ultramicroscopy 2*(1): 121.

Calvert, H. E., and Peters, G. A. (1981). *New Phytol. 89*(2): 327.

Carraro, S. (1983). *Etude de l'influence de différents regimes alimentaires a base d'azolla sur la croissance de Sarotherodon niloticus*. Mémoire présente en vue de l'obtention du grade de Licencié en Sciences. Université Catholique de Louvain, Louvain-la-Neuve, Belgium.

Cassani, J. R. (1981). *J. Fish Biol. 18*(2): 127.

Castillo, L. S., Gerpacio, A. L., and Pascual, A. L. (1981). Exploratory studies on azolla and fermented rice hulls in broiler diets. In *Proceedings of the Philippines Society of Animal Science. Annual Convention, Nov. 13-14, 1981.* College, Laguna, Philippines.

Chen, J. M., and Li, S. L. (1980). *Acta Pedol. Sin. 17*(4): 390 (in Chinese).

Cheng, J. F., Xu, S. X., and Wang, M. Y. (1980). *Jiangsu Nongye Keji. 4*: 16 (in Chinese).

Demalsy, P. (1953). *Cellule 56*: 5.

Dolberg, F. Saadullah, M., and Haque. M. (1981). *Trop. Anim. Health Prod. 6*(4): 322.

Experiment Station, Liaoguan Production Brigade and Soil and Fertilizer Branch, Agricultural Research Institute, Baoshan County (1978). *Yunan Nongye Keji 2*: 24.

FAO (1978). *FAO Soils Bull. Rome 41*: 81.

Feng, Y. Z. (1977). *Nongye Kexue Changshi 7*: 11 (in Chinese).

Ferrera-Cerrato, R., and Miranda, R. A. (1982). Propagation of an *Azolla* sp. and its potential as a green manure for corn in Mexico, in *Biological Nitrogen Fixation Technologies for Tropical Agriculture*, P. H. Graham and S. C. Harris, eds. Centro Internacional de Agricultura Tropical, Cali, Colombia, pp. 561-564.

Fiore, M. de Fatima (1984). *Notas Cientificas Pesq. Agropec. Bras.*, Brasilia *19*(3): 387.

Fowler, K., and Stennett-Willson, J. (1978). *Fern Gaz. 11*(6): 405.

Haq, N., and Rosh, D. (1984). *Int. Rice Res. Newsl. 9*(3): 24.

He, G. F., and Lin, Y. C. (1981). *Sci. Agric. Sin. 3*: 54 (in Chinese).

He, F. G., Lin, Y. C., and Ke, Y. S. (1982). *Sci. Agric. Sin. 1*: 28 (in Chinese).

IAEA (1985). The role of isotopes in studies on nitrogen fixation and nitrogen cycling by blue-green algae and the *Azolla-Anabaena azollae* association. Proceedings Consultants Meeting, *IAEA-TecDoc-325.* International Atomic Energy Agency, Vienna.

IRRI (1979). *Nitrogen and Rice.* International Rice Research Institute, Manila, Philippines.

Jaeger, E. C. (1978). *A Source-Book of Biological Names and Terms*, 3rd ed. Charles C Thomas, Publisher, Springfield, Ill., p. 32.

Janiya, J. D., and Moody, K. (1981). *Int. Pest Control 23*: 136.

Kannaiyan, S., Thangaraju, M., and Oblisami, G. (1983). *Sci. Cult. 49*(7): 217.

Kaushik, B. D., and Venkataraman, G. S. (1981). *Int. Rice Res. Newsl. 6*(6): 24.

Kawamatu, S. (1963). *Cytologia 28*: 12.

Ke, Y. S., Zhang, Z. T., Liu, X. L., and Ling, D. Q. (1981). *Guangdong Nongye Kexue 3*: 29 (in Chinese).

Kulasooriya, S. A., Hirimburegama, W. K., and De Silva, R. S. Y. (1980). *Oecol. Plant.* *1*(4): 355.
Li, J. M. (1977). *Hubei Nongye Kexue 11*: 30 (in Chinese).
Lu, S. Y., Chen, K. A., Zhou, Z. B., and Shen, Z. H. (1965). *Sci. Agric. Sin. 5*: 35 (in Chinese).
Lumpkin, T. A. (1981). *Acta Bot. Sin. 23*(6): 492.
Lumpkin, T. A. (1984). *Int. Rice Res. Newsl. 33*(1): 30.
Lumpkin, T. A. (1985). *Proc. R. Soc. Edinburgh 86*: 161.
Lumpkin, T. A., and Bartholomew, D. P. (1986). *Crop Sci. 26*: 107.
Lumpkin, T. A., and Plucknett, D. L. (1980). *Econ. Bot. 34*(2): 111.
Lumpkin, T. A., and Plucknett, D. L. (1982). *Azolla as a Green Manure: Use and Management in Crop Produciton.* Westview Press, Inc., Boulder, Colo., 230pp.
Ma, J. Y., and Xu, G. M. (1962). *Zhongguo Nongbao 6*: 13.
Mandal, B. K., and Bharati, A. K. (1983). *Indian J. Agric. Sci. 53*(6): 472.
Moore, A. W. (1969). *Bot. Rev. 35*: 17.
Morton, C. V., and Wiggins, I. L. (1971). *Flora of the Galapagos Islands.* Stanford University Press, Stanford, Calif., pp. 61-63.
National Azolla Action Program (1982). *Introduction to Azolla and Its Uses.* Publication 1. Ministry of Agriculture and the University of the Philippines at Los Banos, College, Laguna, Philippines, 16pp.
Nickell, L. G. (1961). *Phyton 17*: 49.
Ott, F. D., and Petrik-Ott, A. J. (1973). *Brit. Fern Gaz. 10*: 305.
Partohardjono, S., Hendrick, V., and Bastaman, M. (1983). *Contrib. Cent. Res. Inst. Food Crops Bogor 69*: 11.
Perkins, S. K., Peters, G. A., Lumpkin, T. A., and Calvert, H. E. (1985). *Scanning Electron Microsc. 4*: 1719.
Peters, G. A. (1977). The *Azolla-Anabaena azollae* symbiosis. In *Genetic Engineering for Nitrogen Fixation*, A. Hollaender, ed. Plenum Press, New York, pp. 231-258.
Peters, G. A., Toia, R. E., Jr., Raveed, D., and Levine, N. J. (1978). *New Phytol. 80*: 583.
Peters, G. A., Toia, R. E., Jr., Evans, W. R., Crist, D. K., Mayne, B. C., and Poole, R. E. (1980). *Plant Cell Environ. 3*: 261.
Pfeiffer, W. M. (1907). *Bot. Gaz. 44*: 445.
Rains, D. W., and Talley, S. N. (1979). Uses of azolla in North America. In *Nitrogen and Rice.* International Rice Research Institute, Philippines, Manila, pp. 419-431.
Reed, C. F. (1954). *Bot. Soc. Broteriana Ser. 2a 28*: 5.
Roger, P. A., and Reynaud, P. A. (1979). *Oecol. Plant. 14*(1): 75.
Satapathy, K. B., and Singh, P. K. (1985). *J. Aquat. Plant Manage. 23*: 40.

Scharpenssel, H. W., Menke, K. H., Goetz, D., Meyer-Spasche, H., and Dorffling, K. (1982). *Landwirtsch. Forsch. 35*(3-4): 200.

Sculthorpe, C. D. (1967). *The Biology of Aquatic Vascular Plants.* St. Martin's Press, Inc., New York.

Shi, S. L., Wen, Q. X., and Liao, H. Q. (1980). *Acta Pedol. Sin. 17*(3): 240 (in Chinese).

Srinivasan, S. (1981). *Int. Rice Res. Newsl. 6*(3): 22

Strasburger, E. (1873). *Ueber Azolla.* Verlag von Ambr. Abel. Jena, Leipzig.

Subei Agricultural College (1971). Green Azolla. In *Cultivation of Water Crops.* People's Literature Publishing House, Shanghai, pp. 109-134.

Subudhi, B. P. R., and Singh, P. K. (1983). *Indian J. Agric. Sci. 53*(5): 320.

Subudhi, B. P. R., and Watanabe, I. (1981). *Soil Sci. Plant Nutr. 27*(2): 237.

Svenson, H. K. (1944). *Am. Fern J. 34*: 69.

Swatdee, P., and Seetanum, W. (1979). *Int. Rice Res. Newsl. 4*(5): 24.

Sweet, A., and Hills, L. V. (1971). *Fern J. 61*: 1.

Talley, S. N., and Rains, D. W. (1980). *Agron. J. 72*(1): 11.

Tank, J. M., Wen, Y. H., and Xiong, W. (1984). *Acta Pedol. Sin. 21*(2): 217 (in Chinese).

Tuzimura, K., Ikeda, F., and Tukamoto, K. (1957). *J. Sci. Soil Manure 28*: 17.

Van Hove, C., Diara, H. F., and Godard, P. (1983). *Azolla en Afrique de l'ouest in West Africa.* E. Oleffe., Court-St.-Etienne., Belgium.

Vo, M. K., and Tran, Q. T. (1970). *Beo Hoa Dau (Azolla).* Nha Xuat Ban Nong Thon, Hanoi (in Vietnamese).

Waha Baillonville, T. (1983). *Etude de la valeur alimentaire d'azolla pour les ruminants.* Mémoire présente en vue de l'obtention du grade de Licencié en Sciences. Université Catholique de Louvain, Louvain-la-Neuve, Belgium.

Wang, Q. P., Chen, J., and Wu, M. H. (1981). *Fujian Nongye Keji 3*: 21 (in Chinese).

Watanabe, I. (1984). *Outlook Agric. 13*(4): 166.

Watanabe, I., and Berja, N. S. (1983). *Aquat. Bot. 15*: 175.

Watanabe, I., and Ramirez, C. (1984). *Soil Sci. Plant Nutr., 30*(4): 595.

Watanabe, I., Subudhi, B. P. R., and Aziz, T. (1981). *Curr. Sci. 50*(21): 937.

Win, S. U. (1984). Effect of azolla application on rice yield (unpublished). Paper presented at the Workshop to Assess the Potential of Azolla Use in Tropical Asia, January 22-28, Bangkok, Thailand.

Wu, L. H. (1978). *Nongye Kexue Changshi 11*: 15 (in Chinese).

Xiang, Z. A., and Li, Z. F. (1981). *Zhejiang Nongye Kexue 6*: 281 (in Chinese).

Yatazawa, M., Tomomatsu, N., Hosoda, N., and Nunome, K. (1980). *Soil Sci. Plant Nutr. 26*(3): 415.

Ye, L. S., and Wu, C. L. (1964). *Zhejiang Nongye Kexue 12*: 585 (in Chinese).

Yuan, Z. P. (1964). *Zhejiang Nongye Kexue 12*: 590 (in Chinese).

4

Collection, Isolation, Cultivation, and Maintenance of Associative N_2-Fixing Bacteria

DAVID A. ZUBERER *Texas Agricultural Experiment Station, Texas A&M University, College Station, Texas*

I. INTRODUCTION

Considerable attention has been focused on the role of free-living N_2-fixing bacteria (diazotrophs) in natural and agricultural environments during the past decade. It is now widely recognized that a variety of free-living diazotrophs associate with the roots of many plant species, including many nonlegumes as well as some legume species, and enter into a "loose" associative symbiosis. The degree of intimacy of the association may vary from rhizosphere colonization to more closely coupled associations in which the diazotrophs are thought to reside within the outer cell layers of the root tissue. It is generally accepted that the diazotrophs acquire their carbon and energy supplies from the roots as exudates, secretions, lysates, and sloughed cell debris. As a result of root and rhizosphere colonization, dinitrogen is fixed and eventually made available for plant growth. The nature of the colonization, the quantity of nitrogen "fixed," and the contribution of "associative nitrogen fixation" to plant growth remain subjects of intensive investigation and have been reviewed elsewhere (Neyra and Dobereiner, 1977; Dobereiner and DePolli, 1980; Patriquin, 1982; van Berkum and Bohlool, 1980).

It should be remembered that associative nitrogen-fixing bacteria are primarily free-living diazotrophs which are capable of associating with higher plants as a means of obtaining carbon for growth and N_2 fixation. Their contribution to the nitrogen content of a given soil in the absence of "host plants" can still be considerable if exogenous carbon sources such as plant residues are incorporated into suitably moist soils.

Research aimed at gaining a better understanding of the ecology of free-living diazotrophs and their contribution to the nitrogen economy of various ecosystems constitutes a small but nevertheless important part of the overall research efforts for enhancing biological nitrogen fixation for food and fiber production. Although there remain many obstacles to the successful use of free-living diazotrophs to increase plant production, the study of these microorganisms will continue to receive attention for many different reasons. The materials given in this chapter are intended to provide the investigator with enough information to conduct basic studies which necessitate the isolation, cultivation, and maintenance of associative nitrogen-fixing bacteria. The intent of this chapter is to provide information useful to investigators wishing to study associative symbioses as well as to lend a degree of completeness to the coverage (in this volume) of the challenging research area of enhancement of biological nitrogen fixation.

II. COLLECTION OF ASSOCIATIVE NITROGEN-FIXING BACTERIA

The collection of associative diazotrophs is not unlike the collection of bacteria for any other purpose. All collection procedures begin with a judicious choice of samples suspected to contain the bacteria which are the objects of the search. For associative diazotrophs this usually means obtaining root samples in some form, although the idea of isolating from soil populations should not be discounted, since bacteria capable of survival in the absence of a host plant (a useful attribute) can be obtained in this way.

A. Collection of Soil Samples

There are few precautions necessary when collecting soil samples as sources of bacteria (see Wollum, 1982). The primary aim should be to maintain the sample in as natural a state as possible to maximize survival prior to processing for isolation of desired bacteria. In general, if excessive temperature and moisture fluctuations are avoided, the soil samples will remain suitable for subsequent isolations for several weeks. Longer storage can be achieved by keeping samples sealed in plastic bags in the refrigerator (5°C). Whenever possible, samples should be processed on the same day of collection to avoid marked changes in populations. This is particularly important if the collection of samples is for the purpose of enumerating populations of diazotrophs. If same-day processing is not possible, the samples should be kept in a refrigerator for handling on the following day.

Soil samples can be obtained with spades or trowels, or a variety of coring devices. In general, samples are obtained from the surface horizon to a depth of 10 to 20 cm. If sample-to-sample contamination is considered a problem, sufficient equipment must be available for the required number of samples or equipment can be disinfected using common disinfectants or by flaming the tools with a portable torch. The degree of caution regarding cross-contamination clearly depends on the purpose for which the samples are collected. If the colleciton is intended to show the distribution of a given group of organisms, due caution should be used. If, however, the purpose of the collection is simply to obtain the diazotrophs in a field soil, little caution is required. In either case, subsequent handling of the samples should be done so as to maximize survival of bacteria in the samples.

B. Collection of Root-Rhizosphere Samples

Since the extent of the rhizosphere cannot always be carefully quantitated, most investigators sample the soil that remains adherent to the root following its careful removal from the soil as "rhizosphere" soil. Coring devices consisting of a series of coring tubes of uniform length and diameter attached side by side (Papavizas and Davey, 1961; Johnson and Curl, 1972) have been used successfully to show the decline in microbial numbers as distance from the root increases. Such a device is not necessary when the intent is merely to sample soils closely associated with roots. For this purpose, the technique of collecting the soil from unearthed roots is usually satisfactory.

Root samples can be obtained in several ways. If the soil is of light texture (i.e., sandy), plants can be excavated with a spade and the soil gently shaken from the roots. Roots exposed in this manner can be removed from the plant and placed in plastic bags containing a moistened paper towel to prevent undue drying. Once collected, the roots should be processed quickly since changes in microbial populations occur when the roots are excised and stored for lengthy periods.

An alternative method for collecting roots is to obtain intact plant-soil cores using any of several coring devices. Choice of the device should take into account the nature (i.e., physical properties) of the soils from which cores are to be extracted. The soil-root core is usually placed in a plastic bag or some other suitable container (canning jars of various size are inexpensive and convenient) for return to the laboratory. The use of cores has the advantage of leaving the roots in a relatively undisturbed condition until they can be removed from the soil at a later time. Another advantage of obtaining intact plant-soil cores is that they can be placed in a greenhouse for continued growth of the plants and later studies of the root-associated bacteria. Investigators have frequently chosen robust plants growing in soils of low nutrient status as possible sources of root-associated diazotrophic bacteria and in many cases such a selection yields fruitful returns (Wright and Weaver, 1981).

Roots can be washed free of soil by using a gentle spray from a hose or by holding the core in a pail of water and removing the soil by hand. If possible, fresh water should be used for each sample and the container rinsed thoroughly between samples. Neither of these methods preserves the rhizosphere soil. Washed roots should be processed for isolation or enrichment procedures as soon as possible after they are washed free of soil. Prolonged storage of roots should be avoided if the aim of the study is an attempt to demonstrate the most abundant diazotrophs associated with roots under conditions of plant growth. Dobereiner (1980) discusses ways to determine which part of the diazotrophic microflora is responsible for the nitrogenase activity associated with roots.

III. ENRICHMENT AND ISOLATION OF ASSOCIATIVE DIAZOTROPHS

For convenience, the material that follows will focus on isolation of diazotrophic bacteria from roots. In general, the procedures can be adapted to many different starting materials, including soils as well as plant parts other than roots. For further information on the isolation and cultivation of diazotrophs, the reader should consult the reviews by Dalton (1980) and Dobereiner (1980).

Techniques for enrichment of a wide variety of bacteria are detailed by Aaronson (1970), Veldkamp (1970), and Krieg (1981). An extensive list of culture media for the maintenance and preservation of bacteria including diazotrophs is given by Lapage et al. (1970a) and by the American Type Culture Collection (ATCC, 1982). Excellent coverage of general bacteriological methods is provided in the series *Methods in Microbiology* (Academic Press, New York) and in the *Manual of Methods for General Bacteriology* [P. Gerhardt (ed.), American Society for Microbiology, Washington, D.C., 1981].

A. Enrichment Techniques

Enrichment techniques for associative diazotrophs almost universally rely on the use of a wide variety of liquid, semisolid, or solid culture media lacking combined nitrogen (e.g., ammonium, nitrate, amino acids, etc.). These nitrogen-free media take advantage of the unique ability of diazotrophs to use N_2 as a sole nitrogen source, thereby at least temporarily precluding the growth of the nondiazotrophic microflora. Further selectivity is built into these nitrogen-free media by varying such factors as carbon source, pH, and the oxygen tension at which the cultures are incubated. These are discussed in more detail below.

Some information as to the degree of intimacy of the association between the bacteria and roots can be gained from using washed roots and surface-disinfected washed roots as starting materials for enrichment procedures (Dobereiner, 1980; Patriquin, 1982). The time of exposure to the disinfectant (e.g., 1% chloramine-T; 5% Chlorox) should be varied to determine a suitable time for adequate disinfection, and the disinfectant should be rinsed from the roots with several changes of distilled water or suitable buffer before transferring them to the enrichment medium in tubes or plates. The presence of specific diazotrophs following surface disinfection implies a close association between the bacteria and root. Caution is urged, however, in drawing any conclusions regarding the actual sites of specific diazotrophic bacteria within the roots. For studies of this type, the investigator should resort to electron microscopy and the use of sophisticated immunolabeling techniques, such as the use of colloidal gold coupled to specific immune sera for localization of bacteria in root tissues.

Aerobic Diazotrophs

Aerobic diazotrophs are perhaps the easiest to obtain because they require less effort and equipment in terms of incubation conditions and other factors. For the enrichment of aerobic diazotrophs, the inoculum (root pieces, etc.) is placed into a suitable nitrogen-free medium (see below) and incubated under aerobic conditions (cotton plug or culture tube closure). The vessel should allow gas exchange and should provide a surface area-to-depth ratio for the medium near unity (Dobereiner, 1980). Shaking liquid cultures favors the development of diazotrophs most tolerant of oxygen, while unshaken liquid and semisolid cultures offer the best chances for obtaining microaerophilic diazotrophs such as *Azospirillum* sp. It should be remembered that most aerobic diazotrophs are better able to fix N_2 under conditions of reduced pO_2. This is true in culture (Okon et al., 1977; Dalton and Postgate, 1969) and in root associations (Dobereiner et al., 1972; Zuberer and Alexander, 1986). A major advantage of the use of semisolid enrichment media is that they allow the development of organisms in a zone at which the pO_2 is most favorable for N_2 fixation. Thus *Azospirillum* species frequently form a distinctive narrow band of growth (often referred to as a "pellicle") in semisolid agar deeps. Such a band is not always proof of diazotrophs, as many cultures exhibiting this feature do not yield positive results for acetylene reduction assays or subsequent isolation of diazotrophs (Tyler et al., 1979). A disadvantage of the use of semisolid agar deeps arises when conditions in the culture tube allow the proliferation of facultative anaerobes and some of the less stringent anaerobes. For enrichment purposes this may not be a problem; however, in the enumeration of aerobic diazotrophs by the most probable number technique it can lead to confusion.

The use of an array of different sugars and sugar alcohols in liquid media of neutral to alkaline pH incubated aerobically favors the enrichment of members of the genus *Azotobacter*, while more acid media (pH 5.0) tend to favor the enrichment of *Beijerinkia* sp. (Thompson and Skerman, 1979; Becking, 1961). A detailed account of the enrichment and isolation of the Azotobacteraceae can be found in the comprehensive treatment of this subject by Thompson and Skerman (1979). Claus and Hempel (1970) suggest that media containing resorcinol, ethylene glycol, or glutaric acid can be used for the enrichment of *Azotobacter vinelandii* in the presence of other species of *Azotobacter*. The use of enrichment media containing malic acid, succinic acid, or lactic acid is helpful in selecting for species of *Azospirillum* (Dobereiner, 1980), particularly if low concentrations of agar (2.0 to 5.0 g/liter) are added to impede the diffusion of O_2 into the media. The enrichment of *Derxia* is more difficult. Thompson and Skerman (1979) suggest using media containing a range of alcohols utilized by *Derxia* or media containing 1,4-butylene glycol, 1,3-butylene glycol, or 1,2-propylene glycol as carbon sources.

Facultative Anaerobes

The facultatively anaerobic diazotrophs include *Bacillus polymyxa*, *Bacillus macerans*, and many members of the Enterobacteriaceae, including species of the following genera: *Klebsiella*, *Enterobacter*,

Erwinia, and *Citrobacter* (Hino and Wilson, 1958; Nielson and Sparell, 1976). With the exception of *Klebsiella pneumoniae* (Klucas, 1972; Hill, 1975, 1976) these bacteria are capable of fixing N_2 only under anaerobic conditions. Nitrogen-free enrichment media are, therefore, incubated under anaerobic conditions. The anaerobic incubations can be accomplished in several ways. These include the use of Pankhurst tubes (Pankhurst, 1967; Campbell and Evans, 1969; Raju et al., 1972), the use of gas-impermeable bags (Hill, 1973; Wright and Weaver, 1981), and placement of cultures in a variety of chambers, including anaerobe jars, vacuum desiccators, and serum bottles. The Hungate roll-tube technique (Hungate, 1969) is widely used for the culture of strict anaerobes.

Anaerobic conditions are most frequently achieved by replacing the atmosphere inside the incubation chambers (vacuum desiccators, etc.) with an atmosphere lacking O_2 (e.g., O_2-free N_2 gas). The chambers should be evacuated (50 cm Hg) and flushed several times before incubating at the desired temperature. Tubed media as well as plates can be incubated successfully under these conditions. The anaerobic atmosphere in flexible gas-impermeable bags is achieved by constant flushing with a gas mixture devoid of O_2 (Hill, 1973), or by the use of O_2-scavenging systems such as alkaline pyrogallol (Kaspar and Tiedje, 1982) or steel wool treated with acidified copper sulfate (Attebery and Finegold, 1970; Wright and Weaver, 1981). Commercial systems (e.g., GasPak; BBL Microbiology Systems, Cockeysville, Maryland) for the cold catalytic removal of O_2 and production of CO_2 and H_2 for anaerobic atmospheres are also available but can be costly. Campbell and Evans (1969) used alkaline pyrogallol in the small sidearm of Pankhurst tubes to provide the anaerobic atmosphere necessary to culture facultative and anaerobic diazotrophs for acetylene reduction assays.

Facultatively anaerobic bacteria have been obtained from semisolid agar deeps containing the combined carbon medium of Rennie (1981; see Section III.B) incubated under aerobic conditions (Zuberer, unpublished data). Enteric bacteria from these tubes were recovered on Violet Red Bile agar (Difco), purified on Tryptic Soy agar, and confirmed as diazotrophs by acetylene reduction assay under anaerobic conditions. The aerobic microflora consumes sufficient O_2 from the agar deeps to allow the proliferation of the facultative anaerobes. Knowles (1982) reported using a nitrogen-deficient Endo-type agar (a selective medium for enteric bacteria) for enumeration of diazotrophic Enterobacteriaceae. The plates were incubated in an anaerobe jar (24 to 48 hr) and rose-colored, convex, mucoid colonies and colonies with a metallic sheen (both 75 mm in diameter) were counted.

Many different media have been employed for the enrichment of the diazotrophic Enterobacteriaceae. Most employ sucrose as the carbon source; hence they have little specificity for various enteric bacteria. Nielson and Sparell (1976) isolated diazotrophic enterobacteria from pulp mill effluents with a medium containing galactitol as the carbon source, and Talbot and Seidler (1979) indicated that Klebsiellae from botanical environments utilize cyclitols such as inositol, sequoyitol, and pinitol. They suggested that the ability to use these carbon sources might provide some competitive advantage to these organisms in competing for nutrients and colonizing botanical environments. Members of the Klebsiella-Enterobacter group have also been isolated (with the medium of Hino and Wilson; see Section III.B) from the surfaces of root nodules and roots of white clover plants (Evans et al., 1972).

Another approach for obtaining diazotrophic Enterobacteriaceae is to use a selective medium such as m-Brilliant Green broth (Difco; see Knowles, 1982) for a typical most probable number (MPN) enumeration. Following this first enumeration, aliquots from positive tubes of the Brilliant Green broth are transferred to a nitrogen-free medium such as Hino and Wilson's (1958) medium and incubated under anaerobic conditions until turbidity develops. At this point 0.25 ml of a 20% solution of sucrose is added and 5% of the gas phase is replaced with acetylene. After another 24 hr of incubation the tubes are assayed for ethylene production and any tubes showing more ethylene than an uninoculated blank are scored as positive. The most probably number is derived from the appropriate tables. Wright and Weaver (1981) obtained diazotrophic enteric bacteria using MacConkey agar, but there was no advantage in using the medium since all colonies had to be tested individually to determine if they were nitrogenase positive. Twenty-six percent of the isolates from MacConkey agar were nitrogenase positive.

Facultative diazotrophs of the genus *Bacillus* have frequently been obtained with the medium of Hino and Wilson (Hino and Wilson, 1958; Neal and Larson, 1976). When attempting to obtain N_2-fixing *Bacillus* sp., the resistance of endospores to "pasteurization" should not be overlooked as a means for selection. Thus inoculating enrichment media with heat-treated (80°C, 10 min) soil suspensions or macerated roots followed by incubation of nitrogen-free enrichment media under anaerobic conditions could significantly increase the ease of isolation of the spore-forming *Bacillus* species. *Bacillus* species can be differentiated from clostridia by streaking the nitrogenase-positive enrichment culture onto a general-purpose medium (e.g., Tryptic Soy agar) incubated under aerobic conditions. Single-colony isolates should be examined for nitrogenase activity in a nitrogen-free medium under anaerobic conditions.

Anaerobic Bacteria

Detailed accounts of methods for enrichment and isolation of anaerobic bacteria, including diazotrophs, are given by Skinner (1971), Holdeman et al. (1972), Costilow (1981), Kaspar and Tiedje (1982), and Knowles (1982). The procedures for incubating media under anaerobic conditions, described briefly in the preceding section, can be applied to the culture of the anaerobic bacteria as well. Some further precautions may be necessary for strictly anaerobic bacteria. These precautions include use of prereduced, deoxygenated media and performance of all manipulations in an anaerobic environment such as a glove bag or glove box. Deoxygenation can be achieved by heating the media (100°C) followed by rapid cooling just prior to inoculation or by sparging the medium with sterile (glass fiber or micropore filter) O_2-free nitrogen gas (Kaspar and Tiedje, 1982). Reducing agents such as sodium thioglycollate (E_0' < -100 mV, 0.05% final concentration) and cysteine hydrochloride (E_0' -210 mV, 0.025% final concentration), as well as others (see Costilow, 1981; Kaspar and Tiedje, 1982), can be added to poise the redox potential of the media. This is necessary for the culture of the sulfate-reducing bacteria (*Desulfovibrio* and *Desulfotomaculum*) which might be associated with soils and roots of rice and aquatic plants. Culture vessels should be flushed with N_2 gas and tightly stoppered following inoculation or they should be placed in a convenient anaerobic chamber (anaerobe jar, vacuum desiccator, etc.).

For the culture of nonstringent anaerobes (e.g., most clostridia) most operations for media preparation and transfer of bacteria can be conducted in a normal atmosphere, taking care to minimize exposure of media, and so on, to oxygen. Costilow (1981) and Kaspar and Tiedje (1982) list useful procedures for handling nonstringent and stringent anaerobic bacteria.

B. Media for Isolation

Formulas for many nitrogen-free culture media can be found in the literature dating back to the original works of Beijerinck (ca. 1901-1908) and Winogradski (ca. 1938-1949). For the sake of brevity, only a few of the more commonly used media for the enrichment and isolation of various root-associated and soil bacteria will be given here. Responsibility for any glaring omissions rests solely with the author. However, the media listed should provide the researcher with at least a "good start" down the road toward isolation and purification of a variety of diazotrophic bacteria.

Media for the enrichment, isolation, and culture of diazotrophs are commonly phosphate-buffered solutions containing a variety of essential mineral nutrients and a carbon source, and lacking a source of combined nitrogen such as ammonium ions, nitrate ions, and organic nitrogen-containing compounds. Iron and molybdenum must be provided for the synthesis of nitrogenase. Truly nitrogen-free solid media can be prepared with silica gel (see Dobereiner, 1980). Media containing agar will usually contain trace amounts of combined nitrogen. Skinner (1971) reported the nitrogen content of several commercially prepared agars to range from 0.06 to 0.098% nitrogen. He calculated that a medium containing 1.5% w/v of agar with 0.098% nitrogen would contain 14.7 mg N/liter. He considered this amount too low to interfere with nitrogenase activity.

Since the phosphates in many nitrogen-free media form precipitates with other salts upon autoclaving, the phosphate salts are usually autoclaved separately and added to the bulk of the medium when it has cooled. Autoclaving the phosphate separately from the sugars also prevents the formation of toxic products and the caramelization of the sugar. This is especially true for glucose-containing media (Skinner, 1971; Brown et al., 1962).

Media for Azotobacter *spp.*

Burk's medium (Burk and Lineweaver, 1930; Rennie, 1981)

Sucrose	20.0 g
K_2HPO_4	0.08 g
KH_2PO_4	0.20 g
$MgSO_4 \cdot 7H_2O$	0.41 g
$CaCl_2$	0.30 g
$Na_2MoO_4 \cdot 7H_2O$	0.10 mg
$FeSO_4 \cdot 7H_2O$	0.70 mg
Distilled water	1000 ml
pH 7.3	

Glucose agar (Brown et al., 1962)

Glucose	5.0 g
$MgSO_4 \cdot 7H_2O$	0.2 g
$FeSO_4 \cdot 7H_2O$	0.04 g
$Na_2MoO_4 \cdot 2H_2O$	0.005 g
$CaCl_2$	0.15 g
Agar	15.0 g
Distilled water	1000 ml

The ingredients above are autoclaved together, and after cooling to ≅42°C, 0.8 g/liter K_2HPO_4 is added from a separately sterilized solution and the medium is distributed aseptically into tubes or plates. The final pH is 6.8 to 7.0.

Modified Burk's medium (mannitol-B_6 medium; Dalton and Postgate, 1969; Dalton, 1980)

Mannitol	10.0 g
K_2HPO_4	0.64 g
KH_2PO_4	0.16 g

NaCl	0.20 g
$MgSO_4 \cdot 7H_2O$	0.20 g
$CaCl_2$	0.10 g
Trace elements	
$FeSO_4 \cdot 7H_2O$	2.5 mg
H_3BO_3	2.9 mg
$CoSO_4 \cdot 7H_2O$	1.2 mg
$CuSO_4 \cdot 7H_2O$	0.1 mg
$MnCl_2 \cdot 5H_2O$	0.1 mg
$Na_2MoO_4 \cdot 2H_2O$	2.5 mg
$ZnSO_4 \cdot 7H_2O$	2.1 mg
Nitrilotriacetic acid	100.0 mg
Distilled water	1000 ml

Nitrilotriacetic acid is added to a solution of trace elements and adjusted to pH 7.0 with NaOH prior to addition to the remaining medium. A precipitate formed during autoclaving can be eliminated by autoclaving the phosphates separately or by allowing the medium to stand for several days. The medium can be solidified by adding agar (20 g/liter). Final pH is 7.4 ± 0.2.

Media for Beijerinckia *spp.*

Enrichment medium for *Beijerinckia* (Becking, 1961)

Glucose	20.0 g
KH_2PO_4	1.0 g
$MgSO_4 \cdot 7H_2O$	0.5 g
Distilled water	1000 ml
Adjust pH to 5.0	

The liquid medium is poured in thin layers (2 to 3 mm) into petri dishes. Approximately 0.5 g of soil is used as inoculum for each plate. The plates are incubated for 2 to 3 weeks at 30°C and then examined microscopically for the presence of cells resembling *Beijerinckia*. In an advanced stage of development the entire medium changes into a viscous slimy mass which can be drawn out into threads.

Plating medium for *Beijerinckia* (Becking, 1961)

Glucose	0.1 g
KH_2PO_4	1.0 g
$MgSO_4 \cdot 7H_2O$	0.5 g
$FeCl_3 \cdot 6H_2O$	0.1 g
$Na_2MoO_4 \cdot 2H_2O$	0.02 g
Agar	20.0 g

Distilled water	1000 ml
pH 5.0	

The medium is prepared as two solutions:

Solution A. 20.0 g glucose is dissolved in 500 ml of distilled water.
Solution B. The salts are dissolved in 500 ml of distilled water, adjusted to pH 5.0, and agar is then added.

Solutions A and B are autoclaved (121°C, 15 min) and mixed after cooling to approximately 42°C.

Liquid medium for *Beijerinckia* (Becking, 1974)

Glucose	20.0 g
K_2HPO_4	0.8 g
KH_2PO_4	0.2 g
$MgSO_4 \cdot 7H_2O$	0.5 g
$FeCl_3 \cdot 6H_2O$	0.05 g
$CaCl_2 \cdot 2H_2O$	0.05 g
$MnSO_4 \cdot 2H_2O$	0.005 g
$ZnSO_4 \cdot 6H_2O$	0.005 g
$CuSO_4 \cdot 5H_2O$	0.004 g
$Na_2MoO_4 \cdot 2H_2O$	0.002 g
Distilled water	1000 ml
Adjust pH to 6.9	

All strains of *Beijerinckia* were reported to grow well in this medium at pH values down to 4.0.

Media for Azospirillum

Nitrogen-free malate medium (medium NFb; Dobereiner, 1980)

Malic acid	5.0 g
K_2HPO_4	0.5 g
$MgSO_4 \cdot 7H_2O$	0.2 g
NaCl	0.1 g
$CaCl_2$	0.02 g
$Na_2MoO_4 \cdot 2H_2O$	0.002 g
$MnSO_4 \cdot H_2O$	0.01 g
KOH	4.5 g
Biotin	0.1 mg
Fe-EDTA (1.64% w/v, aqueous)	4.0 ml
Bromothymol blue (0.5% w/v in ethanol)	3.0 ml
Distilled water	1000 ml

Adjust pH to 6.8 with NaOH, then add agar. The medium is green at its initial pH. For semisolid medium add 1.75 g of agar; for solid medium add 15 g of agar and 20 mg of yeast extract.

Combined carbon medium (Rennie, 1981; see "Multipurpose Nitrogen-Free Media" below). *Azospirillum* species can be grown on the combined carbon medium of Rennie (1981) in semisolid agar deeps or on agar plates.

Medium of Okon et al. (1977).

K_2HPO_4	6.0 g
KH_2PO_4	4.0 g
$MgSO_4 \cdot 7H_2O$	0.2 g
NaCl	0.1 g
$CaCl_2$	0.02 g
$FeCl_3$	0.01 g
$Na_2MoO_4 \cdot 2H_2O$	0.002 g
Distilled water	1000 ml
Adjust pH to 6.8 with NaOH	

Succinate, malate, or lactate (0.5%) are sterilized separately in 10-fold concentration and added to the cooled medium.

The medium was devised primarily for the laboratory cultivation of cells. The medium can be made semisolid by adding 0.5% agar, which will encourage growth under microaerophilic conditions. The nitrogen-free malate medium of Dobereiner (above) is generally used for isolation of *Azospirillum* from roots and soils.

RC medium (rojo Congo or Congo red; Rodríquez-Cáceres, 1982)

K_2HPO_4	0.5 g
$MgSO_4 \cdot 7H_2O$	0.2 g
NaCl	0.1 g
Yeast extract	0.5 g
$FeCl_3 6H_2O$	0.015 g
DL-Malic acid	5.0 g
KOH	4.8 g
Agar	20.0 g
Distilled water	1000 ml

Adjust pH to 7.0 with 0.1 *N* KOH and autoclave at 121°C for 20 min. After autoclaving add 15 ml of a 1:400 aqueous solution of Congo red (autoclaved separately) to each liter of the melted medium.

Colonies of *Azospirillum* (96-hr incubation, 37°C) are recognized by a characteristic scarlet coloration in addition to other typical colony characteristics, including "dry consistency, diameter of 1.5 to 2.0 mm, round or irregular form, undulate edge and rugose surface with ridges radiating from the center" (Rodríquez-Cáceres, 1982). The medium is of practical value when sufficient experience in recognizing the characteristic colonies of azospirilla has been acquired (J. R. Milam, University of Florida, personal communication). It is apparently more useful in working with samples with abundant azospirilla.

Media for Facultative Anaerobes Including Bacillaceae *and* Enterobactericeae

Hino and Wilson medium (Hino and Wilson, 1958): The medium is prepared from two solutions:

Solution A:

Sucrose	20.0 g
$MgSO_4 \cdot 7H_2O$	0.5 g
NaCl	0.01 g
$FeSO_4 \cdot 7H_2O$	0.015 g
$Na_2MoO_4 \cdot 2H_2O$	0.005 g
$CaCO_3$	10.0 g
Distilled water	500 ml

Solution B:

K_2HPO_4-KH_2PO_4 buffer, 0.1 *M*, pH 7.7	500 ml
p-Aminobenzoic acid	10 μg
Biotin	5 μg

The solutions are autoclaved separately and combined when cool. If a clear medium is desired, the $CaCO_3$ can be replaced by $CaCl_2$ (0.05 g/liter). The medium can be made semisolid by addition of 1.75 g/liter agar or solidified by the addition of 2% agar. The medium was originally used to isolate diazotrophic *Bacillus* species but many investigators have successfully used it to culture diazotrophic *Enterobacteriaceae*. The medium can also be used for the large-scale cultivation of *Klebsiella* for investigations requiring large quantities of cells (Dalton, 1980).

Medium for *Klebsiella pneumoniae* (Hill, 1976)

Sucrose	20.0 g
Na_2HPO_4	10.4 g
KH_2PO_4	3.4 g
$CaCl_2 \cdot 2H_2O$	26 mg
$MgSO_4$	30 mg
$MnSO_4$	0.3 mg
Ferric citrate	36 mg
$Na_2MoO_4 \cdot 2H_2O$	7.6 mg
p-Aminobenzoic acid	10 μg
Biotin	5 μg
Distilled water	1000 ml

The phosphates are sterilized separately and added to the medium when cool. The buffering capacity maintains a pH of 6.7 ± 0.1. The medium was used for growth of *Klebsiella* in a chemostat with limiting concentrations of glucose (15 to 30 m*M*; Hill, 1976).

Burk's medium (see "Media for *Azotobacter* spp." above): Burk's medium has been used to culture *Klebsiella* (Bergersen and Hipsley, 1970) and *Bacillus* sp. (Nelson et al., 1976). The latter species was obtained through the use of Pankhurst tubes to enumerate diazotrophs associated with grass roots in Oregon.

Media for Clostridium *spp.*

Nitrogen-free medium for *Clostridium pasteurianum* (Daesch and Mortenson, 1972; Smibert and Krieg, 1981)

$MgSO_4 \cdot 7H_2O$	0.0493 g
$FeCl_3 \cdot 6H_2O$	0.0541 g
$MnSO_4 \cdot H_2O$	0.0034 g
$Na_2MoO_4 \cdot 2H_2O$	0.0048 g
$ZnSO_4 \cdot 7H_2O$	0.00058 g
$CuSO_4 \cdot 5H_2O$	0.00050 g
$CoCl_2 \cdot 6H_2O$	0.00048 g
K_2HPO_4	1.132 g
Biotin	Trace
Sucrose	20.0 g
$CaCO_3$	3.0 g
Distilled water	1000 ml

All ingredients are combined and autoclaved. The $CaCO_3$ can be increased to 30 g for increased buffering. The medium should be made anaerobic by steaming or bubbling with O_2-free N_2 gas prior to inoculation.

Medium containing a reducing agent and indicator (Brouzes et al., 1971)

Glucose	20.0 g
Sodium acetate	2.0 g
$CaCO_3$	0.50 g
$Na_2MoO_4 \cdot 2H_2O$	0.01 g
KH_2PO_4	0.50 g
Sodium thioglycollate	0.5 g
Agar	0.75 g
Phenosafranin (0.2% ag.)	8 ml
Soil extract	10 ml

The medium is dispensed in culture tubes or Pankhurst tubes containing inverted Durham tubes, and sterilized at 121°C for 15 min. The medium should be heated to 100°C before use unless used immediately following sterilization. The cooled tubes are inoculated and sealed with sterile rubber stoppers. The possible criteria for positive tubes are production of gas, reduction of the indicator, and acetylene reduction activity (Knowles, 1982).

Modified Burk's medium for N_2-fixing clostridia (Skinner, 1971)

Glucose (or sucrose)	10.0 g
K_2HPO_4	0.8 g
KH_2PO_4	0.2 g
$MgSO_4 \cdot 7H_2O$	0.2 g
NaCl	0.2 g
$FeSO_4 \cdot 7H_2O$	0.01 g
$MnSO_4 \cdot 7H_2O$	0.01 g
$CaCl_2$	0.01 g
$Na_2MoO_4 \cdot 2H_2O$	0.025 mg
Yeast extract	1-10 mg

Trace elements: 1 ml of a solution containing the following (g/liter):

$NaBO_2$	0.05 g
$[Co(NO_3)_2 \cdot 6H_2O]$	0.05 g
$CdSO_4$	0.05 g
$CuSO_4$	0.05 g
$ZnSO_4$	0.05 g
Soil extract	10 ml
Sodium thioglycollate	1.0 g
Distilled water	1000 ml
Adjust pH to 7.2	

Since facultative *Bacillus* sp. will grow under the same conditions used to obtain clostridia, isolates obtained from enrichment cultures should be checked for growth under aerobic conditions on a general-purpose medium such as nutrient agar to detect the facultative anaerobes. Pasteurization (80°C, 10 min) of samples at any of several points in isolation procedures is useful for purifying cultures that have sporulated.

Media for Derxia

Jensen's medium for *Derxia* (medium B; Jensen et al., 1960)

Glucose	20.0 g
K_2HPO_4	1.0 g
$MgSO_4 \cdot 7H_2O$	0.2 g
$CaSO_4$	0.2 g
$FeSO_4 \cdot 7H_2O$	0.2 g
$Na_2MoO_4 \cdot 2H_2O$	0.5 mg
Distilled water	1000 ml
Agar (when needed)	20.0 g
Adjust pH to 6.8-7.2	

Nitrogen-containing medium for *Derxia* (ATCC medium No. 165)

Glucose	20.0 g
K_2HPO_4	1.0 g
$MgSO_4 \cdot 7H_2O$	0.2 g
$CaSO_4$	5.0 mg
$FeSO_4 \cdot 7H_2O$	5.0 mg
$Na_2MoO_4 \cdot 2H_2O$	0.5 mg
NH_4Cl	2.0 g
Distilled water	1000 ml
Agar (when needed)	20.0 g

This medium is employed by the American Type Culture Collection for the culture and maintenance of *Derxia*.

Multipurpose N-Free Media

Combined carbon medium of Rennie (1981): The medium is prepared in two solutions containing the following:

Solution 1:

K_2HPO_4	0.8 g
KH_2PO_4	0.2 g
NaCl	0.1 g
$Na_2FeEDTA$	28.0 mg
$Na_2MoO_4 \cdot 2H_2O$	25.0 mg
Yeast extract	100.0 mg
Mannitol	5.0 g
Sucrose	5.0 g
Sodium lactate	0.5 ml (60% v/v)
Distilled water	900 ml

Solution 2:

$MgSO_4 \cdot 7H_2O$	0.2 g
$CaCl_2$	0.06 g
Distilled water	100 ml

The solutions are autoclaved separately and mixed when cool. Add 5.0 μg of biotin and 10.0 μg of *p*-aminobenzoic acid (from filter-sterilized solutions) for each liter of medium. The final pH is adjusted to 7.0. The medium is reported to be useful for the enrichment, enumeration, and isolation of a variety of free-living diazotrophs.

Low-nitrogen medium of Watanabe and Barraquio (1979): This medium is a semisolid glucose-yeast extract medium containing the following:

Glucose	5.0 g
Yeast extract	0.1 g
K_2HPO_4	0.5 g

$FeSO_4 \cdot 7H_2O$	0.04 g
$MgSO_4 \cdot 7H_2O$	0.2 g
$CaCl_2$	0.2 g
$Na_2MoO_4 \cdot 2H_2O$	5.9 mg
H_3BO_3	0.15 mg
$MnCl_2 \cdot 4H_2O$	0.004 mg
$ZnSO_4 \cdot 7H_2O$	0.11 mg
$CoSo_4 \cdot 7H_2O$	0.07 mg
$CuSo_4 \cdot 7H_2O$	0.005 mg
Noble agar (Difco)	1.75 g
Distilled water	1000 ml
Adjust pH to 7.0	

Watanabe and Barraquio (1979) reported that a large percentage of isolates from rice roots can grow in this medium. The same isolates are unable to grow and reduce acetylene in the same medium lacking yeast extract. Casamino acids (0.1 g/liter) and glutamic acid (0.1 g/liter) also promote growth and acetylene reduction. The authors suggested that use of media containing low levels of combined nitrogen can reveal a much higher incidence of diazotrophs among the root-associated microflora than was previously suspected.

C. General Procedures for Isolation

Once a source material (soil, washed roots, etc.) has been chosen as inoculum, there are several routes (i.e., solid or liquid enrichment media) which can be followed to obtain pure cultures of diazotrophs. The techniques are essentially the same whether aerobic, facultative, or anaerobic bacteria are sought. If strict anaerobes are the object of the search, the special precautions for working with anaerobes should be employed. Otherwise, conventional procedures for anaerobic bacteria can be employed (see Kaspar and Tiedje, 1982).

Many of the Azotobacteraceae can be obtained from plates of nitrogen-free media which have been inoculated by sprinkling soil onto the agar surface. Large, usually mucoid, raised colonies developing on the plates are examined microscopically for the presence of cells resembling *Azotobacter* [large ovoid cells with PHB (poly-β-hydroxybutyric acid) granules], and cells from these colonies are streaked onto the same medium to obtain single-colony isolates. Frequent restreaking (e.g., at daily intervals) helps to eliminate nondiazotrophs which can adhere to the slime produced by the azotobacters. Similar procedures have been used for other species of diazotrophs.

A convenient method to obtain root-associated diazotrophs is simply to press washed roots into the surface of the nitrogen-free medium in petri plates. After incubation under suitable conditions, the occurrence of large mucoid colonies along the root is presumptive evidence for diazotrophs. Material from these colonies can be purified by repeated streaking as above. It may also be helpful to suspend a small portion of the growth from enrichment plates in distilled water or buffer and mixing thoroughly to dislodge nondiazotrophs associated with the slime of the diazotrophs. The suspension (or dilutions thereof) is then streaked onto fresh solid medium.

Isolation of diazotrophs from soils and roots can also be initiated using liquid nitrogen-free enrichment media particularly suitable for a specific group of bacteria. Separation of the diazotrophs is best accomplished by frequent transfer (1 to 3 days) of the enrichment culture to fresh medium. This minimizes the development of nondiazotrophs utilizing nitrogenous excretions in the enrichment culture. The enrichment culture is finally streaked onto a nitrogen-free medium (preferably of the same composition) to obtain single-colony isolates. All isolates developing on nitrogen-free agar plates should be tested for ability to reduce acetylene in semisolid deeps, and positive tubes should be streaked onto tryptic soy agar to check purity.

Semisolid agar (1.75 to 4.0 g/liter) deeps are widely used in the enrichment and enumeration (by MPN techniques) of diazotrophs. For enrichment techniques the semisolid medium is usually inoculated with a small piece (0.5 to 1.0 cm) of washed or surface-sterilized root or with dilutions of soil samples as in most probable number techniques. The root should be crushed with a foreceps as it is placed into the agar deep. For facultative and anaerobic bacteria the tubes can be incubated in an anaerobic chamber (vacuum desiccator, etc.) or they can be flushed individually with N_2 and capped with a serum bottle stopper. The latter procedure facilitates the assessment of acetylene reduction following a suitable incubation period. Growth of aerobic (microaerophilic) bacteria usually develops as a narrow distinct band of cells situated at a favorable oxygen tension which may move toward the agar surface as the abundance of cells increases. Anaerobes usually grow at the lower depths in the agar and the growth tends to be more diffuse. Isolation from the semisolid enrichment cultures or from MPN tubes which are positive for acetylene reduction is accomplished by streaking solid media of the same composition and incubating under an atmosphere appropriate for the group of bacteria being sought. Alternatively, the culture may be transferred one to several times to fresh semisolid deeps prior to streaking for isolation. As indicated earlier, the use of semisolid media can allow for the development of facultative and anaerobic bacteria in tubes incubated "aerobically." This is not so much a problem in enrichment as it is in enumeration procedures for aerobic diazotrophs since acetylene reduction in tubes incubated aerobically may be mediated by facultative or even anaerobic diazotrophs.

The incorporation of small amounts of combined nitrogen into solid media often facilitates the development of aerobic diazotrophs on culture plates incubated in air. It is believed that this "starter" nitrogen is helpful in establishing cells and contributes to the synthesis of an active nitrogenase system. It is also possible that the greater number of cells formed in the presence of small amounts of nitrogen helps to lower the oxygen concentration in colonies, leading to a more active synthesis of nitrogenase. Watanabe and Barraquio (1979) reported obtaining greater numbers of diazotrophs from rice roots when they used media containing yeast extract or casamino acids (100 mg/liter). Rennie (1981) also incorporated yeast extract (100 mg/liter) into his combined carbon medium, and Dobereiner (1980) recommended the incorporation of 20 mg/liter of yeast extract when using the malate medium for *Azospirillum* in culture plates.

Another approach to the enrichment of root-associated diazotrophs, the spermosphere model, has recently been described by Thomas-Bauzon et al. (1982). They described the technique for isolating diazotrophs associated with rice roots; however, it should be suitable for other plant species as well. The technique utilizes young gnotobiotic seedlings grown in a carbon- and nitrogen-free medium for several days. Following their establishment (coleoptiles 1 cm tall) and confirmation of sterility the tubes containing the seedlings are inoculated with a soil suspension or other suitable inoculum. Diazotrophic bacteria capable of using seed or radicle exudates develop in these cultures and following confirmation of acetylene reduction the enrichment culture is streaked onto solid media for isolation. The technique has the advantages of selecting those organisms that proliferate on plant-derived carbon sources and of maintaining very low concentrations of combined nitrogen through uptake by the growing seedling. Using this model system with rice seedlings, approximately 78% of the isolates obtained were members of the Enterobacteriaceae; the remainder were *Pseudomonas*-like organisms and *Azospirillum*.

The purification of cultures of diazotrophs is frequently complicated by the production of copious quantities of slime on nitrogen-free media. Contaminating microorganisms become embedded in the slime and are difficult to separate from the diazotroph. This can be overcome, in part, by suspending the culture in a suitable diluent prior to streaking (as above) and also by purification on media such as nutrient or tryptic soy agar. Slime production is greatly reduced on these media and purification is made less complicated. Development of distinct colonial morphologies [e.g., the pinkish, often wrinkled colonies of *Azospirillum* on Tryptic Soy agar (Difco)] is also facilitated on these general-purpose media. If several colony types emerge during this phase of purification, each should be tested for its ability to reduce acetylene under aerobic and anaerobic conditions. If all isolates do not reduce acetylene separately, combinations may be inoculated into nitrogen-free agar deeps to check for acetylene reduction by mixed cultures. In our laboratory we have had the experience of individual isolates failing to reduce actylene while recombined mixed cultures of the same organisms demonstrate acetylene reduction. These interesting microbial associations deserve further attention. They also serve to illustrate an important point about purification from enrichment cultures—that cultures become more difficult to grow under conditions for nitrogen fixation as purification (simplification of the mixed microflora) proceeds away from the initial enrichment (Lindberg and Granhall, 1984; Tyler et al., 1979). It may become necessary to use more exacting methods during later stages of isolation schemes. This is particularly true for the isolation of the anaerobic diazotrophs.

IV. MAINTENANCE AND PRESERVATION OF DIAZOTROPHS

The major goal of culture preservation is to keep the organism alive, free of contamination, and without variation due to mutation. There are many methods for culture preservation; however, not all methods achieve equal results with different organisms. In general, available methods for culture preservation fall into two broad categories: short-term preservation and long-term preservation. Short-term methods include routine serial subculturing, storage under mineral oil, conventional freezing, and assorted drying techniques. Long-term methods include freeze-drying (lyophilization) and storage at ultralow temperatures (e.g., in liquid nitrogen; -196°C). Detailed accounts of methods and problems involved in maintenance and preservation are given by Lapage et al. (1970b), Lapage and Redway (1973), Bridges (1966), and Gherna (1981).

The decision of which method to use is usually based on the size of the culture collection to be maintained, the equipment available, and the duration of preservation required.

A. Short-Term Methods

Subculturing

Probably the simplest and most widely used method for culture maintenance is that of serial subculturing in which the organism is periodically transferred to fresh medium and stored at room temperature or in a refrigerator. The maintenance medium should preferably be a minimal medium to reduce the metabolic activity of the culture. Culture tubes should be sealed with airtight closures to reduce or eliminate evaporation and subsequent dehydration of the medium. Screw caps with rubber or Teflon liners are suitable for this purpose, and extra protection can be provided by wrapping the tops of the tubes with Parafilm. Storage in the refrigerator (4 to 8°C) further reduces the metabolic rate of the cells and lengthens the period of storage.

Many bacteria can be maintained for periods of 3 to 5 months in this manner (Gherna, 1981).

The transfer interval is variable and prior experience is frequently the best reference. Most free-living diazotrophs in culture collections are not highly sensitive and will tolerate storage periods of one to several months. Thompson and Skerman (1979) maintained *Azotobacter* species on nitrogen-free sucrose-containing media with transfers at 1- to 2- month intervales. Tyler et al. (1979) maintained strains of *Azospirillum brasilense* and *Azospirillum lipoferum* on Tryptic Soy agar (Difco) with monthly transfers. Tarrand et al. (1978) maintained stock cultures of *Azospirillum* on the semisolid malate medium of Day and Dobereiner (1976) containing 0.005% yeast extract. Strains requiring yeast extract were transferred biweekly; all others were transferred monthly. We have maintained diazotrophic strains of *Klebsiella* and *Enterobacter* on slants of Tryptic Soy agar for periods of up to 6 months at room temperature or in the refrigerator. These cultures grow overnight when transferred to fresh media and nitrogenase activity is observed when the cultures are transferred to nitrogen-free media incubated under N_2 (Zuberer, unpublished).

It is wise to keep two tubes of each culture in the stock collection. One tube is used as a working stock to prepare subcultures for distribution; the other is entered only if the other tube becomes unusable for some reason (i.e., contamination or loss of viability). If the reserve culture must be opened, it should be used immediately to initiate two new tubes: one to serve as a new reserve and one to serve as a working culture. When subculturing it is advisable to sample from the whole slant or from 50 to 100 colonies from a plate. This reduces the chances for genotypic and phenotypic variations in the culture (Lapage et al., 1970b). The disadvantages of the serial transfer method are the risks of contamination, mislabeling of cultures, the selection of variants, possible loss of cultures, and storage space required (Gherna, 1981).

Preservation Under Oil

The longevity of cultures of many bacteria can be extended by placing them under sterile medicinal-grade mineral oil (paraffin oil of specific gravity of 0.865 to 0.890) (Gherna, 1981). Once the culture has been grown out on an agar slant or agar deep it is covered with at least 2.0 cm of sterile mineral oil (sterilized in an oven 170°C for 1 to 2 hr). In the case of agar slants the oil should extend at least 1.5 cm above the upper end of the slope. Culutres can be stored at room temperature or, preferably, in a refrigerator at 0 to 5°C. The oil further reduces the metabolic activity of the culture, and cultures of many bacteria survive for periods of several months to years.

The same disadvantages as those of serial transfer apply. In addition, the oil can be messy and care should be used when flaming the needle or loop following transfer from a culture with oil, as the oil will spatter, causing contamination of lab surfaces and personnel (Gherna, 1981).

Conventional Freezing

Survival of bacteria at normal freezer temperatures (0 to -20°C) is highly variable and depends largely on the organism. Some organisms will last for several years under these conditions. Damage to organisms results from the concentration of salts due to freezing and in general the method is not recommended.

Short-Term Preservation Through Drying Procedures

Methods for preservation of cultures through drying include storage in sterile air-dried soil, on sterile filter paper disks or strips, and in dried gelatin droplets. Spore-forming bacteria are especially amenable to storage in air-dried soil. A well-sporulated culture (1 ml) is added to sterile soil, left at room temperature until visibly dry, then sealed with an airtight closure and stored in the refrigerator. Cultures have been stored for years with this method (Gordon and Rynearson, 1962). The culture can be rejuvenated by aseptically retrieving a pinch of the soil as inoculum for fresh media.

Preservation of cultures through drying on sterile filter paper is easy and inexpensive. A bacterial suspension containing approximately 10^8 cells/ml is allowed to saturate the paper strips or disks, which are then dried in a vacuum desiccator. The dried samples are then sealed in tubes or between clear plastic sheets (Coe and Clark, 1966) and stored in a desiccator in the refrigerator. The cultures are rejuvenated by aseptically transferring a paper strip or disk to a fresh liquid medium. Gherna (1981) indicates that members of the Enterobacteriaceae are successfully preserved for several years in this manner.

Another simple drying procedure is the gelatin disk method (Stamp, 1947), which requires no specialized equipment and can be carried out in almost any laboratory (see Lapage et al., 1970b). Bacteria are suspended in a mixture of nutrient broth, gelatin (10% w/v), and ascorbic acid (0.25% w/v). Small drops of the cell suspension are placed in sterile plastic petri dishes and transferred to a vacuum desiccator containing phosphorus pentoxide. A vacuum of 10 to 30 cm Hg is applied to the desiccator and the drops allowed to dry at room temperature. The dried disks are picked from the petri dish and stored at 0 to 5°C in a sterile vial containing silica gel. The culture is rejuvenated by transferring a disk to a broth medium. Stamp (1947) found a wide range of organisms to survive up to 4 years in this manner.

Another method which may be suitable for many free-living diazotrophs is that of drying on unglazed porcelain beads (Hunt et al., 1958; Norris, 1963). Small screw-capped vials (15 ml capacity) containing indicating silica gel covered with glass wool and capped with aluminum foil are sterilized in an oven (170°C, 1 to 2 hr). The caps are autoclaved separately to preserve rubber liners. About 20 to 30 beads are soaked in a cell suspension, drained, and transferred to the sterile vial, which is then tightly sealed. Vials are stored at room temperature or in the refrigerator and should last for several years. The culture is reactivated by placing a bead into a fresh liquid medium. One source for the beads is McDaniel Refractory Porcelain Co., Beaver Falls, Pennsylvania.

B. Long-Term Methods

Freeze-Drying (Lyophilization)

Freeze-drying is generally cited as the method of choice for long-term preservation of cultures (Lapage and Redway, 1973). It is economical and effective for storage of large numbers of cultures in a minimum of space. There are several methods for freeze-drying bacteria for long-term preservation. Excellent treatments of this subject are given by Lapage et al. (1970b) and Gherna (1981); the reader is urged to consult these references for detailed information.

Freeze-drying involves the removal of water from frozen cell suspensions by sublimation under vacuum; that is, water passes from the solid to the vapor phase without going through the liquid state. Cells dried in this manner can survive for extended periods and can be conveniently rehydrated and restored to their previous state.

Successful freeze-drying depends on starting with healthy cells grown under optimum conditions and harvested during the late logarithmic or early stationary phases of growth (Gherna, 1981). Lapage et al. (1970a) have presented an extensive listing of bacterial genera and the media used in the maintenance, preservation, and preparation for freeze-drying, including suspending media for the freeze-drying process. The list includes many diazotrophic bacteria. According to Lapage and Redway (1973), suspending media should contain a substance capable of maintaining the residual moisture content at an optimal level [1 to 2%; e.g., carbohydrates such as glucose, sucrose, or sorbitol (some carbohydrates are deleterious)]; a compound that contains amino groups (e.g., peptone or glutamic acid) which neutralize carbonyl groups said to be toxic; minimal electrolytes; and a substance that forms a cake when dried to serve as mechanical protection of the dried organisms, to aid the sublimation of liquid, and to prevent dispersal of the organism when the vial is opened. The suspending medium usually includes a cryoprotective agent such as skim milk (20%)

or sucrose (12% final concentration) (Gherna, 1981). Other cryoprotectants include 10% dextran, horse serum, inositol, and other chemicals.

The equipment for freeze-drying can be quite simple and reasonably inexpensive or complex and expensive (e.g., commercial freeze-driers). Good results can be obtained with simple systems composed of a high-vacuum pump, a condenser, and a manifold. Gherna (1981) describes equipment and methodology for freeze-drying using double-vial and single-vial manifold methods. Following drying, vials are sealed under vacuum using a dual-tipped air/gas torch.

Freeze-dried cultures are best maintained below 5°C and storage at -30 to -69°C can be expected to increase shelf life. Many bacteria (*Bacillus, Klebsiella, Enterobacter, Clostridium*) can be expected to remain viable in the freeze-dried state for periods exceeding 30 years.

Revival of freeze-dried cultures is accomplished by opening the sealed vial under aseptic conditions. After scoring the neck of the vial with a file, it should be swabbed with alcohol (70%) to remove contamination. To break the vial, place it in sterile gauze and apply pressure to the scored mark. Using sterile gauze minimizes the possibility for contamination by airborne microorganisms when air rushes into the evacuated vial and can help contain the escape of pathogens. The culture is rehydrated by adding 0.3 to 0.5 ml of a suitable sterile broth medium to dissolve the cake or pellet. The culture is then transferred to a culture tube of the same medium and following thorough mixing, 0.2 ml is transferred to an agar slant or semisolid medium of the same composition. A purity check is performed by streaking the serially diluted rehydrated culture onto a solid medium. Tubes and plates are incubated under optimal conditions, and transferred to a fresh medium as soon as growth occurs and purity is confirmed. Ample time should be allowed for growth, as lyophilized cultures frequently exhibit long lag periods during reconstitution.

Ultralow-Temperature Freezing

A second method for long-term preservation of bacteria is that of ultralow-temperature freezing, in which cultures are maintained from -70 to -196°C. This is frequently suitable for organisms that do not withstand freeze drying and is said to overcome some of the undesirable aspects of freeze-drying (Bridges, 1966). The metabolic rate of the cells is reduced to virtually nil and many bacteria can be returned to the thawed condition with 90 to 100% survival. This affords the advantage of quite rapid recovery from the frozen state with almost immediate resumption of metabolic activity. Bridges (1966) suggested that frozen cultures could serve as sources of uniform inoculum for subsequent experiments to be carried out over extended time intervals.

As with freeze-drying, ultralow-temperature freezing requires the use of cryoprotectants in the suspending medium. The cryoprotective agents include glycerol and dimethyl sulfoxide (DMSO), which appears to provide intracellular and extracellular protection against freezing; and agents such as sucrose, lactose, glucose, mannitol, sorbitol, dextran, polyvinylpyrrolidone, and polyglycol, which appear to exert a protective effect extrenal to the cell membrane. Glycerol and DMSO are generally most effective in preserving a wide range of bacteria and are used at concentrations of 10 and 5% (v/v), respectively, in an appropriate medium (Gherna, 1981).

Cells to be frozen are harvested during the mid-to-late logarithmic phase of growth. Broth cultures are harvested by centrifugation with due regard to avoidance of contamination and the cells resuspended in fresh medium containing the cryoprotectant. Cells from agar cultures can be harvested by washing them from the agar surface using fresh liquid medium containing the cryoprotectant. A small volume (0.4 ml) of a cell suspension containing at least 10^8 cells/ml is transferred to a suitable glass or plastic vial which is then sealed and readied for freezing.

The rate of freezing is critical, and best survival and recovery is obtained by lowering the temperature 1°C/min to -30°C and 15 to 30°C/min to -150°C. This can be achieved with commercially available programmed freezers (e.g., Linde BF3-2; Union Carbide Corp.), or if such equipment is not available, satisfactory results can be achieved by placing the filled vials in a stainless steel pan on the floor of a Revco freezer at -60°C for 1 hr and then plunging the vials into liquid nitrogen for 5 min. The rate of cooling to -60°C in this manner is about 1.5°C/min (Gherna, 1981).

Thawing of frozen cultures should be rapid for greatest recovery of most microorganisms. Cultures can be thawed by immersion in a 37°C water bath with agitation until the ice melts (about 1 to 2 min depending on whether glass or plastic vials are used). The vials should be swabbed with 70% ethanol to disinfect the surface, opened, and the culture transferred to fresh medium. Use caution when bringing glass vials out of liquid nitrogen. Sealed vials may explode during warming.

Many bacteria (including many diazotrophs) can be preserved for periods in excess of 30 years in liquid nitrogen. The rapid recovery of frozen cultures is a major advantage of the method. The disadvantages of ultralow-temperature freezing are primarily those of expense, although moderately priced liquid nitrogen storage systems are available. The cost of liquid nitrogen is moderate and if the institution has a facility (e.g., cyclotron or electron microscopy center) using large quantities of liquid nitrogen, it is usually not too difficult to obtain a constant supply at a reasonable cost.

V. CONCLUSIONS

The techniques and culture media described herein should enable the researcher to obtain isolates of a wide variety of diazotrophic bacteria. The use of nitrogen-free media has been the mainstay for isolating diazotrophs; the lack of combined nitrogen conferring selectivity for diazotrophs. Further experimentation with selective agents such as dyes, antibiotics, and bile salts should be encouraged since such work may lead to the formulation of media with higher degrees of selectivity. Such an approach might prove worthwhile in the isolation and differentiation of diazotrophic enterobacteriaceae.

The characterization of diazotrophic isolates can be conducted with routine bacteriological techniques and should include tests for as many distinguishing features as possible. If the purpose of characterization is to establish that an isolate represents a new genus or species, description of the isolate should be exhaustive including, when possible, an analysis of quanine plus cytosine ratios and nucleic acid hybridization studies. The possibility of obtaining help from cooperative laboratories engaged in such procedures should not be overlooked. If, on the other hand, the purpose of characterization is simply to determine the major groups of diazotrophs in a given system, the investigator may wish to identify isolates to the genus level only.

The classification of isolates must assume added importance if such isolates are to be employed as inoculants in greenhouse or field soils where their effect on plant growth is to be assessed. Such investigations should include methods for monitoring the establishment of the organism on roots of a given host plant. This can be accomplished by reisolating the bacterium from inoculated plants or through the use of fluorescent antibody techniques to detect the organism on the roots. Plants from uninoculated treatments must also be observed to establish presence or absence of the bacterium used as inoculum. Prior to initiating inoculation experiments in field soils a thorough investigation should be made to estimate existing numbers of diazotrophs, especially those closely related to organisms employed as inoculants. Emphasis should be placed on determining that the inoculum establishes on the roots, remains throughout much of the growth period of the host, and can be recovered at the termination (e.g., harvest) of the experiment. If possible, any observed increases in N_2 fixation (acetylene reduction), dry matter production, nitrogen content, and so on, should be correlated with numbers of the bacteria used as inoculants before conclusions regarding the effects of inoculation are drawn.

Finally, any chapter of this nature would be remiss in not pointing out that in addition to taxonomic descriptions, *Bergey's Manual of Systematic Bacteriology* [Krieg and Hold (eds.), Vol. 1, Williams & Wilkins, Baltimore, Md., 1984] contains a wealth of information dealing with ecology, enrichment and isolation, maintenance, and preservation of bacteria. Valuable information on many nitrogen-fixing bacteria is available in the articles on pertinent genera.

ACKNOWLEDGMENTS

The author wishes to thank D. Sen and D. B. Alexander for their helpful suggestions regarding the manuscript, and Nancy Lee for skilled assistance in word processing of the manuscript.

REFERENCES

Aaronson, S. (1970). *Experimental Microbial Ecology.* Academic Press, Inc., New York, 236 pp.

ATCC (1982). *American Type Culture Collection Catalogue of Strains I*, 15th ed. American Type Culture Collection, Rockville, Md.

Attebery, H. R., and Finegold, S. M. (1970). *Am. J. Clin. Pathol. 53*: 383.

Becking, J. H. (1961). *Plant Soil 14*: 49.

Becking, J. H. (1974). *Soil Sci. 118*: 196.

Bergersen, F. J., and Hipsley, E. H. (1970). *J. Gen. Microbiol. 60*: 61.

Bridges, B. A. (1966). *Lab. Pract. 15*: 418.

Brouzes, R., Mayfield, C. I., and Knowles, R. (1971). *Plant Soil, Spec. Vol.*: 481.

Brown, M. E., Burlingham, S. K., and Jackson, R. M. (1962). *Plant Soil 17*: 309

Burk, D., and Lineweaver, H. (1930). *J. Bacteriol. 19*: 389.

Campbell, N. E. R., and Evans, H. J. (1969). *Can. J. Microbiol. 15*: 1342.

Claus, D., and Hempel, W. (1970). *Arch. Mikrobiol. 73*: 90.

Coe, A. W., and Clark, S. P. (1966). *Mon. Bull. Min. Health Public Health Lab. Serv. 25*: 97.

Costilow, R. N. (1981). In *Manual of Methods for General Bacteriology*, P. Gerhardt, ed. American Society for Microbiology, Washington, D.C.

Daesch, G., and Mortenson, L. E. (1972). *J. Bacteriol. 110*: 103.

Dalton, H. (1980). In *Methods for Evaluating Biological Nitrogen Fixation,* F. J. Bergersen, ed. John Wiley & Sons, Inc., New York.

Dalton, H., and Postgate, J. R. (1969). *J. Gen. Microbiol. 56*: 307.

Day, J. M., and Dobereiner, J. (1976). *Soil Biol. Biochem. 8*: 45.

Dobereiner, J. (1980). In *Methods for Evaluating Biological Nitrogen Fixation*, F. J. Bergersen, ed. John Wiley & Sons, Inc., New York.

Dobereiner, J., and DePolli, H. (1980). In *Nitrogen Fixation*, W. D. P. Stewart, ed. Academic Press, Inc., New York, p. 301.
Dobereiner, J., Day, J., and Dart, P. (1972). *J. Gen. Microbiol.* *71*: 103.
Evans, H. J., Campbell, N. E. R., and Hill S. (1972). *Can. J. Microbiol.* *18*: 13.
Gherna, R. L. (1981). In *Manual of Methods for General Bacteriology*, Philipp Gerhardt, ed. American Society for Microbiology, Washington, D.C.
Gordon, R. E., and Rynearson, T. K. (1962). In *Culture Collections: Perspectives and Problems*, S. M. Martin, ed. University of Toronto Press, Toronto, p. 118.
Hill, S. (1973). *Lab. Pract.* *22*: 193.
Hill, S. (1975). *J. Gen. Microbiol.* *91*: 207.
Hill, S. (1976). *J. Gen. Microbiol.* *93*: 335.
Hino, S., and Wilson, P. W. (1958). *J. Bacteriol.* *75*: 403.
Holdeman, L. V., Cato, E. P., and Moore, W. E. C. (1972). *Anaerobe Laboratory Manual*, 4th ed. Virginia Polytechnic Institute and State University, Blacksburg, Va.
Hungate, R. E. (1969). In *Methods in Microbiology*, Vol. 3B, J. R. Norris and D. W. Ribbons, eds. Academic Press, Inc., New York, p. 117.
Hunt, G. A., Gourevitch, A., and Lein, J. (1958). *J. Bacteriol.* *76*: 453.
Jensen, H. L., Petersen, E. J., De, P. K., and Bhattacharya, R. (1960). *Arch. Mikrobiol.* *36*: 182.
Johnson, L. F., and Curl, E. A. (1972). *Methods for Research on the Ecology of Soil-Borne Plant Pathogens.* Burgess Publishing Company, Minneapolis, Minn., 247pp.
Kaspar, H. F., and Tiedje, J. M. (1982). In *Methods of Soil Analysis, Part 2, Chemical and Microbiological Properties*, Agronomy Monograph 9, 2nd ed., A. L. Page, ed. ASA-SSSA, Madison, Wis., p. 989.
Klucas, R. (1972). *Can. J. Microbiol.* *18*: 1845.
Knowles, R. (1982). In *Methods of Soil Analysis,Part 2, Chemical and Microbiological Properties*, Agronomy Monograph 9, 2nd ed., A. L. Page, ed. ASA-SSSA, Madison, Wis., p. 1071.
Krieg, N. R. (1981). In *Manual of Methods for General Bacteriology*, P. Gerhardt, ed. American Society for Microbiology, Washington, D.C., p. 112.
Krieg, N. R., and Holt, J. G. eds. (1984). *Bergey's Manual of Systematic Bacteriology*, Vol. 1. Williams & Wilkins, Baltimore, Md., p. 964.
Lapage, S. P., and Redway, K. F. (1973). In *Handbook of Microbiology*, Vol. 1, *Organismic Microbiology*, A. J. Laskin and H. A. Lechevalier, eds. CRC Press, Inc., Cleveland, Ohio, p. 713.

Lapage, S. P., Shelton, J. E., and Mitchell, T. G. (1970a). In *Methods in Microbiology*, Vol. 3A, J. R. Norris and D. W. Ribbons, eds. Academic Press, Inc., New York, p. 1.

Lapage, S. P., Shelton, J. E., Mitchell, T. G., and MacKenzie, A. R. (1970b). In *Methods in Microbiology*, Vol. 3A, J. R. Norris and D. W. Ribbons, eds. Academic Press, Inc., New York, p. 135.

Lindberg, T., and Granhall, U. (1984). *Appl. Environ. Microbiol. 48*: 683.

Neal, J. L., Jr., and Larson, R. L. (1976). *Soil Biol. Biochem. 8*: 151.

Nelson, A. D., Barber, L. E., Tjepkema, J., Russell, S. A., Powelson, R., Evans, H. J., and Seidler, R. J. (1976). *Can. J. Microbiol. 22*: 523.

Neyra, C. A., and Dobereiner, J. (1977). *Adv. Agron. 29*: 1.

Nielson, A. H., and Sparell, L. (1976). *Appl. Environ. Microbiol. 32*: 197.

Norris, D. O. (1963). *Emp. J. Exp. Agric. 31*: 255.

Okon, Y., Albrecht, S. L., and Burris, R. H. (1977). *Appl. Environ. Microbiol. 33*: 85.

Pankhurst, E. S. (1967). *Lab. Pract. 16*: 58.

Papavizas, G. C., and Davey, C. B. (1961). *Plant Soil 14*: 215.

Patriquin, D. G. (1982). In *Advances in Agricultural Microbiology*, N. S. Subba Rao, ed. Butterworth & Company Publishers Ltd., London, p. 139.

Raju, P. N., Evans, H. J., and Seidler, R. J. (1972). *Proc. Nat. Acad. Sci. USA 69*: 3474.

Rennie, R. J. (1981). *Can. J. Microbiol. 27*: 8.

Rodríquez-Cáceres, E. A. (1982). *Appl. Environ. Microbiol. 44*: 990.

Skinner, F. A. (1971). In *Isolation of Anaerobes*, D. A. Shapton and R. G. Board, eds. Academic Press, Inc., New York, p. 57.

Smibert, R. M., and Krieg, N. R. (1981). In *Manual of Methods for General Bacteriology*, Philipp Gerhardt, ed. American Society for Microbiology, Washington, D.C., p. 411.

Stamp, L. (1947). *J. Gen. Microbiol. 1*: 215.

Talbot, H. W., and Seidler, R. J. (1979). *Appl. Environ. Microbiol. 37*: 909.

Tarrand, J. J., Krieg, N. R., and Dobereiner, J. (1978). *Can. J. Microbiol. 37*: 1016.

Thomas-Bauzon, D., Weinhard, P., Villecourt, P., and Balandreau, J. (1982). *Can. J. Microbiol. 28*: 922.

Thompson, J. P., and Skerman, V. B. D. (1979). *Azotobacteraceae: The Taxonomy and Ecology of the Aerobic Nitrogen-Fixing Bacteria.* Academic Press, Inc. London Ltd., London, 417 pp.

Tyler, M. E., Milam, J. R., Smith, R. L., Schank, S. C., and Zuberer, D. A. (1979). *Can. J. Microbiol. 25*: 693.

van Berkum, P., and Bohlool, B. B. (1980). *Microbiol. Rev. 44*: 491.

Veldkamp, H. (1970). In *Methods in Microbiology*, Vol. 3A, J. R. Norris and D. W. Ribbons, eds. Academic Press, Inc., New York, p. 305.

Watanabe, I., and Barraquio, W. L. (1979). *Nature 277*: 565.

Wollum, A. G. (1982). In *Methods of Soil Analysis*, Part 2, *Chemical and Microbiological Properties*, Agronomy Monogrpah 9, 2nd ed., A. L. Page, ed. ASA-SSA, Madison, Wis., p. 981.

Wright, S. F., and Weaver, R. W. (1981). *Appl. Environ. Microbiol. 42*: 97.

Zuberer, D. A., and Alexander, D. B. (1986). *Plant Soil 90*: 47-58.

5

Fluorescence Methods for Study of *Rhizobium* in Culture and in Situ

B. BEN BOHLOOL *NifTAL Project, University of Hawaii, Paia, Hawaii*

I. INTRODUCTION

Direct examination of small portions of the natural environment under the microscope reveals interesting relationships between microorganisms, plants, and the physical environment. Conventional staining procedures can aid in improving the clarity of microscopic observations, but it is often difficult to distinguish microbial cells and filaments from organic and inorganic particles and fibers. Certain fluorescent dyes that selectively stain components of biological origin can be used to stain microbial cells and distinguish them from the environmental background. Dyes such as acridine orange and fluorescein isothiocyanate, which react with nucleic acids and proteins, respectively, have been used to stain and enumerate microorganisms directly in the natural habitat (see Parkinson et al., 1971; and Ramsey, 1978, for detailed references).

While fascinating physical relationships among different organisms can be discerned with such approaches, it is virtually impossible to identify the various components of microbial consortia in situ. For example, Figure 1 shows a complex microbial assemblage on the surface of azolla roots as revealed by acridine orange staining. But the small size and the nondescript morphology of the microbial partners impose serious limitations on the value of this type of approach.

Although microbes are, for the most part, nondescript in morphology, they do have unique chemical features that can be used to identify genera, species, and even strains of a particular microorganism. Serological techniques rely on differences in the chemical nature of particulate and soluble antigens of cells and provide a powerful tool for mi-

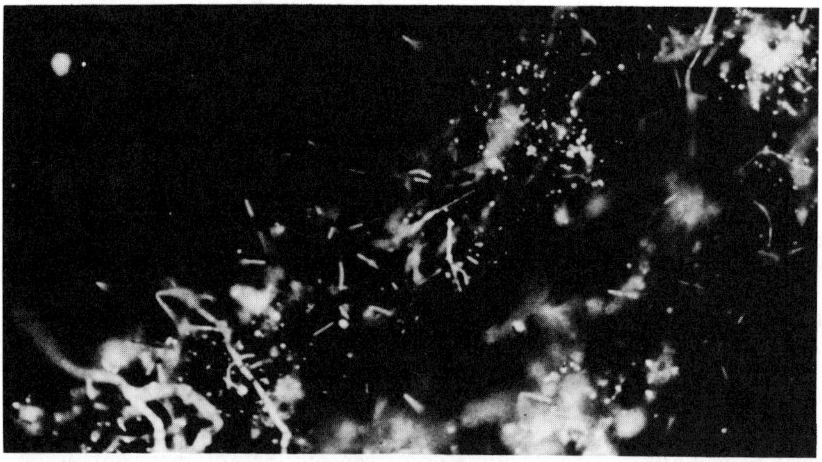

Figure 1 Acridine orange staining of the root surface of azolla showing a variety of shapes and sizes of colonizing microorganisms (1 cm = 15 μm.)

crobial differentiation. Immunofluorescence combines the specificity of immunological reactions with sensitivity and precision of fluorescence microscopy to facilitate simultaneous detection and identification of specific microorganisms in situ. A chapter by Schmidt and Paul (1982) gives a comprehensive coverage of microscopic methods for soil observations.

In Section II, I outline two acridine orange procedures for staining of *Rhizobium* cells: (1) on root (and other) surfaces, and (b) inside infection threads of clover. The techniques are only useful for the studies in gnotobiotic systems, because the nonselective staining procedure does not allow for differentiation of *Rhizobium* from other rod-shaped bacteria in complex microbial assemblages. In Section III, I describe more specific staining procedures, immunofluorescence and fluorescent lectin labeling, that relay on unique chemical features of rhizobia.

II. NONSELECTIVE FLUORESCENT STAINS: ACRIDINE ORANGE STAINING

A. Rhizobia on Root Surfaces

Most conventional staining procedures have the disadvantage that the excess stain must be removed by thorough washing. This often results in removal and/or distortion of the microbial arrangements. With acridine orange staining there is no need for any washing of the excess stain at all, if the concentration of the dye in solution and the volume of the stain applied are carefully adjusted.

Materials Needed

1. Acridine orange stock solution, 0.035% acridine orange in 0.1 *M* sodium phosphate buffer, pH 8.0.
2. Staining solution. Dilute stock solution 1:10 in water (or solutions isotonic with sample to be used); add merthiolate (Thiomersal) to 0.01% final concentration and filter through a 0.2-μm membrane filter to remove bacteria and debris.
3. Pasteur pipettes, microscope slides, coverslip No. 1 (22 mm × 60 mm), and blotting paper.
4. Fluorescence microscope equipment for reflected-light fluorescence observations (see Section IV).

Procedure

1. Place root (or other surfaces) to be stained on a glass slide and cover with a small amount of stain solution.

2. Mount the slide with the coverslip directly in acridine orange and wipe the edges with blotting paper.
3. Observe with reflected-light fluorescence. If background is too intense, rinse the space between slide and coverslip with a gentle flow of 0.01 *M* phosphate buffer, pH 8.0.

Application

Figure 2 shows *Rhizobium trifolii* colonizing root hairs of clover grown aseptically in Fahreus slides and inoculated with a strain of *R. trifolii* 5 days prior to staining. Notice selective massive colonization, with perpendicular orientation, of a single root hair in the same region of the root where other root hairs are only sparsely colonized.

B. *R. trifolii* Inside Clover Infection Threads

Acridine orange can be used to stain bacteria inside cells and infection threads. However, interference from background fluorescence of host

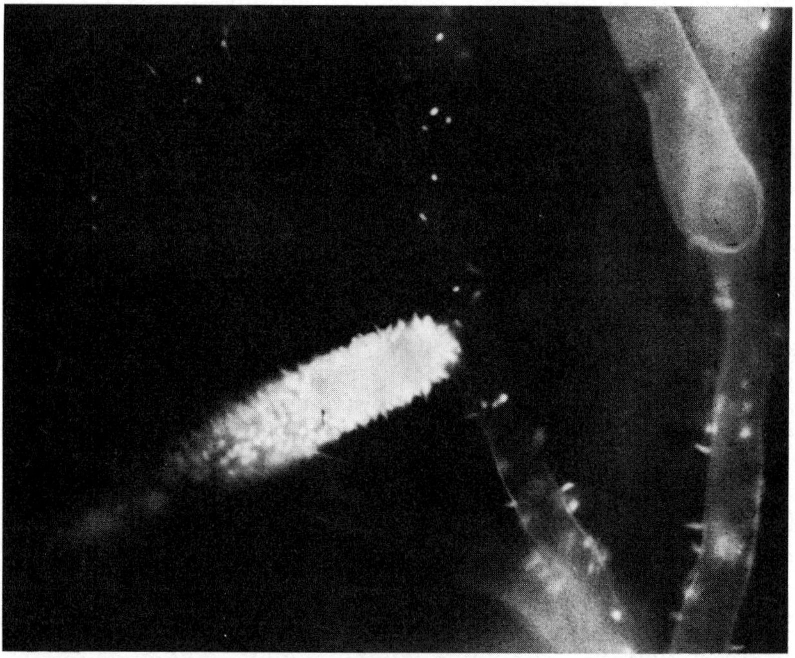

Figure 2 Colonization of root hairs of clover by *R. trifolii* as revealed by acridine orange staining: Massive colonization of one root hair, with others only sparsely colonized (1 cm = 15 μm.)

cell's cytoplasm and organelles must be suppressed. Furthermore, the host tissue must be rendered permeable so that the stain can reach the bacteria effectively.

Materials Needed

1. One of the following root pretreatment solutions:
   a. 10% alkali extract of bark powder from *Pinus radiata*. Bring 100 ml of 0.5 *N* NaOH to near boiling. Add 10 g of bark powder and mix in a blender for 5 min. Allow estraction to proceed for 24 hr, centrifuge, and filter solution through Whatman No. 2 filters. Store in tight bottle in the dark. Use straight or dilute with 0.5 *N* NaOH to obtain desired background.
   b. 2 to 5% tannic acid, catechin, or catechol in 0.5 *N* NaOH.
   c. 6 *M* urea in 0.01 *M* Tris, 1 m*M* EDTA, and 0.1 *M* NaCl, pH 7.5.
2. Acridine orange solution and other materials as in Section II.A.

Procedure

Pretreatment: Clover seedlings are grown in Fahreus slides and inoculated with the desired strain of *R. trifolii*. For most suitable orientation of root hairs, the slides are incubated at a 30° angle inside a deep petri plate. After the desired period of incubation, the seedlings are removed from the chamber and placed in pretreatment solution (in bark extract, tannic acid, catechin, catechol solutions for 24 hr or until roots are colored dark brown; in urea solution for 30 to 60 min, depending on size of root). The roots are washed thoroughly and soaked in water for several hours with frequent changes. Roots from the urea treatment must be counterstained before acridine orange staining. Eriochrome black (see Section III.A), in particular, provides a pleasant dull-red background (see Section III.A).

Staining: Pretreated roots are placed in the working solution of acridine orange for 15 to 30 sec, rinsed, and soaked in water for at least 15 min, but longer if the background is too intense. Stained roots are rinsed in phosphate-buffered saline (PBS), pH 8.0, and mounted in PBS on glass slides for observation.

Application

These procedures work exceptionally well with several species of clover, especially white, red, and strawberry clover. However, the author has not been successful visualizing infection threads on soybean, part of the reason being the extensive convolutions of the infected soybean root hair and the intense background fluorescence. The bark extract-treated roots provide the best background for fluorescence observation of bacteria in infection threads of clover. This material, however, is not readily available and effective prestaining sometimes takes as long as 3 to 4 days. Figure 3a shows the rod-shaped cells of *R. trifolii* lined up along the infection thread inside a clover root hair, and Figure 3b, the continuation of the same infection thread, at a higher magnification, entering the cortical layer. It is difficult to see the infection thread beyond the first cortical layer without sectioning of the root. However, extensive ramification of infection thread inside epidermal cells can be detected in localized regions of the root, as in Figure 3c. These preparations were made with roots pretreated with bark extract. Other pretreatments are also effective, but the contrast between the cells and the background is not as pronounced.

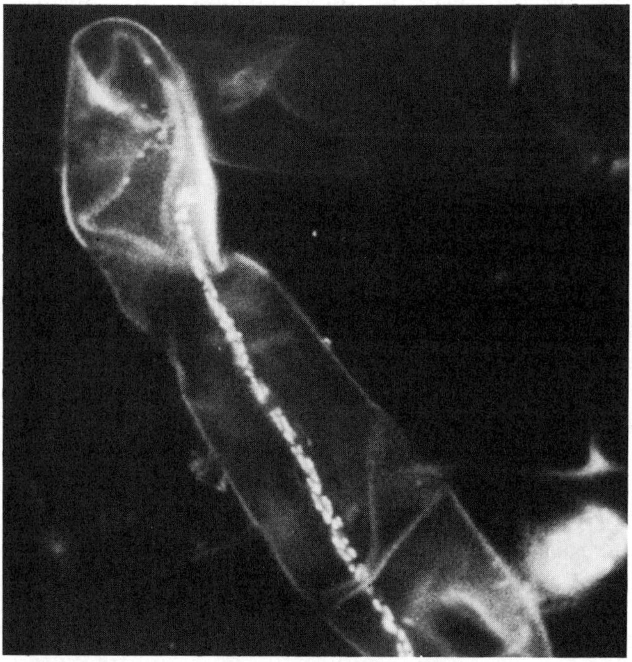

(a)

Figure 3 Acridine orange staining of *R. trifolii* inside infection threads on root hairs of clover: (a) infection thread extending from point of entry toward the root (1 cm = 15 μm); (b) same infection thread as in (a) entering the cortical region (1 cm = 15 μm); (c) ramification of an infection thread in the epidermal layer (1 cm = 10 μm).

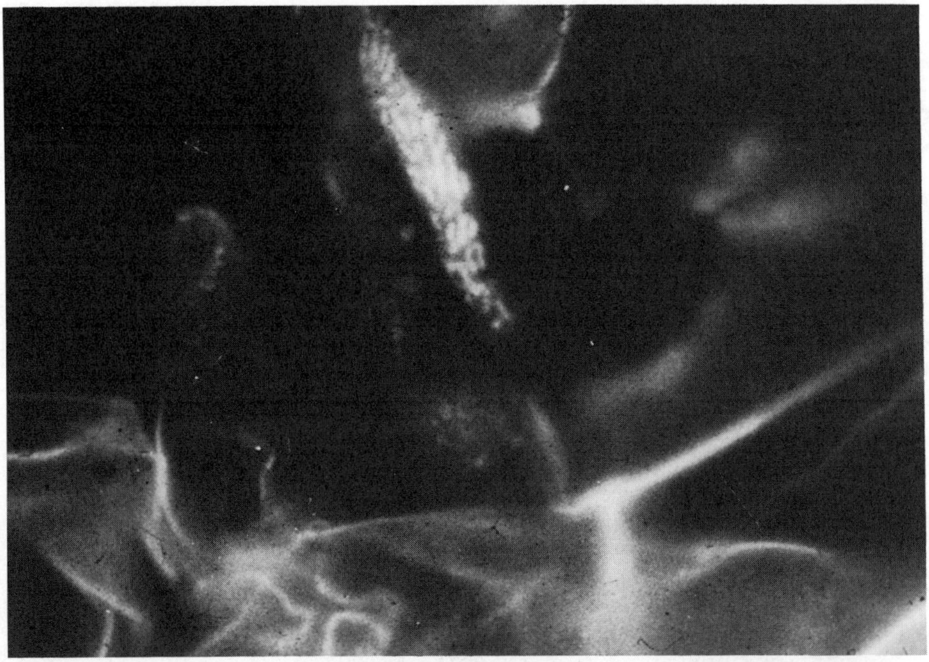

(b)

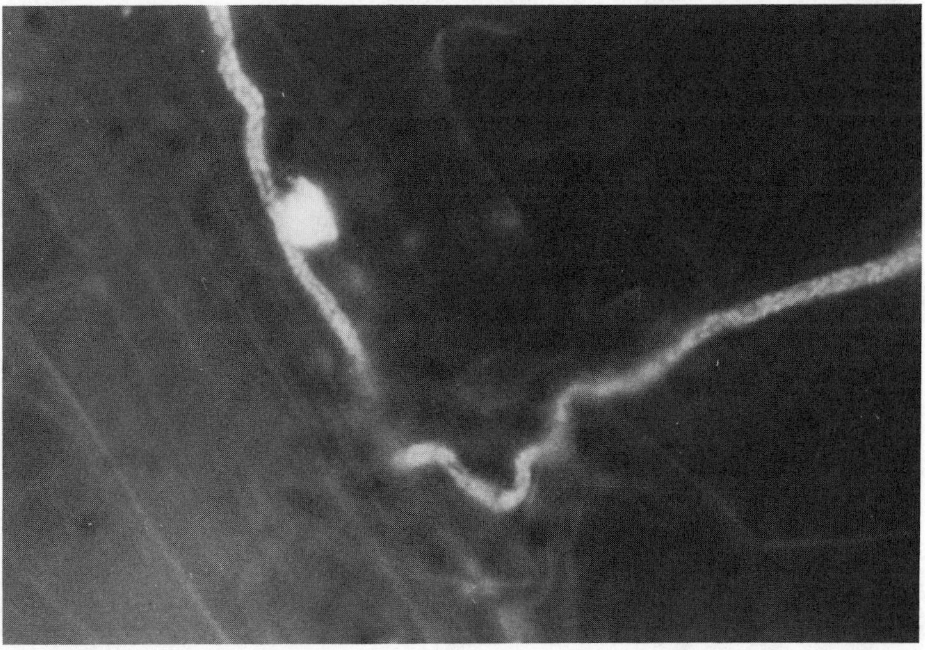

(c)

C. Limitations and Precautions

The direct mounting of specimens in acridine orange under the coverslips has the advantage that it allows for staining with minimum of disturbances. However, if the layer of acridine between the slide and coverslip is excessive, intense background will interfere with effective observation. It is, therefore, important to determine the optimum amount of the dye to be used, and carefully blot and rinse the excess.

Acridine dyes react almost exclusively with nucleic acids: DNA and RNA. Fluorescence emission is primarily concentration dependent but is also significantly affected by pH. At the concentration recommended in this chapter, plant nucleus and bacteria normally appear apple green, with cytoplasm and its organelles, yellow to red. It is therefore important to adjust the concentration of dye and the pH of the stain to the values recommended above.

Acridine orange stains cells that have been killed by a variety of means (i.e., formalin, ethanol, and autoclaving). It is essential, therefore, to remove bacteria, by membrane filtration, from all diluent and wash solutions. Autoclaving of solutions is not adequate to eliminate unwanted staining of extraneous bacteria.

By virtue of their strong affinity for nucleic acids, acridine dyes are potent mutagens and must be handled with care and caution. Mouth pipetting should be avoided, and surgical gloves must be worn when handling strong solutions of acridine orange.

III. SPECIFIC FLUORESCENT STAINS: FLUORESCEIN-CONJUGATED PROTEINS FOR LABELING RHIZOBIA

Fluorescent dyes (fluorochromes) when coupled to specific proteins, such as antibodies, enzymes, and lectins, provide a highly sensitive probe for visualization of the material of interest. Many fluorochromes, with different excitation and emission spectra, that readily bind to proteins are available. Fluorescein isothiocyante (FITC) is, however, the most commonly used fluorochrome for binding to proteins, and thus the only one that would be discussed here. Fluorescein has absorption wavelengths of 436 nm and 489 nm, and an emission wavelength of 511 nm. Highly efficient filter packs combining interference excitation filters with barrier filters and dichroic mirrors are now available for FITC (and other fluorochromes) from several major manufacturers.

A. Immunofluorescence Staining of Rhizobia from Culture and in Situ

The fluorescent antibody method (FA), or immunofluorescence (IF), is still the only effective technique available that provides the means of recognizing specific microorganisms directly in their natural environment.

A number of useful references on theory and application of immunofluorescence are available and should be consulted. The treatises on the subject by Goldman (1968) and Nairn (1975) and the more comprehensive text on experimental immunology by Weir (1978) are excellent sources of information. Applications related to microbial ecology have been reviewed by Bohlool and Schmidt (1980).

Materials Needed

Fractionation of serum globulins

1. High-titered antibody samples (agglutination titer: 1280 or greater).
2. Ammonium sulfate, 3.9 *M*.
3. Saline, 0.85% NaCl.
4. Dialysis tubing.
5. Barium chloride, saturated.

Conjugation of FITC to globulin

1. Sodium phosphate solutions, 0.1 *M*, pH 8.0 and pH 9.0.
2. Sodium hydroxide, 0.1 *N*.
3. Fluorescein isothiocyanate, FITC, Isomer 1 (Baltimore Biological Labs, Cockeysville, Maryland).
4. Protein determination: chemical method (biuret, Lowry, Biorad) or spectrophotometric (260/280 nm).
5. Merthiolate stock solution 1.0% in water.
6. Sephadex G-25 (coarse).
7. Phosphate (0.1 *M*) buffered saline (0.85%), pH 7.2.

Testing of FA

1. Fresh broth cultures of desired organism.
2. Microscope slides (1 mm thick) and coverslips (No. 1).
3. Nonfluorescent mounting fluid (Difco) or buffer (carbonate/bicarbonate, pH 9.0): glycerol mixture at 1:9 proportion, respectively.
4. Fluorescence microscope (see Section IV).

Immunofluorescence staining of nodule smears, soil contact slides, and root surfaces

1. General materials and supplies, as above.
2. Gelatin-rhodamine conjugate (after Bohlool and Schmidt, 1968). Hydrolyze a 2% solution of gelatin by adjusting the pH to 10.5 with 1 *M* NaOH and autoclaving for 10 min. After cooling, add enough rhodamine isothiocyanate (RhITC; Sigma Chemical), dissolved in a minimum amount of acetone, to provide 0.04 mg

of dye per milligram of gelatin used. Allow conjugation to proceed in the dark for 12 hr at 25°C. Dialyze at room temperature against PBS, pH 7.2 (with 0.01% merthiolate) until no further color is detected in the dialysate. Store as a freeze-dried powder in a desiccator, or as a liquid in small screw-capped tubes at -20°C.

3. Eriochrome black (modified from Goldman, 1968). Dissolve 9 mg of dye in 1 ml of dimethyl sulfoxide (DMSO) and add 5 ml of chelating solution (50 ml of DMSO; 20 ml of H_2O; 10 ml of 0.1 *M* aluminum chloride; 10 ml of 1.0 *M* acetic acid, with pH adjusted to 5.2 with 1 *M* NaOH and the volume to 100 ml with distilled H_2O).
4. Crystal violet, 100 ppm (after Hockenhull, 1979).

Enumeration of *Rhizobium* from soil, peat, and rhizosphere

1. Hydrolyzed gelatin extractant, for releasing bacteria from soils (after Kingsley and Bohlool, 1981). Hydrolyze a 1% solution of gelatin at pH 10.5, as above. Dilute 1:10 with 0.1 *M* $(NH_4)_2HPO_4$.
2. Sodium hydroxide, 0.2 *N* NaOH.
3. Membrane filters, nucleopore, prestained with Irgalan Black (Ciba-Geigy) (after Hobbie et al., 1977).

Procedure

Fluorescent antibody preparation (after Lewis et al., 1964): Numerous protocols are available for obtaining highly purified immunoglobulins from serum (see Weir, 1978). For most routine applications, however, high levels of purity are not essential and a partially fractionated preparation is adequate.

1. To a measure volume (i.e., 10 ml) of undiluted serum add an equal volume of cold 3.9 *M* ammonium sulfate. Keep serum cold and slowly add ammonium sulfate dropwise with constant stirring of the mixture.
2. Allow the cloudy mixture to stand in the cold for at least 1 hr, and separate globulins by centrifugation (10,000 × g, for 30 min).
3. Decant supernatant and dissolve precipitate in saline to the original volume.
4. Repeat ammonium sulfate precipitation at least two more times or until the globulin precipitate is completely white and free of hemoglobin.
5. Dissolve the final precipitate in a minimum volume of water and dialyze against saline until sulfate is no longer detectable in the dialysate. The presence of sulfate can be determined by mixing equal volumes of dialysate and saturated barium

chloride ($BaCl_2$). If no precipitate ($BaSO_4$) develops, the solution can be considered free of sulfate, and the dialysis complete.

6. Determine the protein content of the solution inside the dialysis tubing by any of the available methods (biuret, Biorad, Lowry, of 260/280-nm spectrophotometric).
7. Adjust a sample of the solution to 1.0% (10 mg/ml) protein by the addition of saline.
8. Add 4.0 ml of 0.1 *M* sodium phosphate (pH 9.0) to 10 ml of the 1.0% globulin solution.
9. Follow by adding 4.0 ml of 0.1 *M* sodium phosphate (pH 8.0) that contains sufficient freshly dissolved FITC (BBL) to provide 0.05 mg of FITC per milligram of protein (5.0 mg of FITC in this case).
10. Adjust the pH to 9.5, with 0.1 *N* NaOH, and the volume to 20.0 ml, with saline. Take care that the pH does not exceed 9.5.
11. Add 0.2 ml of merthiolate stock (final concentration 0.01%) and allow conjugation to proceed in the dark at room temperature and with gentle stirring, for 6 hr.
12. Separate the conjugated protein from unreacted dye by Sephadex chromotography or dialysis, using merthiolated PBS, pH 7.2.
13. Incubate FA at 4°C for 12 to 24 hr; centrifuge (10,000 × *g*, 10 min) and membrane-filter (0.45 μm) to remove precipitate.
14. Test FA for reactivity with homologous antigen, as described in the following section.
15. If satisfactory, distribute the FA in small screw-capped tubes (1.0 ml) and store at -20°C.

Initial testing of FA" The FA should be tested against its homologous antigen to ensure its quality. In addition, the optimum working dilution of the FA should be assessed. Make at least six smears from fresh broth culture(s) of the homologous organism, air dry, and heat fix. Make twofold serial dilutions of the FA to a final dilution of 1:32. Test each dilution in the following manner:

1. Cover each smear with a drop of a particular dilution of the FA on separate slides to avoid mixing.
2. Allow staining to proceed for 20 min.
3. Remove excess FA by washing in PBS for 15 min. This is best accomplished by first rinsing off the FA with about 0.5 ml of PBS, then placing the slide in a slide staining dish containing PBS.
4. Dip each slide in distilled water, air dry, and mount in nonfluorescent mounting fluid.
5. Observe each slide using a fluorescence microscope.

Controls: Before a new FA can be used in ecological studies it must stain its homologous antigen brightly. In addition, many control checks are necessary. The specificity of the FA must be tested; thus other related strains should be stained to ensure that the FA is strain specific. In addition, unrelated FAs (or preimmune fluorescent globulin) should be used to test the staining characteristic of the organism. If the FA is to be used to identify the organism of interest in a complex environment such as soil, the FA should be tested against unrelated microorganisms from the same soil. For details of controls pertinent to ecological applications, see Bohlool and Schmidt (1980).

Applications

Detection of rhizobia in legume root nodules: Immunofluorescence can be used to identify the bacteria that form nodules on roots of leguminous platns. The nodule tissue can either be sectioned or homogenized for staining. The following methods are quite suitable:

1. Roots are washed in distilled water containing 0.1% Tween-80 and rinsed four times. Nodules are removed and crushed with alcohol-flamed forceps; each nodule is then touched to the surface of glass slides (as many as needed; e.g., if two FAs are to be tested against the nodule smears, two to six separate slides should be made for each nodule to be used for the FAs, and each control). As many as 12 different nodule impressions can be made onto the same slide, but no more than one FA should be applied, to avoid mixing of FAs on slide. Smears are air dried, heat fixed, and 1 drop of RhITC-conjugate is applied to each smear. The RhITC is allowed to dry as a film in a drying oven (60°C). After cooling, FA staining proceeds according to the protocol above.
2. Nodulated roots are washed as above. Nodules are placed in weels of microtiter plates. Each nodule is crushed in a drop of sterile water. The homogenate is then used to make smears on glass slides with a microbiological loop. FA staining proceeds as above. Figure 4 shows bacteroids of *B. leguminosarum* from lentil nodules stained by homologous strain-specific FA, with this procedure.

Staining of soil contact (buried) slides: Rossi-Cholodney soil contact slides can be stained with fluorescent antibodies. In most soils it is necessary to treat the slide with gelatin-rhodamine conjugate (Bohlool and Schmidt, 1968) to prevent the adsorption of FA to soil particles on the slides and to provide a suitable dim orange (rhodamine) background against which the apple green (FITC) bacteria can be observed more clearly.

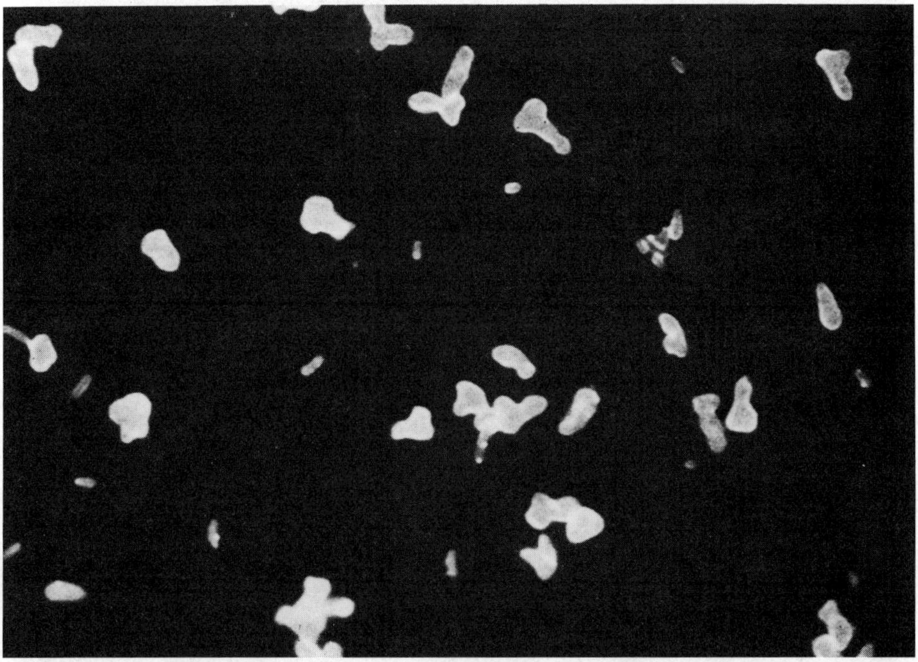

Figure 4 Immunofluorescence staining of *R. leguminosarum* bacteroids from lentil nodules using strain-specific FA (1 cm = 5 μm). (From S. N. May and B. B. Bohlool, unpublished.)

Proceed as follows. Bury the slides in contact with soil (or other material) for various lengths of time, either in flasks or in the field. Shake off the excess soil, air dry, and heat fix. Add enough of the gelatin-rhodamine solution to cover the surface of the slide (about 0.5 to 1.0 ml) and allow to dry as a film in a 60°C drying oven. After cooling, cover the surface with an appropriate dilution of the FA (depending on the titer of the FA) and stain in a moist chamber for 30 to 45 min. Wash as above. Air-dried slides can then be coverslipped with FA mounting fluid and observed as before. Figure 5a shows fluorescent cells of *Rhizobium japonicum* strain USDA 31 on contact slides from sterile peat, and Figure 5b, the same strain growing in sterile soil and colonizing the filaments of a fungus co-cultured in the same soil, stained by this procedure.

Staining bacteria colonizing root surfaces: Roots to be stained are first counterstained in eriochrome black solution until the root is com-

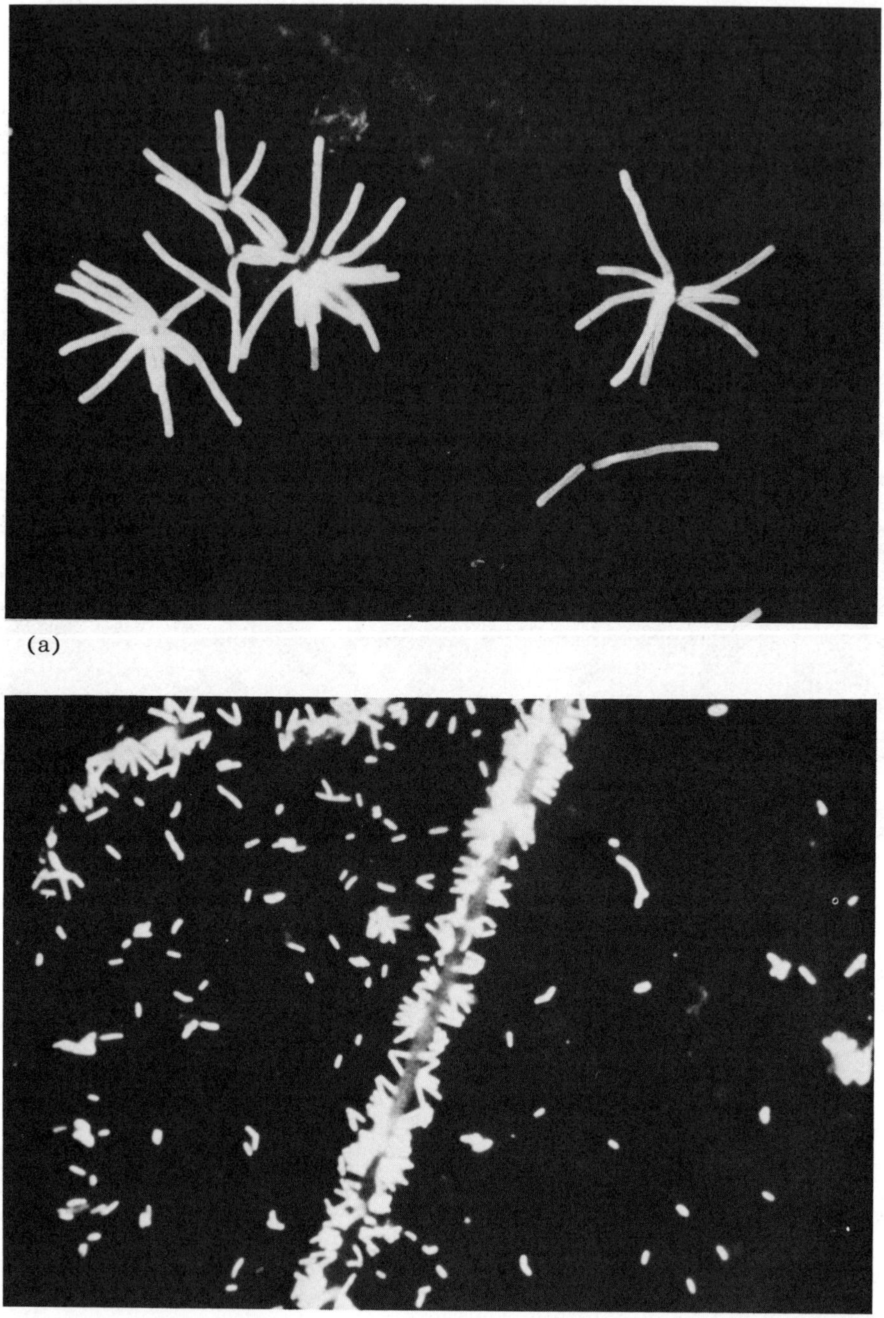

(a)

(b)

Figure 5 Immunofluorescence detection of *R. japonicum*, USDA 31, using strain-specific FA: (a) growing in sterilized peat on the surface of contact slides (1 cm = 10 μm); (b) growing in association with a fungus in sterile soil (1 cm = 10 μm). (From Bohlool and Schmidt, unpublished.)

pletely colored (pink to dark red). The excess stain is removed by thorough washing and soaking in saline until no further color comes out of root. Pieces from known region of the root are cut (2 to 3 cm) and placed on glass slides. Gelatin-rhodamine is applied to cover the root pieces completely. The solution is allowed to dry at 65°C. The appropriate dilution of FA is applied and staining and washing is continued as above. Root pieces are mounted under a cover glass in mounting fluid and observed.

An alternative method for staining bacteria in plant tissue is a modification of that developed by Hockenhull (1979). This does not involve eriochrome black prestaining, but the roots are dipped in crystal violet stain after FA treatment. A combination of rhodamine-gelatin pre-FA treatment and crystal violet post-FA counterstaining gives excellent results.

Quantitative estimation of number of *Rhizobium* in soil and rhizosphere: Enumeration of bacteria in water samples presents no special difficulties. An appropriate volume of the sample is filtered through a non-fluorescent membrane filter (e.g., Nucleopore) and washed through with 0.9% saline. The rest of the staining procedure is the same as that for smears. The enumeration of bacteria in soil, on the other hand, is complicated by the colloidal and particulate nature of the sample. The protocol for the removal of these fractions and the recovery of the specific bacteria is given below. For data concerning the efficiency of recovery by this method, see Bohlool and Schmidt (1973), Schmidt (1974), and Kingsley and Bohlool (1981).

Suspend 10.0 g of soil in 95.0 ml of hydrolyzed gelatin extractant and shake for 5 min on wrist-action shaker. Remove colloidal material either by mild centrifugation (700 × *g*, 5 min) or by flocculation with 0.7 g of $Ca(OH)_2$ and $MgCO_3$ in a 2:5 mixture (the most suitable procedure varies for different soils). Filter a known volume (from 1 to 5 ml) of supernatant through Irgalan-treated Nucleopore filters and wash through with prefiltered saline. Transfer the filter to a glass slide and apply 0.5 ml of gelatin-rhodamine solution directly on the effective filtering area. Allow the solution to dry on the filter at 65°C. Apply appropriate dilution of FA (prefiltered through the membrane filter, 0.45 μm) and incubate in a moist chamber for 30 min. Carefully transfer the filter to a suction unit and wash through with prefiltered saline (normally, 200 to 300 ml of slaine wash is required to provide a dark background). Mount the filter on a glass slide under mounting fluid. Count the fluorescent bacteria per field of microscope and extrapolate to numbers per gram of dry soil, taking into account the area of the microscopic field, area of effective filtering surface, the volume and dilution of the soil filtered, and the moisture content of the soil.

B. Fluorescent-Lectin Binding to Rhizobia

Lectins, or phytohemagglutinins, are glycoproteins distinguished by their ability to recognize and bind to specific sugars and sugar moieties in carbohydrates found on cell surfaces. They are quite ubiquitous in nature and are thought to be involved in sugar transport as well as in cell-to-cell recognition. Lectins might play a role in host-symbiont recognition and in the defense mechanisms of plants against phytopathogenic microroganisms (for reviews, see Schmidt, 1979; and Graham, 1980). Many legume seeds contain high concentrations of lectins, some of which (e.g., soybean lectin) are quite specific, while others (e.g., Con-A from *Canavalia*) react with a wide range of different sugars. The binding of soybean lectins to the soybean rhizobia, *Rhizobium japonicum*, can be visualized under the fluorescence microscope if the lectin is first conjugated to a fluorochrome. The reaction is quite rapid and stable. However, mounting the stained smears in buffered glycerol (FA mounting fluid) causes the dissociation of the complex in some systems. For this reason, PBS is used under the cover glass.

Materials Needed

1. Fluorescent soybean lectin. Highly pruified material can be purchased from several biological supply companies. These preparations are quite expensive and not at all necessary for demonstration of lectin binding to rhizobia. Crude extracts of soybean seeds, partially purified with ammonium sulfate (after Liener and Pallansch, 1952), are quite effective for this purpose.

   Suspend 100 g of hexane-defatted soybean flour in 1 liter of H_2O. Adjust the pH to 6.7 and remove the residue by centrifugation. Adjust the pH of the supernatant to 4.2 and remove the precipitate by centrifugation. To 1 liter of the supernatant, slowly add 350 g of $(NH_4)_2SO_4$ while stirring. Harvest the precipitate by centrifugation, dissolve in 150 ml of saline, adjust the pH to 6.8, and dialyze against saline (pH 6.8) until no more $(NH_4)_2SO_4$ is detectable in the dialysate. Proceed with FITC conjugation as described for FA (Section III.A).
2. *Rhizobium japonicum* cultures. Cultures should be grown for 5 to 7 days in media with a high C:N ratio [e.g., yeast extract-mannitol (YEM)] (Vincent, 1970). Several strains have been shown to be suitable for demonstration. Strains USDA 110, 123, 136, and 138 produce copious amounts of lectin-binding material; while USDA 31, 62, and 117 produce very little, if any.
3. PBS, pH 7.2, as before.
4. Other materials and supplies, as before.

Procedure

1. Place a drop of the culture on a glass slide and *gently* spread with a microbiological loop.
2. Air dry and heat fix the smear.
3. Apply a few drops of a dilution (1:4 or 1:8) of the fluorescent lectin solution on smear and incubate in moist chamber for 15 min.
4. Rinse off the excess stain with PBS and soak in PBS for 15 min.
5. Gently remove the slide from the buffer, drain the surface, wipe the back, and mount a cover glass directly in the remaining PBS.
6. Observe under a fluorescence scope as before.

Application

A double-light microscope system using reflected fluorescence and transmitted white light will allow visualization of the cells with a fluorescent halo of lectin attached to the polysaccharides. Figure 6 shows the reaciton of *R. japonicum* USDA 110 under reflected light fluorescence.

C. Limitations and Precautions

For all techniques there are intrinsic limitations and possible sources of error with which the user should be familiar. With immunofluorescence, this is particularly crucial because the results are often based on subjective assessment of microscopic observations. Controls are therefore essential in all applications of immunofluorescence.

The most critical factors that limit the usefulness of fluorescent antibodies are the specificity of the reagent (too much, in some cases, and not enough, in others), the background interference due to autofluorescence and/or nonspecific adsorption of FA (which in most cases can be controlled), the stability of the microbial antigens in different environments, and the inability to differentiate between live and dead cells. For further discussion of these problems, see the review by Bohlool and Schmidt (1980).

If only fluorescent illumination is used for observation of lectin binding, with some cultures the reactive material appears as intensely fluorescing debris. It is therefore desirable to use double-light systems to visualize the cells with the transmitted white light. Some *Rhizobium* polysaccharide-lectin complexes dissociate in glycerol. It is important to mount the stained slides on the same buffered saline in which the smears were washed.

Production of lectin-binding polysaccharide in *Rhizobium* is dependent on the carbon and nitrogen status of the grown medium. The same

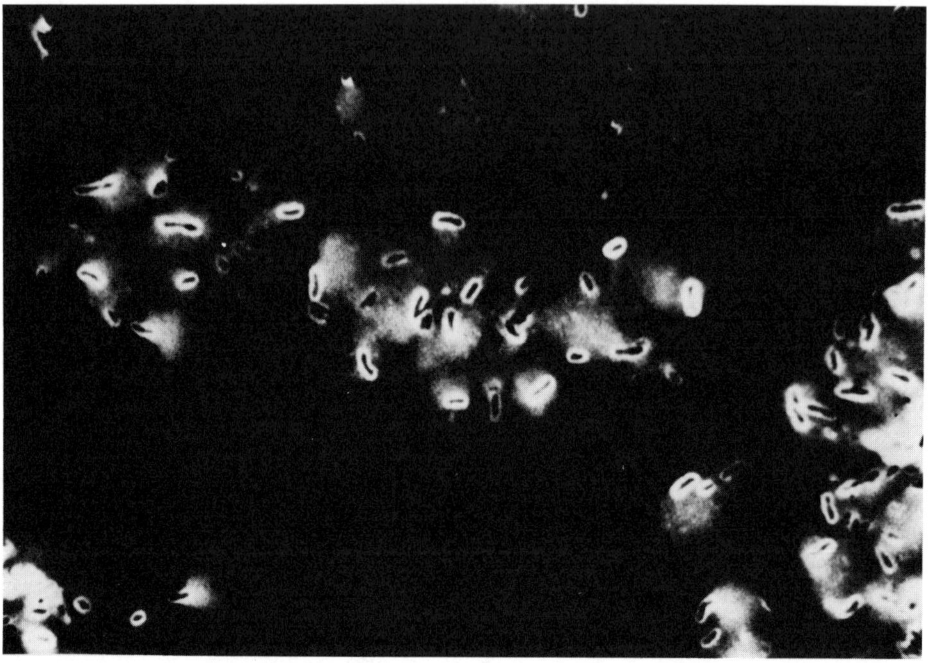

Figure 6 Fluorescent lectin staining of *R. japonicum*, USDA 110, from a 7-day culture in YEM medium (1 cm = 5 μm). (From Bohlool and Schmidt, 1974.)

strains that produce copious amounts of the material in a high-carbon low-nitrogen medium will be totally negative in a high-nitrogen medium. The best medium for demonstration of lectin binding in *Rhizobium* is a basal medium containing 0.05% yeast extract and 1.0% of the suitable sugar (e.g., mannitol) as the carbon source. The age of the culture also has an effect on production and consistency of the lectin-binding polysaccharide. Cultures in late logarithmic or early stationary phases generally give the best results.

Fluorescent samples mounted in the glycerol mounting medium fade within 10 to 20 sec when observed under the microscope. A semi-permanent mounting medium based on polyvinyl alcohol (Elvanol, Gelvatol) prolongs the fluorescence of mounted speciment (Rodriques and Deinhardt, 1960). An improved version of this was reported by Lennette (1978).

IV. FLUORESCENCE MICROSCOPY

The fluorescence microscope is a conventional compound microscope equipped with a special light source of suitable wavelength and intensity and the proper combination of objective and ocular lenses and filters. Most manufacturers now make versatile fluorescence equipment that is reasonably priced and readily available in most countries. The exact model to purchase depends entirely on the needs of the investigator. Information about the appropriate model and the correct combination of components for fluorescence microscopy is readily available from technical representatives for major manufacturers or through most scientific supply houses.

With new advances in microscope optics, it is no longer necessary to obtain the superpressure mercury arc (HBO-200) or xenon (XBO 150). These models are much more expensive, have a shorter life, and produce a large amount of heat. The author has used the HBO-50 mercury assembly in several microscope models with good results. Also for routine work, the 12-V 100-W quartz-halogen lamp is quite adequate for fluorescence microscopy. It has the added advantage that it can also be used for white-light observations.

If double-light illumination is desired, a combination of HBO-50 (or 12 V 100 W quartz-halogen) for reflected-light fluorescence and 12 V 100 W quartz-halogen for transmitted white light is highly recommended. For transmitted-light observations, a condenser must be used. The oil-immersion dark-field "ultracondenser" or the phase-contract condenser is suitable for this purpose. Specialized objectives, however, are needed for use with these condensers (see below).

Most fluorescence scopes can now be fitted with a wide variety of filter combinations. However, for the work described in this chapter, the "FITC filter set," consisting of an interference excitation filter (suitable for FITC as well as acridine orange), a dichroic mirror for selective reflection of the correct wavelengths, and the corresponding barrier filter, is entirely adequate. Most microscope models offer this filter set (and other sets for rhodamine, quanacrine) as an option.

High-quality objective lenses are important requirements for effective fluorescence microscopy. The top of the line in objective lenses are the Plan-apo series lenses, which are available for vairous magnifications, with or without phase-contrast capability. These are, however, extremely expensive; a set of them often costs more than the price of the entire microscope. Many of the other new objectives (e.g., Neofluars) are quite acceptable for routine fluorescence observations. If double-light microscopy is to be used, the objective must have a phase capability (pH objective) or be equipped with an iris diaphragm for dark-field illumintation.

In general, the quality of eyepiece (ocular) lenses is not as crucial as that for objectives. The medium-power eyepieces (e.g., 8× or 10×), however, give the best balance of brightness and resolution.

ACKNOWLEDGMENTS

This chapter is dedicated to my mentor, Dr. E. L. Schmidt, in recognition of his numerous contributions to this and other areas of microbial ecology. The research was supported in part by grants from the U.S. Agency for International Development (AID).

REFERENCES

Bohlool, B. B., and Schmidt, E. L. (1968). *Science 162*: 1012.

Bohlool, B. B., and Schmidt, E. L. (1973). A fluorescent antibody technique for determination of growth rates of bacteria in soil. In *Modern Methods in the Study of Microbial Ecology*, T. Rosswall, ed., Bull. Ecol. Res. Comm. (Stockholm) 17: 336-338.

Bohlool, B. B., and Schmidt, E. L. (1974). *Science 185*: 269.

Bohlool, B. B., and Schmidt, E. L. (1976). *J. Bacteriol. 125*: 1188.

Bohlool, B. B., and Schmidt, E. L. (1980). The immunofluorescence approach in microbial ecology. In *Advances in Microbial Ecology*, Vol. 4, M. Alexander, ed. Plenum Publishing Corporation, New York, pp. 203-241.

Goldman, M. (1968). *Fluorescent Antibody Methods*. Academic Press, Inc., New York.

Graham, T. L. (1980). Recognition in the *Rhizobium*-legume symbiosis. In *Biology of the Rhizobiaceae*, A. Atherly and K. Giler, eds. Academic Press, New York.

Hobbie, J. E., Daley, R. J., and Jasper, S. (1977). *Appl. Environ. Microbiol. 33*, 1225.

Hockenhull, J. (1979). *K. Vet. Landbohoejsk. Arsskr. 1979*: 1.

Kingsley, M. T., and Bohlool, B. B. (1981). *Appl. Environ. Microbiol. 42*: 241.

Lennette, D. A. (1978). *Am. J. Clin. Pathol. 69*: 647.

Lewis, V. J., Jones, W. L., Brooks, J. B., and Cherry, W. B. (1964). *Appl. Microbiol. 12*: 343.

Liener, I. E., and Pallansch, M. J. (1952). *J. Biol. Chem. 197*: 29.

Nairn, R. C. (1975). *Fluorescent Protein Tracing*, 4th ed. Churchhill Livingstone, Edinburgh.

Parkinson, D., Gray, T. R. G., and Williams, S. T. (1971). *Methods for Studying the Ecology of Soil Microorganisms* (IBP Handbook 19). Blackwell Scientific Publications Ltd., Oxford, Chap. 4, pp. 19-35.

Ramsey, A. J. (1978). *N.Z. J. Mar. Freshwater Res. 12*: 265.
Rodríques, J., and Deinhardt, F. (1960). *Virology 12*: 316.
Schmidt, E. L. (1974). *Soil Sci. 118*: 141.
Schmidt, E. L. (1979). *Annu. Rev. Microbiol. 33*: 355.
Schmidt, E. L., and Paul, E. A. (1982). Microscopic methods for soil microorganisms. In *Methods of Soil Analysis*, Part 2, A. L. Page, ed., *Agronomy 9*: 803.
Vincent, J. M. (1970). *A Manual for the Practical Study of Root-Nodule Bacteria* (IBP Handbook 15). Blackwell Scientific Publications Ltd., Oxford.
Weir, D. M. (1978). *Handbook of Experimental Immunology*, 3rd ed. Blackwell Scientific Publications Ltd., Oxford.

6

Serological Techniques for *Bradyrhizobium* and *Rhizobium* Identification

ARTHUR G. WOLLUM II *North Carolina State University, Raleigh, North Carolina*

I. INTRODUCTION

The bacterial cell possesses a variety of specific antigenic determinants that may be used to identify specific strains. Of the two basic types, somatic and flagellar antigens, the former have proven very useful for

---

Paper No. 10950 of the Journal Series of the North Carolina Agricultural Research Service, Raleigh, NC 27695-7601

routine identification of bacterial strains (Means and Johnson, 1964; Williams and Chase, 1977).

One of the most widely used immunological procedures for strain identification is the agglutination reaction. Wide acceptance of this procedure probably results from its technical simplicity. However, to utilize the procedure, one must have a supply of antiserum, specific for the strain being tested. Specific polyclonal antibodies can be conveniently produced in a variety of animals. Although most investigators use rabbits because of the ease of handling and the amount of serum that can be produced, other animals can be used, including mice, rats, and goats. Unfortunately for those investigators interested in rhizobia or bradyrhizobia, commercial antisera are not available. However, antisera for specific strains may be available on a contractual basis for some rhizobial strains from individuals who custom make antisera. Because of the specific nature of the test, most investigators find it more convenient to produce their own specific antiserum. I will not provide detailed instructions for antisera production because methodology is available in numerous sources (e.g., Carpenter, 1975; Vincent, 1970). Rather, it is the purpose of this chapter to provide detailed information on the use of the agglutination procedure as a diagnostic tool for ecological studies involving rhizobia or bradyrhizobia.

II. ANTIGEN PREPARATION

A. Background

Bacteroids from nodules or bacterial cultures may be used as the test antigen. When working with nodule antigens, most investigators prefer to use fresh nodules. However, a recent report by Somasegaran et al. (1983) is noteworthy. These investigators demonstrated that dried nodules worked as well as freshly picked nodules as the antigen source. By using dried nodules, large numbers of nodules can be accumulated for analysis as a unit at some future date. Nodules may be frozen as well and stored for future analysis without a loss of antigenic properties. This process is the procedure preferred by many investigators.

B. Culture Antigens

Inoculate an agar slant of yeast extract-mannitol (YEM) medium (Weaver and Frederick, 1982) with the desired organism. Culture under conditions giving optimum growth. When abundant growth is visible (typically 2 to 3 days for rhizobia and 5 to 7 days for bradyrhizobia), add a small amount of saline (0.85% NaCl) to the culture tube and with a wire loop scrape the growth off the agar surface. Pipette an aliquot of the cell suspension to a clean test tube, steam the tube and its contents at 100°C for 15 min, and then cool to ambient temperature. Adjust the optical density (OD) at 600 nm of the steamed suspension to an absorbance reading of about 0.5 (0.4 to 0.6).

C. Nodule Antigens

Gently wash nodules under a steam of water. A small brush aids cleaning nodule surfaces. Place individual nodules into separate test tubes containing 1 to 2 ml of 0.85% saline (NaCl). Reduce the suspending volume for small nodules. Crush the nodule with a sterile glass rod. Use a fresh rod or alcohol-flamed rod for each nodule. Alternatively, soak the rod in a hot detergent solution and rinse in a large volume of hot water before crushing each nodule. This procedure works well by using two rods and alternating the rods, soaking one, while using the other to crush a nodule. After crushing the nodule, steam the tube and its contents for 15 min at 100°C to destroy the heat-sensitive flagellar antigen and then cool to ambient temperature. Adjust the optical density (OD; 600 nm) of the steamed suspension to an absorbance reading of about 0.5 (0.4 to 0.6).

If oven-dried nodules are used, they should be rehydrated first. To accomplish this, pipette 1 to 2 ml of 0.85% saline in a test tube and place individual nodules into separate tubes. Allow the nodule to imbibe water. This can be done conveniently by placing the test tube in the refrigerator overnight. Somasegaran et al. (1983) notes that nodules allowed to imbibe moisture overnight in the refrigerator are easier to crush than nodules that have imbibed moisture for 1 hr at ambient temperatures. Furthermore, oven-dried nodules do not need to be steamed after crushing.

III. DETERMINATION OF SERUM TITER

Agglutination reactions can be conducted in several different containers. Standard test tubes are readily available in most laboratories and have been commonly used. Other laboratories have used the polystyrene microtiter U-shaped well plate (Model 0010102201, Dynatech Lab., Inc., Alexandria, Virginia)* Another type of test container is a disposable tray (Disposo Tray, Model 240-U-CS clear; Linbro Chemical Co., Inc., New Haven, Connecticut). These trays have 24 shallow wells. Basically, the procedures for setting up the containers for

---

*The use of trade names in this publication does not imply endorsement by the North Carolina Agricultural Research Service or criticism of similar ones not mentioned.

agglutination tests are similar, regardless of the container used. Exceptions will be noted as necessary.

Dispense a known volume of 0.85% saline into each tube or well. For test tubes (10 mm × 75 mm or 11 mm × 100 mm), pipette 0.9 ml into the first tube and 0.5 ml into each remaining tube. For the microtiter and disposable tray plates, dispense 0.025 and 0.25 ml, respectively, into each well. Carefully pipette 0.1 ml of antiserum into the first test tube containing 0.9 ml of saline. Mix the contents of the tube by aspirating most of the suspension into the pipette and blowing out from three to four times. This tube now contains a 1:10 dilution of the serum. Next transfer 0.5 ml from the first tube to the next tube in the series. Repeat the mixing procedure outlined previously. Continue to make serial dilutions to succeeding tubes. Each tube of the series now contains one-half the concentration of the antiserum in the tube preceding (1:10, 1:20, 1:40, etc.). In the last tube remove 0.5 ml of the diluted serum and discard. Include in the series a saline control containing only 0.5 ml of saline.

For the microtiter plates pipette 0.025 ml of antigen into the first well and mix the contents. Dynatech Lab., Inc. manufactures a unique set of dilutors that are convenient to use. Refer to the manufacturer's specifications for instructions. This well now contains a 1:2 dilution of the serum. Initially, the serum can be diluted with saline to any specific dilution desired. Subsequent determination of dilutions must take into account the original dilution. Continue the dilution process by transferring 0.025 ml to succeeding wells, reserving one well as a saline control. For the disposable trays use a 0.25-ml volume for dilution. For this type of dilution and these kinds of plates, each well of the series now contains one-half the concentration of the antiserum in the tube preceding (1:4, 1:8, 1:16, etc.).

Carefully pipette aliquots of the diluted antigen (OD = 0.5) to each tube or well, including the saline control. For the test tubes use 0.1 and 0.025, or 0.25 ml, respectively, for the microtiter or disposable tray plates. Carefully mix the contents of the tubes or wells by shaking them back and forth.

Manufacturer recommendations suggest that microtiter plates be covered with plastic film tape. Cover each plate and roll the surface with a roller until each well is airtight. Working on a moistened towel prevents the buildup of static electricity.

Incubate the tubes in a water bath at 52°C for 5 hr or overnight at 37°C. The water level should be about halfway up the level of the suspension in the tube. Plates are normally floated on top of the water in the water bath. A water bath cover prevents drying of the wells of the disposable tray plates.

Determine which dilutions responded positively for agglutination following the instructions in Section IV. Identify the highest dilution to give a positive agglutination reaction. For subsequent analyses of unknown antigens use the dilution that was one less than the highest titer giving a positive reaction. For example, if 1:512 was the titer giving a positive reaction, use 1:256 as the working dilution for the antiserum. Because of individual strain by animal interactions, each antiserum will have its own unique titer. Frequently, the strongest agglutination reactions occur with serum dilutions somewhat less than the highest dilution or titer giving a positive reaction.

IV. READING AGGLUTINATION TESTS

Good lighting is essential for reading the agglutination tests. For reading the reactions in the test tubes and the disposable-tray plates a dark backgorund and an incandescent light have worked well. To read the U-shaped wells of the microtiter plates, a stereoscope fitted with substage incandescent light is helpful.

After incubation remove the tubes or plates from the water bath, wipe dry, and allow to set 5 to 10 min to equilibriate with ambient temperature. Examine each tube or well for the characteristic agglutination reaction. Generally, a strong reaction is one in which all the cells are clumped in the bottom of the tube or well. The clumps may appear as distinct granules, flakes, or flocculent masses. Where a strong reaction has occurred, the suspending fluid is usually clear. When plates are used as the reaction vessel, the subsequent agglutination reaction may take on a ropy or knarled appearance.

Those wells or tubes in which no reaction has occurred exhibit no clumping except for a central compact button of sedimented cells. Generally, a decreasing gradation in the concentration of cells away from the central button can be seen. Also, the suspending fluid tends to remain quite turbid. Between the strong reaction and the negative reaction are varying degrees of agglutination. Inclusion of known controls helps inexperienced operators distinguish between weak positive and negative reactions.

V. STRAIN IDENTIFICATION

Distribute aliquots of each antigen into test tubes or wells of the desired plates. For the test tube procedure, use 0.5 ml of antigen suspension, while for the microtiter and disposable tray plates use 0.025 and 0.25 ml, respectively. Add an equal volume of antiserum diluted to the optimum titer for the test. Incubate the test tubes or plates according to the procedures outlined for the determination of serum titers (Section III). Read the reactions according to the specifications outlined in Section IV.

VI. ADSORBED ANTISERA

Two or more isolates may share common antigenic determinants. In these instances utilization of a common antiserum would not permit recognition of the two individual strains. While sharing common determinants, similar isolates also may have distinctive antigenic determinants that could be utilized for identification purposes. For isolates sharing common determinants as well as distinctive determinants, an adsorbed antiserum may be prepared to take advantage of the differences between the isolates.

Prepare a suspension of the test antigen previously described and adjust the cell concentration to log 9 to log 10 per milliliter. Pipette 2.5 ml of the antigen into 10 ml of a 1:20 dilution of the serum to be used. Suspend the cells in the serum by gentle mixing and incubate in a water bath at 37°C for approximately 1 hr or place in the refrigerator overnight. Centrifuge the tubes and their contents, conserving the supernatant for future use. Test the specificity of the serum using the appropriate antigens and the procedures previously outlined. Repeat the adsorption process if the specificity of the serum has not been enhanced to a satisfactory level. Two to three adsorption cycles result in a satisfactorily adsorbed serum.

VII. COMMENTS

The apparent simplicity of the agglutination test belies the inherent difficulties associated with the reaction. Despite problems in reading tubes or plates, the agglutination procedure can be routinely used with good results if a few precautions are noted.

Prior familiarization with the test strains is essential. Agglutination reactions tend to be strain and serum specific. While all positive reactions tend to be alike, individual strain agglutination patterns tend to differ. Therefore, each strain should be observed for unique agglutination reactions prior to any large test of unknowns.

When conducting agglutination tests, investigators should include known antigen controls of the type being tested. For example, if nodule antigens are being evaluated, include known nodule antigens as positive standards. By including positive standards as controls, the efficacy of the test can be verified. Furthermore, these standards provide a reference for agglutination to compare unknown samples which may show some deviation from the "standard" reaction.

One trained person can type from 300 to 500 antigens per week. Increasing the number of support personnel, of course, would increase the number of samples that could be read. When dealing with similar specimens from indigenous populations, similarity of antigen reactions to a common antiserum suggests some sharing of common antigenic determinants. By using adsorbed antiserum, the specificity of the reaction can be improved in some, but not all cases.

The agglutination procedure of antigen identification is reliable. Positive reactions admittedly are interpreted differently by different people. However, identification agreement generally occurs for more than 90% of the antigens studied. No substitute is available for experience when reading agglutination reactions. If possible, researchers should work with someone acquainted with the procedure prior to embarking on a large-scale testing program.

REFERENCES

Carpenter, P. L. (1975). *Immunology and Serology*, 3rd ed. W. B. Saunders Company, Philadelphia.

Means, U. M., and Johnson, H. W. (1964). *Appl. Microbiol. 16*: 203.

Somasegaran, R., Woolfenden, R., and Halliday, J. (1983). *J. Appl. Bacteirol. 55*: 253.

Vincent, J. M. (1970). *A Manual for the Practical Study of Root-nodule Bacteria* (IBP Handbook 15). Blackwell Scientific Publications Ltd., Oxford.

Weaver, R. W., and Frederick, L. R. (1982). *Rhizobium.* In *Methods of Soil Analysis*, Part 2, *Chemical and Microbiological Properties*, American Society of Agronomy, Madison, Wis., pp. 1041-1070.

Williams, C. A., and Chase, M. W., eds. (1977). *Methods in Immunology and Immunochemistry*, Vol. 4, *Agglutination, Complementation, Neutralization, and Inhibition.* Academic Press, Inc., New York, 468pp.

7

Enzyme-Linked Immunosorbent Assay (ELISA) for the Detection and Identification of *Rhizobium* Strains

B. DOV KISHINEVSKY *The Volcani Center, Bet Dagan, Israel*

DAVID GARETH JONES *University College of Wales, Aberystwyth, Dyfed, Wales*

I. INTRODUCTION

Almost all aspects of research into the legume-*Rhizobium* symbiosis have benefited from the development of several methods for the precise identification of strains of the microsymbiont. Such marker techniques have proved invaluable in ecological studies, especially in respect of the competitive ability of strains, host plant preference for strains, and in studies of saprophytic competence, a factor of great practical importance in terms of the longevity of strains used as inoculants in various soils.

Marker techniques can be conveniently grouped into three categories (a) those based on serological differences, (b) genetic markers, and (c) other natural markers. The whole topic of *Rhizobium* strain identification has been received recently (Schwinghamer and Dudman, 1980; Jones, 1983). It will be appreciated that to facilitate autecological studies, a marker technique that is both rapid and reliable is essential. The method must allow the identification of the marked strain in a population of closely related strains of the same cross-inoculation group, and this must involve the soil situation if the method is to be adopted extensively. In addition, wherever possible, the marked strain should be "wild type," as the introduction of other markers may produce changes that do not reflect the parental strain.

There can be no doubt that serological markers--the use of the antibody-antigen reaction—form the basis of one of the most precise methods of identifying bacteria, whether in medical pathology or agricultural microbiology. Serological methods are frequently favored because of their high specificity and reproducibility, these advantages being facilitated by the fact that rhizobial strains possess a high degree of antigenic heterogeneity and therefore can be separated into many antigenically distinct types. Traditionally, of the serological methods available, agglutination and immunodiffusion have been the most regularly employed in strain identification. Means et al. (1964) were able to identify antigens in single nodules by the agglutination test, but this method is insensitive when small nodules are examined, a limitation that is also a problem with the immunodiffusion technique (Dudman, 1977). In addition, both techniques have the disadvantage of the requirement to isolate and subculture the rhizobia from nodules prior to conducting the test. The same limitation is inherent in other identification techniques using genetic markers such as antibiotic resistance (Obaton, 1971; Schwinghamer and Dudman, 1973; Brockwell et al., 1977; Hardarson and Jones, 1979) and auxotrophy (Johnston and Beringer, 1975).

The immunofluorescence technique (FAT) is as precise as any serologically based test, the limitations being the levels of cross-reactivity of the strains to be used and the requirement for expensive microscopic equipment. It has been used successfully to identify the rhizobia in crushed nodules (Schmidt et al., 1968; Trinick, 1969), in hand-cut sections of nodules (Russell and Jones, 1975), and for the quantitative determination of specific rhizobia in soils (Bohlool and Schmidt, 1970). It is therefore very apparent that a more sensitive technique is necessary when only small amounts of antigen are available, a problem especially relevant to studies using perennial herbage legumes such as white clover.

The introduction of the enzyme-linked immunosorbent assay (ELISA), first outlined by Engvall and Perlmann (1971, 1972) and by Van Weemen and Schuurs (1971), offered a most promising alternative to the conventional serological methods mentioned previously. The ELISA technique combines simplicity of operation with specificity, sensitivity, and rapidity of performance. The enzyme-labeled reagents are relatively cheap to prepare, highly stable, and have a long shelf life if stored correctly. The ELISA technique, which is used extensively for the identification of viral and bacterial plant pathogens (Thresh et al., 1977; Bar-Joseph and Salomon, 1980; Koenig, 1978; Saunders and French, 1983), as well as in clinical bacteriology (Nassau et al., 1976) and immunology, where it was reported to have sensitivity comparable with radioimmunoassay (Maolini and Masseyeff, 1975).

Kishinevsky and Bar-Joseph (1978) were the first to utilize ELISA for the identification of *Rhizobium* strains of the cowpea miscellany isolated from peanuts (*Arachis hypogaea*). Successive workers have modified the technique in attempting to make it even more sensitive (Berger et al., 1979; Morley and Jones, 1980; Fuhrmann and Wollum, 1985). More recently, ELISA has been utilized in studies examining serological diversity of *Rhizobium* strains (Ahmad et al., 1981), their competitive ability (Nambiar et al., 1984; Rice et al., 1984; Martensson and Gustafsson, 1985), and for the quality control of inoculants (Olsen et al., 1983; Nambiar and Anjaiah, 1985). The success of ELISA will still depend on the skill of the operator, but meticulous care in the preparation of antisera, conjugates, and reagents will help in no small way. However, it must be stressed that certain difficulties have been encountered (Tchan, 1982), many have been overcome, but others are still the subject of current investigations, which, hopefully, will not only lead to an increase in ELISA efficiency but also in its usage. This chapter, by including precise details of all aspects of the ELISA technique and its modification and by discussing their advantages and limitations, should provide the reader with a comprehensive practical manual and an insight of its applications.

II. METHODOLOGICAL ASPECTS

Enzyme-linked immunosorbent assay (ELISA) is a term that refers to any procedure involving the use of a solid-phase carrier to facilitate the rapid separation of free antigen and antibody from an antigen-antibody complex and using an enzyme marker to detect an antigen-antibody reaction. The ELISA techniques are based on an important finding that antibodies can be conjugated to an enzyme in such a way that the immunological and enzymatic activity of each moiety is retained. Enzymes are extremely active and one molecule can initiate chain reactions involving many molecules of colorless substrate, resulting in the formation of highly colored products. In this way, enzymes can easily be detected at very low concentrations and therefore facilitate the recognition of binding extremely small amounts of antibodies and antigens which may not be detectable by agglutination and immunodiffusion tests. In principle, ELISA is similar to radioimmunoassay and immunofluorescent methods in which the antigens (or antibodies) are traced by means of radio- or fluorescent labels.

ELISA differs from other serological techniques in that it involves the consecutive immobilization of antigens and antibodies on an insoluble carrier surface followed by the removal of nonrelevant substances from the reaction site. This is an advantage since most of the nonspecifically interfering factors are thus eliminated. In consequence, pure antigen is not an essential requirement and assays may be carried out as effectively with nodule suspensions as with *Rhizobium* culture antigens of equivalent cell concentration.

The ELISA techniques used so far for *Rhizobium* strain identification fall into one of two types: (a) the direct or double-antibody sandwich (DAS) ELISA, and (b) the indirect ELISA. The two techniques differ from each other in the way in which the test antigen is immobilized and by the type of antibodies coupled to the enzyme (Figure 1).

A. Direct "Double-Antibody Sandwich" ELISA

In this method, the enzyme is conjugated directly to strain-specific antirhizobial antibodies. The antigen (cell suspension or nodule crush) is added to a specific antibody which has previously been adsorbed onto the solid inner surface of a well of a polystyrene microtiter plate. The immobilized antigen-antibody complex is then coated with the previously mentioned enzyme-labeled antibody of the same specificity as the sensitizing antibody. Further addition of a suitable enzyme substrate permits the colorimetric detection and measurement of the enzyme-antibody-antigen complex. The reaction product is proportional to the amount of antigen in the test solution. If the antigen is not specific to the antibody, it will be removed by washing steps, no complex will be found, and no colorimetric substrate change will occur.

B. Indirect ELISA

This method employs an enzyme attached to secondary anti-species antibodies (or to protein A as described by Kishinevsky and Maoz,

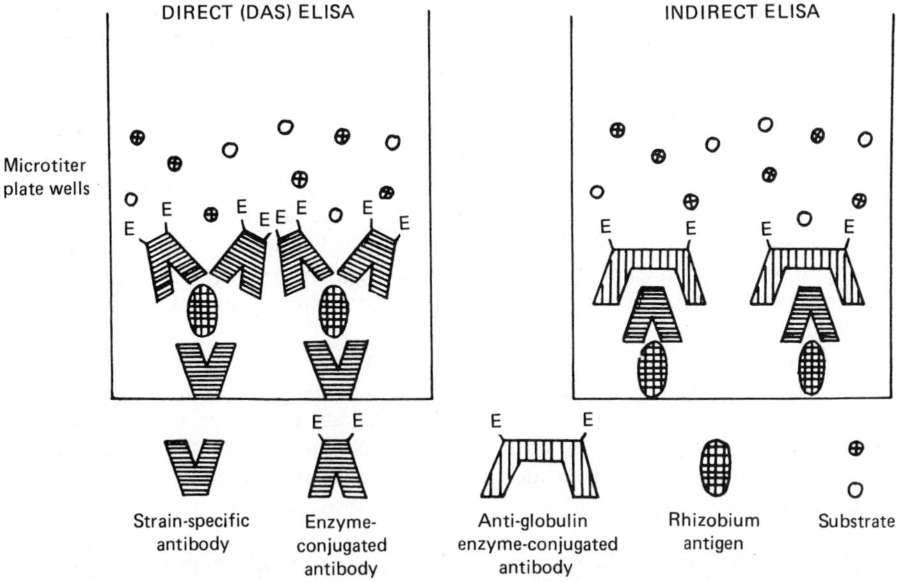

Figure 1 Comparison of reaction mixtures in direct and indirect ELISA.

1983) which are capable of reacting with the strain-specific antibodies. In this technique, the test antigen (cultured cells or crushed nodule suspensions) is nonspecifically bound to glass slides or microtiter plate wells by heat fixation (Olsen and Rice, 1984) or by passive and nonspecific adsorption to the plate wells (Ahmad et al., 1981; Kishinevsky and Maoz, 1983), followed by a specific binding of anti-*Rhizobium* antibodies to the antigen layer. The immobilized antigen-antibody complex is then used to "capture" the enzyme-labeled anti-immunoglobulin molecule. The amount of substrate degraded is proportional to the amount of enzyme present, which, in turn, is indirectly related to the quantity of antigen on the solid surface.

C. Preparation of Antisera

The production of specific antisera with high antibody titers is an essential prerequisite for a successful ELISA test. Although there is no sure way of predicting whether a particular schedule of injection of the animal will provide such quality of antisera, the immunization procedures commonly used in producing antisera for agglutination and immunodiffusion tests have also proved to be adequate for use in ELISA.

Antisera against *Rhizobium* strains are commonly produced by using whole-cell (living or steam-killed) antigens. The problem of low specificity and sensitivity of antibodies produced against purified *Rhizobium* cell wall lipopolysaccharides was encountered by Martensson et al. (1984) in their attempt to use such antibodies for *R. meliloti* strain identification.

Other factors, such as the method of preparing the antigen, the number and sites of injections administered, and the time between primary immunization and bleeding can all be manipulated to enhance the immunoglobulin content in strain-specific antisera. For most practical purposes, it is sufficient to measure the quality of each new antiserum by a tube agglutination test (Kishinevsky and Bar-Joseph, 1978), and only antisera with a high homologous agglutination titer (at least 1280) should be used for the ELISA test.

Since individual animals may differ in their response to identical injections of the same antigen (Chase, 1967) it is good practice to immunize two or more animals at the same time as insurance that at least one of them will produce a satisfactory antiserum. The following procedures have been utilized successfully in the preparation of antisera for ELISA.

Procedure A: Antisera Against Different Strains of *Rhizobium* spp. of Peanut (*Arachis hypogaea L.*) (Kishinevsky and Bar-Joseph, 1978). The *Rhizobium* strains were grown on the defined solid medium of Bergersen (1961) for 7 days at 28°C, after which they were harvested and suspended in a 0.85% solution of NaCl. The antisera were prepared by intramuscular injection of rabbits with 1 ml of emulsion of equal parts of Freund's complete adjuvant (Difco) and a culture suspension containing approximately 2×10^9 cells/ml. Two and three weeks later, two subsequent intravenous injections of 1 ml each were given (no added adjuvant) and the rabbits bled on the fifth day after the last injection.

Procedure B: Antisera Against *R. trifolii* (Morley and Jones, 1980). Whole cells were harvested from 5 to 7-day-old plates with sterile saline (0.85% NaCl) well mixed for a few seconds and then centrifuged at $6000 \times g$ for 30 min at 5°C. The pellet obtained was resuspended in sterile saline and the foregoing process repeated twice more. The resultant homogeneous suspension contained about 10^8 cells/ml. The first two injections (intravenous) were given with 1 ml of culture suspension with a 7-day interval. After a further 7 days an intramuscular injection of 1 ml of emulsion of equal volume of Freund's adjuvant and culture suspension was given and the rabbits were bled on the seventh day after the last injection. Although the foregoing immunization procedure produced good-quality antisera, it has been pointed out (Morley and Jones, personal communication) that the procedure outlined in the preceding paragraph is now adopted.

Procedure C: Antisera Against *R. meliloti* (Olsen et al., 1983). Cells grown in yeast extract-mannitol broth for 5 days at 30°C were harvested with phosphate-buffered saline (PBS; see the Appendix) followed by centrifugation at 6000 × *g* at 4°C. The cells were then washed, pelleted three times in PBS, resuspended, and steamed at 100°C for 30 min. The cells were then immediately pelleted and given three additional PBS washes before being resuspended to 10^{11} cells/ml of which 1 ml was emulsified with 1 ml of Freund's complete adjuvant. The first inection (rabbits) was with 0.1 ml under each rear-foot toe pad and 0.25 ml intramuscularly. Six weeks later, the rabbits were injected intravenously with 0.5 ml of fresh steam-washed cells followed by two intravenous injections of 0.25 ml at week 7 and week 8. The rabbits were bled on the seventh day after the last injection.

D. Purification of Immunoglobulins

Whole antisera often include contaminants that produce "background" reactions that are undesirable and reduce the efficiency of the ELISA test. To overcome this problem, it is necessary to isolate an immunologically pure immunoglobulin (IgG) from the antiserum, which is then conjugated to the enzyme.

Many of the anti-species antisera as well as purified goat (sheep, swine) anti-rabbit γ-globulins which have usually been used as reagents for indirect ELISA studies are now commercially available. These antibodies will have been prepared by accepted procedures and normally give good results.

The following protocols will outline some methods commonly used for the purification of IgG. Familiarity with such techniques as dialysis, centrifugation, and measurement of protein concentration by absorbance at 280 nm is assumed.

Common Materials and Reagents

1. Antiserum (see Section II.C for method of preparation)
2. Deionized water
3. Spectrophotometer (to read absorbance at 280 nm)
4. Glassware, pipettes, beakers, tubes, silicone-treated glass tubes, and so on.

Materials and Reagents for Specific Use

1. PBS containing 0.02% NaN_3 (see the Appendix) (procedures D and E)
2. $(NH_4)_2SO_4$, saturated solution (procedures D and E)
3. DEAE-cellulose, Whatman DE23 (procedure D)
4. LKB Ultragel AcA 34 (procedure E)

5. Column PD-10 (Pharmacia Fine Chemicals) (procedure D)
6. Dialysis tubing (procedure D)
7. Medium-speed centrifuge (procedures D and E)

Methods

Procedure D: preparation of γ-globulins by salt precipitation and DE23 cellulose filtration. This method has been widely adopted in ELISA tests for the identification of plant viruses (Clark and Adams, 1977) and has also been used successfully in DAS procedures for the identification of rhizobial antigens (Kishinevsky and Bar-Joseph, 1978; Morley and Jones, 1980).

1. Dilute 1 ml of rabbit antiserum with 9 ml of deionized water.
2. Add slowly (1 drop per second) saturated $(NH_4)_2SO_4$ solution to give a final concentration of 35% and leave at room temperatures for 30 to 60 min.
3. Centrifuge at 8000 × g to collect precipitate.
4. Dissolve precipitate in 2 ml of half-strength phosphate-buffered saline (PBS).
5. Dialyze against at least three changes of 500 ml of half-strength PBS over a period of 24 hr or filter through a Pharmacia PD-10 column prepacked with Sephadex G-25M (see the Appendix).
6. Filter through 3 to 5 ml of DEAE-cellulose preequilibrated in half-strength PBS (DEAE-cellulose suspended in half-strength PBS was allowed to settle. The buffer is poured off and the washing is repeated five to seven times. A broken 10 ml-pipette can be used as a filter column).
7. Wash γ-globulin through DEAE-cellulose with half-strength PBS collecting the eluate in approximately 2-ml fractions.
8. Monitor effluent at 280 nm and combine all fractions containing first protein peak.
9. Adjust strength of γ-globulin to approximately 1 mg/ml (E_{280} = 1.4).
10. Store in 1-ml amounts in silicone-treated glass tubes at -18°C.

Procedure E: preparation of γ-globulins by using LKB Ultragel AcA34. Rice and Olsen (personal communication) deviated from procedure D by using LKB Ultragel AcA34 as follows:

1. Adsorb antisera extensively with heterologous cross-reacting cells.
2. Slowly add 3 ml of cold saturated $(NH_4)_2SO_4$ to 4 ml of cold adsorbed antisera while lightly vortexing.
3. Let stand at 4°C for 4 hr and centrifuge, decant superantant, and redissolve globulins in 4 ml of PBS.

4. Slowly add 3 ml of cold, saturated $(NH_4)_2SO_4$ to 4 ml of cold globulin in PBS while lightly vortexing. Let stand 4 hr at 4°C, centrifuge, decant and drain supernatant.
5. Redissolve globulins in minimum PBS--usually 1 ml—and pass over AcA34, 2.5 cm by 50 cm. This results in two protein peaks, one over 300,000 mw which is discarded and another of 150,000 to 160,000 mw. Only the most concentrated of the IgG fractions are taken and pooled, ordinarily resulting in about 10 ml of IgG preparation at 1 to 2 mg/ml. Elution buffer is PBS with 0.02% NaN_3 and IgG is stored at 4°C without further concentration or stabilization. This preparation is used for conjugations without any further treatment.

E. Preparation of Enzyme-Labeled Conjugates

The quality of the enzyme-labeled conjugate can be a limiting factor in ELISA because the sensitivity of the assay is directly related to the reactivity of the conjugate yielding visible product molecules. Alkaline phosphatase has usually been used as an enzyme marker for conjugates used in *Rhizobium* antigen-antibody systems. This enzyme is highly reactive, stable, and can be coupled to the antibodies by the "one-step" glutaraldehyde method (Avrameas, 1969), which is easy to perform. *P*-nitrophenyl phosphate, which is usually used as the substrate for this enzyme, yields a strong yellow-colored product, easily assessed by eye or measured spectrophotometrically. For additional sensitivity, especially for use with legumes bearing small nodules, a fluorescent substrate (3-O-methylfluorescein phosphate) can be used (Morley and Jones, 1980), but for the quantitative assessment of results, the fluorescence is measured using a fluorimeter (fluorescence spectrophotometer).

Another enzyme, β-galactosidase, was shown recently (Martensson et al., 1984) to be a good choice for conjugation. Like alkaline phosphatase, this enzyme is very stable and its activity may be sensitively measured using *p*-nitrophenyl-β-D-galactoside or *o*-nitrophenyl-β-D-galactopyranoside.

Since the enzymes used in ELISA are relatively expensive, it is most important that highly titered antiserum be used for conjugation. This enables a significant reduction in the costs of ELISA, by allowing the use of small volumes of conjugates, which may be obtained by one of the several preparation procedures.

The choice of the best conjugation method is very difficult at the present time due to the lack of comparative data on the coupling reactions of different enzymes and on procedures. Furthermore, the reactivity of the conjugates obtained may vary from batch to batch even when the same preparation method is used. However, once prepared, a highly reactive conjugate keeps its activity for at least $1\frac{1}{2}$ years

(Voller et al., 1977), permitting an economical use of this reagent since it can be greatly diluted for the assy. Unfortunately, antibodies produced against *Rhizobium* cell wall components are not as efficient in terms of titer as those that can be prepared against pure protein antigens such as rabbit immunoglobulins. In consequence, enzyme-labeled anti-*Rhizobium* IgG has usually been used at considerably lower dilutions (1/200 to 1/400) than the conjugated goat (sheep, swine) anti-rabbit immunoglobulins (1/2000 to 1/4000).

Common Materials and Reagents

1. Alkaline phosphatase (e.g., Sigma Type VII or EIA Grade Boeringer-Mannheim)
2. Glutaraldehyde, 25%, electron microscope grade (e.g., Sigma Grade I)
3. Bovine serum albumin, Fraction V
4. PBS (see the Appendix)
5. Sodium azide (N.B.: poisonous, handle with care; do not leave in contact with metal surfaces)
6. Distilled water
7. Dialysis tubing
8. Glassware: pipettes, beakers, flasks, tubes, and so on

Materials and Reagents for Specific Use Only

1. Purified IgG (procedure G) (see Section II.D)
2. Protein A (e.g., Sigma type) (procedure H)
3. Medium-speed centrifuge (procedure G)

Methods

Procedure G: conjugation of alkaline phosphatase with γ-globulin by the "one-step" glutaraldehyde method (Avrameas, 1969)

1. Centrifuge 1 ml (= 5 mg) of enzyme precipitate in the cold for 10 min at 1000 rpm. Discard the supernatant liquid.
2. Mix the pellet with 2 ml (= 2 mg) of purified γ-globulin.
3. Dialyze three times against 500 ml of PBS to remove $(NH_4)_2SO_4$.
4. Add fresh glutaraldehyde solution to 0.06% final concentration and mix well.
5. Leave 4 hr at room temperature. A very pale yellow-brown color may develop.
6. Transfer to dialysis tube and dialyze at least three times against 500 ml of PBS to remove glutaraldehyde.
7. If necessary, centrifuge at low speed to remove any precipitate that may have formed.
8. Add bovine serum albumin to about 5 μg/ml and store at 4°C.

A procedure for antibody-enzyme conjugation that allows preparation of an enzyme-labeled antibody without dialysis is given below (Rice and Olsen, personal communication).

1. Add 0.5 mg of IgG (prepared as above) to 0.75 mg of alkaline phosphatase (10 mg/ml, EIA Grade, Boehringer-Mannheim).
2. Dilute mixture to 495 μl.
3. Add 5 μl of 12.5% glutaraldehyde [freshly diluted with H_2O from 25% aqueous purified product (Sigma)].

For example, if the IgG preparation is at 1.3 mg/ml, mix 380 μl of IgG prepared with 0.75 μl of EIA AP; add 40 μl of PBS and 5 μl of 12% glutaraldehyde.

React for 3 hr at room temperature with slight shaking and chromatograph on AcA 34, taking the excluded (first) peak for use in the ELISA. This results in approximately 10 ml of IgG-AP conjugate which can be stabilized with 5 mg/ml BSA. The conjugates are used at concentrations ranging from 0.5 to 10 μl/ml, depending on the particular antibody-antigen reaction being detected.

Procedure H: conjugation of alkaline phosphatase with protein A by the "two-step" glutaraldehyde method (Avrameas and Ternynck, 1971; Engvall, 1978). Protein A, isolated from the cell walls of the bacterium *Staphylococcus aureus*, is known to bind specifically to the Fc region of immunoglobulins for several species of animals, including rabbits (Goudswaard et al., 1978). Therefore, for indirect ELISA detection of rhizobial antigens, it is possible to replace labeled anti-species γ-globulins with labeled protein A.

1. Dialyze 0.5 ml (2.5 mg) of alkaline phosphatase [$(NH_4)_2SO_4$] suspension at least three times against 1 liter of PBS for 24 hr to remove the $(NH_4)_2SO_4$.
2. Transfer the dialysis bag to a beaker with 100 ml of 0.5% glutaraldehyde in PBS and stir overnight at room temperature.
3. Dialyze three times against PBS to remove the excess glutaraldehyde.
4. Dissolve 1 mg of protein A in 0.1 ml of PBS and mix it with the glutaraldehyde-treated enzyme. Leave overnight at room temperature.
5. Store with 0.5% bovine serum albumin at 4°C.

F. Preparation of Antigens and Antibody Adsorption Technique

For preparation of culture antigens the organisms can be grown in analogous way to that used for immunization. When the direct double-antibody sandwich form of ELISA is used, the solution used for extraction of rhizobial antigens (cell suspension or nodules) consisted of phosphate-buffered saline (PBS, pH 7.4) and 0.05% Tween-20. Rhizobial antigen preparation in PBS (without Tween-20) or carbonate buffer (pH 9.6) has been found suitable for use in the indirect ELISA test.

The nodule antigens may be prepared by crushing thoroughly washed nodules individually or using single-strain nodule samples that are collected in bulk. Berger et al. (1979) reported that frozen nodules could be used for ELISA determination of their bacterial content. Similarly, the authors found that the nodules could be tested immediately or stored frozen in 20% (v/v) glycerol at -20°C before use in ELISA (unpublished results).

Antigen preparations used for specific adsorption of antibodies are recommended to be in the form of a heavy suspension (approximately 10^{10} to 10^{11} cells/ml PBS) of whole, washed, and heated (30 min, 100°C) cells. Since monospecific antisera are available for relatively few *Rhizobium* strains, a proper adsorption of antisera should be performed for the removal of heterologous antibodies.

It should be emphasized that there is a difference between adsorption procedures of antisera intended to be used in direct and indirect ELISA tests. The adsorption of IgG after purification is usually acceptable for intermediate antibody used in indirect ELISA but would be unsuitable if the technique being employed were direct ELISA, where the sensitizing antibody should be as pure as possible, as should the antibody to be enzyme conjugated (Olsen and Rice, personal communication).

An adsorption of antibodies with particulate antigens can be conducted by adding 1 ml of heated cells to 1 ml of the antiserum (direct ELISA) or γ-globulin preparation (indirect ELISA) to be adsorbed. The antigen-antiserum (γ-globulin) mixture is left for 1 hr at 37°C, agitated for 2 hr at 4°C on an electrical shaker, and centrifuged for 10 min at 15,000 rpm. The supernatant is then tested for negative ELISA reaction with the absorbing strain and ability to react with the homologous strain. Adsorptions could be repeated until ELISA cross-reactions become negative.

In experiments carried out in soils harboring indigenous rhizobia a situation may arise in which many nodules on uninoculated plants react with antiserum against an inoculant strain. In peanut nodules such cross-reactions were selectively eliminated by crushing 400 mg of bulked nodule material from uninoculated plants in 2 ml of anti-inoculant strain γ-globulin. The antigen γ-globulin mixture was kept at 37°C for 2 hr and stored overnight at 6°C. The suspension was passed through a coarse filter before being centrifuged at 20,000 × *g* for 30 min. The supernatant containing the absorbed antiserum was drawn off with a Pasteur pipette and stored at -20°C (Kishinevsky, unpublished data).

III. THE ELISA TECHNIQUES

Detailed practical instructions for both the direct (DAS) and indirect ELISA techniques as they have been developed and applied in different laboratories for the serological identification of *Rhizobium* strains are given in the following protocols.

The basic ELISA procedure (Engvall and Perlmann, 1971; Voller et al., 1973) remains the same as that applied for the identification of viral plant pathogens (Clark and Adams, 1977), but the experimental conditions, such as incubation times and optimum concentrations, may vary for different antigen-antibody systems. This implies that, before use, each batch of immunoglobulin, antigen, and enzyme-labeled conjugate should be tested over a range of dilutions on microtiter plates, each test to include reference positive and negative controls in every plate (Figure 2). Each step in the procedure and the experimental conditions should be standardized and adhered to rigorously since even small deviations can result in large variations in the sensitivity of the ELISA technique. The level of dilution of reagents giving the best separation of positive and negative test samples should be adopted for the subsequent tests. If conventional ELISA is used, an approximate assessment of the results can be made visually. However, for quantitative assays spectrophotometer readings are necessary.

The discrimination between positive and negative values may present difficulties, especially if the negative samples show a background reaction. As a general rule, it is recommended that the threshold for positive ELISA inactivity be set at two (or three) times above the mean absorbance value of the negative group.

A. Materials and Reagents

1. Antigen
2. Purified immunoglobulin to be used for coating plates (DAS) or as a detecting antibody (indirect tests)
3. Enzyme-labeled conjugates: strain-specific immunoglobulin conjugate (DAS); and anti-species antiglobulin enzyme conjugate (indirect tests)
4. Substrates: 0.6 mg/ml paranitrophenyl phosphate in 10% diethanolamine buffer (if alkaline phosphatase-labeled conjugate

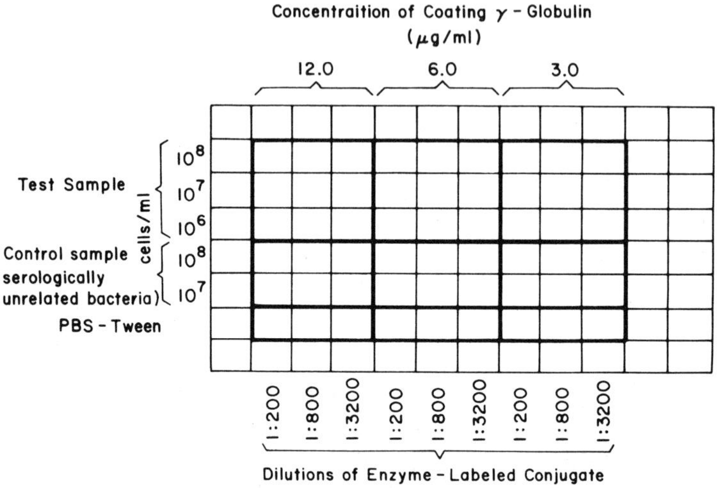

Figure 2 Scheme for determining optimum concentrations of coating and enzyme-labeled γ-globulin. The range of dilutions as they are presented in this scheme have been successfully applied at our laboratories for ELISA determination of rhizobial antigens. The wells of the edge of microtiter plates are usually not used since they sometimes give false-positive results, especially at the corners. (From M. F. Clark and A. N. Adams, Laboratory notes on the ELISA technique for plant viruses, 1977. Reproduced by kind permission of the authors.)

is used); and 3-O-methylfluorescein phosphate (e.g., Sigma) (fluorescent modification of DAS)

5. Bovine serum albumin (BSA)
6. 3.0 *M* NaOH
7. 3.0 *M* NaCl (fluorescent modification of DAS)
8. 0.001 *M* $MgCl_2$
9. Polyvinylpyrrolidone (PVP) (e.g., BDH Chemicals MW 44,000)
10. Buffers (see the Appendix): PBS-Tween; carbonate coating buffer; and diethanolamine buffer
11. Microtiter plates (e.g., Dynatech M29 AR, Linbro, Hamden, Connecticut; Immulon M129A, Dynatech)
12. Plastic plate covers or sandwich wraps
13. Wash bottle for washing plates
14. Incubator at 35°C
15. Light box or photometer for recording results
16. Fluorimeter (fluorescent modification of DAS)

B. Direct Methods

Rhizobial antigen assay using:

A. Double-antibody sandwich ELISA (DAS) (Kishinevsky and Bar-Joseph, 1978; Olsen et al., 1981, 1983)	B. Fluorescent modificaiton of DAS (Morley and Jones, 1980)

1. Add 200 μl of purified strain-specific IgG in coating buffer to each well of the microtiter plate. Cover the plate to prevent evaporation.
2. Incubate at 35°C for 2 to 4 hr.
3. Empty plates and wash by flooding wells in PBS-Tween. Leave soaking at least 2 min. Repeat wash twice, then empty plate and shake out residual liquid.
4. Add 200 μl of test sample (cultured cells or crushed nodule suspensions) diluted in PBS-Tween to duplicate wells.* Cover the plate and incubate at 4°C overnight.
5. Repeat the wash procedure as in step 3.
6. Add to each well 200 μl of enzyme-labeled strain-specific immunoglobulin appropriately diluted in PBS-Tween containing 2% PVP + 0.2% BSA. Cover plate and incubate for 3 hr at 35°C.
7. Repeat the wash procedure as in step 3.

	A	B
8.	Add 200 μl of freshly prepared enzyme substrate solution (*p*-nitrophenyl phosphate) to each well. Incubate at room temperature for 30 min, or as long as positive reference samples reach a desired value (e.g., $A_{405} = 1.0$).	Add 200 μl of freshly prepared 3-O-methylfuorescein phosphate.† Incubate at room temperature for 0.5 to 2 hr.
9.	Stop reaction by adding 50 μl of 3 *M* NaOH to each well.	Stop the reaction by using 50 μl of 3.0 *M* NaCl.
10.	Assess results by visual observation or by measuring absorbance (A) at 405 nm.	Record results by measuring fluorescence (excitation wavelength 468 nm and emission wavelength 512 nm).

---

*For test samples from crushed nodules, PBS + Tween + 2% PVP is usually used since PVP helps to remove any phenolic compounds produced.

†A stock solution of this substrate in cold Tris buffer (pH 8.5, 0.05 *M*) to a concentration of 50 mg/ml is prepared and stored frozen in small amounts. The stock is thawed at 5°C before use and diluted to 10 mg/ml with cold Tris buffer.

C. Indirect Methods

In this variation, two different ways have been used to attach rhizobial cell and nodule antigens to a solid carrier surface where the antigen-antibody reaction takes place.

Assay of Test Antigens Bound to Glass Agglutination Slides by Heat Fixation (Berger et al., 1979)

1. Add 50 μl of test sample (cultured cells or crushed nodule suspensions) diluted in distilled water to duplicate wells and spread evenly over the bottom half surface.
2. Dry the smears under warm air and fix by heating. Slides are used immediately or can be stored at room temperature in vacuo for periods up to 1 week.
3. Wash twice by flooding wells in PBS-Tween containing 0.001 *M* $MgCl_2$. Leave soaking at least 5 min.
4. Add 50 μl of BGG (5 mg/ml) and incubate in a moist chamber at room temperature for 30 min.
5. Repeat the wash procedure three times as in step 3.
6. Add 50 μl of appropriately diluted strain-specific immunoglobulin to each well and incubate in a moist chamber at room temperature for 2 hr.
7. Repeat the wash procedure three times as in step 3.
8. Add 50 μl of antiglobulin enzyme conjugate diluted in PBS-Tween to each well and incubate in a moist chamber at room temperature for 1 hr.
9. Repeat the wash procedure three times as in step 3.
10. Add 50 μl of freshly prepared enzyme substrate solution to each well and reincubate the slides in a moist chamber for 10 to 20 min or as long as positive controls reach a value to be estimated visually and/or measured colorimetrically.
11. Stop the reaciton by adding 50 μl of 3 *M* NaOH. For quantitative measurements the samples should be transferred immediately to tubes containing 3 ml of 0.2 *M* NaOH.
12. Assess the results by visual observation or by measuring absorbance (A) at 405 nm.

In a recently developed modification of this technique (Olsen and Rice, 1984) the glass slides were substituted for by microtiter polystyrene plates followed by evaporation of cell suspension to dryness at 85°C. Steps 4 and 5 were omitted and two successive reactions (steps 6 and 8) were carried out for $2\frac{1}{2}$ and $1\frac{1}{2}$ hr at 37°C.

Assay of Rhizobial Antigens Bound Passively to Microtiter Plates (Ahmad et al., 1981)

1. Add 200 μl of test sample in carbonate buffer pH 9.6 to dupli-

cate wells of the microtiter plate. Cover the plate to prevent evaporation.
2. Incubate at 6°C for 12 to 16 hr.
3. Empty plates and wash by flooding wells in PBS-Tween. Leave soaking for 1 to 2 min. Repeat wash twice, then share out the residual liquid.
4. Add 200 µl of appropriately diluted strain-specific IgG in PBS-Tween to each well of the microtiter plate and incubate at 37°C for 4 hr.
5. Repeat the wash procedure as in step 3.
6. Add 200 µl of antiglobulin enzyme conjugate diluted in PBS-Tween to each well and incubate at 37°C for 3 hr.
7. Repeat the wash procedure as in step 3.
8. Add 200 µl of freshly prepared enzyme substrate solution to each well. Incubate at room temperature for 30 to 60 min or until color has developed to the desired intensity.
9. Terminate reaction with 50 µl of 3 *M* NaOH to each well.
10. Assess the results by visual observation or by measuring absorbance (A) at 405 nm.

Notes

1. A deviation from the standard procedure carrying out the indirect ELISA technique was recently developed at the Volcani Center. By use of enzyme-conjugated goat anti-rabbit globulin as well as enzyme-labeled protein A (Kishinevsky and Maoz, 1983), the following procedure was adopted. The rhizobial antigens (cell suspension or nodules) were extracted and incubated in PBS. The ELISA test was performed at room temperature only. The incubation time for antigen was 3 hr, but for convenience, the antigen preparations were routinely incubated overnight. The incubation time for the detecting antibody was 1 hr and for the conjugate stage, 2-hr instead of 3-hr incubations at 37°C, which are commonly employed in both direct and indirect ELISA procedures.
2. When indirect ELISA with enzyme-labeled protein A is used, the conjugate may be collected and reused up to 10 times for routine tests (Kishinevsky, unpublished data).

IV. APPLICATIONS OF ELISA IN *RHIZOBIUM* STUDIES

The ELISA methods described in Section III have already become accepted as sensitive marker techniques for the identification of *Rhizobium* strains. To date, the tests have been used in (a) the detection of pure culture antigens and the determination of serological relationships between strains, (b) the serological typing of nodule occupants,

(c) strain competition and host plant "preference" studies, and (d) strain quantification in commercial inoculants.

A. Direct ELISA

The earliest report on the use of the direct ELISA was by Kishinevsky and Bar-Joseph (1978), who showed that this technique could be successfully used to identify specific strains of rhizobia (cowpea spp.) in pure culture and in peanut nodules (*Arachis hypogaea*). Homologous isolates were detected with confidence in heat-treated (30 min, 100°C) bacterial suspensions containing 10^4 to 10^5 cells/ml (Figure 3). However, higher cell densities (10^6 to 10^8 cells/ml) were required for positive homologous reactions when unheated culture antigens were used. The suspensions of crushed nodules were not heated before ELISA testing. All the nodules tested gave strong positive reactions with antisera prepared against homologous culture suspension and did not

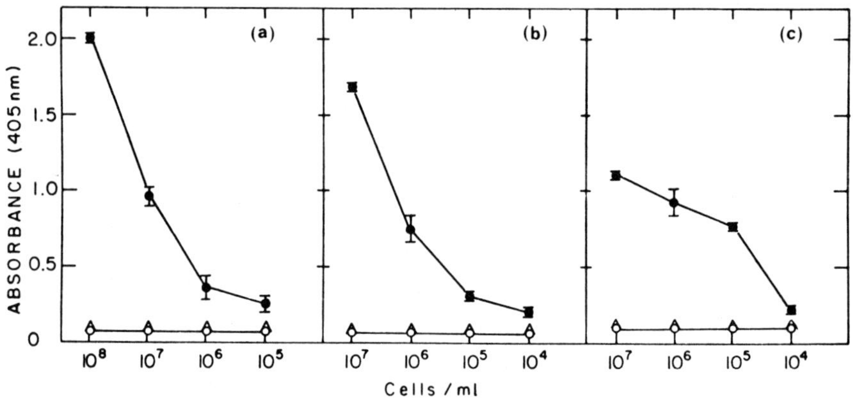

Figure 3 ELISA absorbance values using *Rhizobium* antibodies against a range of concentration of different peanut *Rhizobium* strains. Each point is the mean of four replicates. (a) Antibody to strain 5/70 tested against strains 5/70 (●-●), 3847A (○-○), and 944C (△-△). Plate coated with 5/70 γ-globulin 1 μg/ml. Conjugate 5/70 used at 50 μl/plate (in 20 ml of buffer). (b) Antibody to strain 3847A tested against strains 3847A (●-●), 5/70 (○-○), and 944C (△-△). Plate coated with 3847A-γ globulin 5 μg/ml. Conjugate 3847A used at 50 μl/plate (in 20 ml of buffer). (c) Antibody to strain 944C tested against strains 944C (●-●), 5/70 (○-○), and 3847A (△-△). Plate coated with 944C γ-globulin 1 μg/ml. Conjugate 944C used 50 μl/plate (in 20 ml of buffer). (After Kishinevsky and Bar-Joseph, 1978. Reproduced by permission of Research Journals, National Research Council of Canada.)

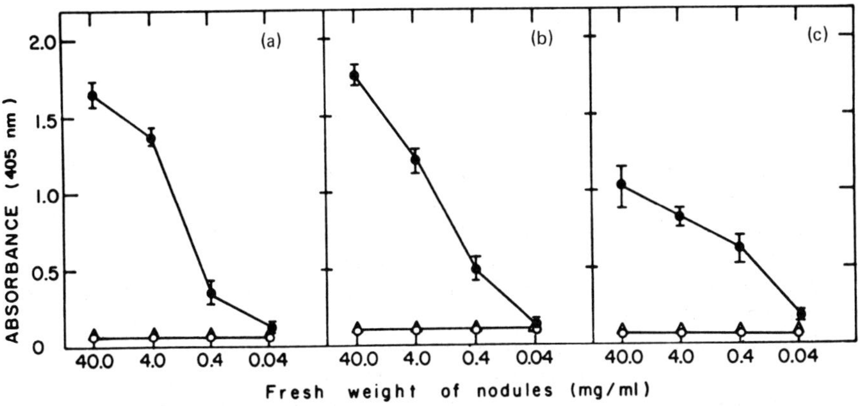

Figure 4 ELISA absorbance values using *Rhizobium* antibodies against various weights of nodule tissue taken from *Arachis hypogaea* plants grown in sterile soil. Each point is the mean of five replicates. Plate coating and conjugate application were as described in Figure 1. (a) Antibody to strain 5/70 tested against nodule antigens of strains 5/70 (●-●), 944C (△-△), and 3847A (○-○). (b) Antibody to strain 3847A tested against nodule antigens of strains 3847A (●-●), 944C (△-△), and 5/70 (○-○). (c) Antibody to strain 944C tested against nodule antigens of strains 944C (●-●), 5/70 (△-△), and 3847A (○-○). (After Kishinevsky and Bar-Joseph, 1978. Reproduced by permission of Research Journals, National Research Council of Canada.)

react with antibodies developed against antigenically distinct strains (Figure 4). Root tissue of uninoculated plants did not interfere with the sensitivity of the ELISA test. It was found that even 0.4 mg fresh weight of nodule tissue crushed in 1 ml of PBS (80 µl per microtiter plate well) was sufficient to give a significant reaction, whereas nodules from plants inoculated with serologically unrelated strains did not react with heterologous antisera. Even the smallest peanut nodule (ca 2 to 8 mg) contained sufficient antigen for the performance of ELISA tests.

Further evaluation of this technique has allowed the identification of rhizobial strains from a number of different taxonomic groups (cross-inoculation groups). Morley and Jones (1980) and Jones and Morley (1981) used the technique with *R. trifolii* and were also able to demonstrate double occupancy of nodules in the latter investigation. Kishinevsky and Gurfel (1980) showed that even 80 µg (per microtiter plate well) of fresh nodules from plants inoculated with *R. lupini*, *R. leguminosarum*, *R. trifolii*, and peanut rhizobia was sufficient for a positive reaction. Olsen et al (1983) further extended the list of applications by using direct ELISA for the identification of *R. meliloti* in commercial inoculants. *R. meliloti* was also the test organism in the studies of Martensson et al. (1984).

One of the main advantages of ELISA tests is the very small requirement for antigen. Pure cultures can be greatly diluted for the assay, and a simple crushed nodule usually provides sufficient antigenic material to be tested against several antisera. This makes the technique very useful in *Rhizobium* competition studies, where a precise identification of singly as well as doubly occupied nodules is of great importance (Kishinevsky and Bar-Joseph, 1978; Jones and Morley, 1981; Rice et al., 1984).

The use of ELISA in the quality control of commercial inoculants provides for both strain identification and, if required, enumeration of cell numbers. Olsen et al. (1981, 1983) used the double-antibody sandwich technique for this purpose in mixed and single-strain commercial inoculants prepared with a nonsterile peat carrier. Each inoculant was also assayed by plant infection, plate count, and agglutination tests. Not only was precise identification facilitated, but the results were available within 5 days, including the time for colony growth on the plate medium. This compared most favorably with the plant infection technique, which was specific only at the *Rhizobium* species level and required 4 to 5 weeks for complete evaluation of the inoculant quality.

Kishinevsky et al. (1982) have recently applied the double-antibody sandwich method to evaluate the quality of commercial inoculants prepared with a sterile peat carrier and designed for peanut (*Arachis hypogea*) inoculation (Figure 5). Uninoculated peat did not interfere with the results of the ELISA test and no color reactions were observed with peat inoculants containing serologically unrelated strains tested with heterologous antisera. Using 3 μg/ml of purified immunoglobulin and 1/400 dilution of stock enzyme-labeled conjugate, it was possible to detect 3.0×10^6 to 1.1×10^7 *Rhizobium* cells per gram of peat without a time-consuming plate or plant infection count. However, difficulties with this technique for the identification of certain *R. meliloti* and *R. leguminosarum* strains have also been encountered. Kishinevsky and Gurfel (1980) obtained erratic DAS ELISA results (often depending on the animal used for antibody production) with 2 of 15 fast and slow-growing *Rhizobium* strains. Negative ELISA results were observed with antisera against these strains, whereas the agglutination titers of the same unfractionated antibodies were reasonably high. A similar failure of direct ELISA to react with one of a group of *R. meliloti* strains was observed by Olsen et al (1983).

Since ELISA enables the detection of *Rhizobium* in suspensions containing 10^4 to 10^5 cells/ml (if the cells are heat treated), it may be considered that the technique has not sufficient sensitivity for legume/*Rhizobium* systems in which the nodules may contain lower cell numbers [e.g., small nodules of white clover (*Trifolium repens* L.)].

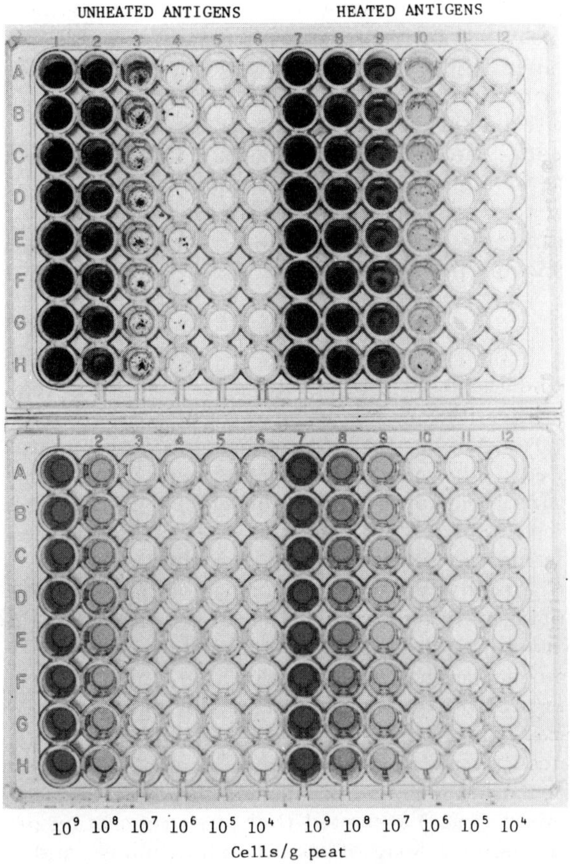

FIGURE 5 ELISA detection of peanut *Rhizobium* cells grown in a peat carrier. Top plate: Inoculant suspensions diluted serially in 10-fold steps. All tests are in eight replicates. Bottom plate: the dark wells indicate positive ELISA results. Note the difference in ELISA reactivity in accordance with heat treatment (30 min, 100°C) and antigen concentration.

Indeed, Morley and Jones (1980) encountered some problems with *Trifolium repens* nodules, which stimulated them to develop a more sensitive modified ELISA technique in which the *p*-nitrophenyl phosphate substrate of the conventional ELISA technique was substituted by 3-O-methylfluorescein phosphate, which degrades to a fluorescent product and can be measured precisely using a spectrofluorimeter (see Section II). This method has proved extremely successful and has been used in several ecological experiments (Jones and Morley, 1981).

Renwick and Jones (1985) compared the fluorescent ELISA and antibiotic resistance identification techniques for use in ecological experiments with *R. trifolii*. Overall, there was significant correlation between the two techniques, but in treatments where an ineffective strain dominated the nodules, there was no correlation. The reason was that 85% of these ineffective nodules were very small, with a volume less than 0.1 mm^3 with bacterial numbers obviously below the threshold for detection using the serologically based method. However, identification of the strain occupying these nodules could be achieved using the antibiotic-resistant marker. Fluorescent ELISA was also shown to be comparable with the antibiotic resistance technique for the enumeration of rhizobia in soils.

B. Indirect ELISA

Although the authors found the DAS ELISA to be a very valuable tool for *Rhizobium* strain identification, it should be recognized that this method necessitates the preparation of a different antibody conjugate for each strain to be tested. This may be a disadvantage, particularly in cases where many strains have to be studied or when it is desirable to test antigens against heterologous antisera. One way to overcome this problem is to use the indirect ELISA method, which allows the detection of different antigens by using a single antiglobulin enzyme conjugate, thereby requiring only one preparation of enzyme-conjugated second antibody or enzyme-conjugated protein A (see procedure H). This device makes the indirect method less complex than the DAS procedure and eliminates the problem of preparing and storing many different conjugated antisera. Furthermore, the costs of indirect ELISA can be significantly reduced by using small amounts of conjugate, which have usually been obtained after a minimum of manipulations. For example, Kishinevsky and Maoz (1983) reported that one preparation of enzyme-labeled protein A, using 2.5 mg of alkaline phosphatase and 1 mg of protein A, contained enough reagent for approximately 15,000 tests (one per well), even though the conjugate was used at moderate (1:4000) dilution. It will be recalled that the commonly used dilutions of conjugated anti-rhizobial globulin in double-antibody sandwich ELISA are 1:200 to 1:400.

Berger et al. (1979) utilized the indirect method to identify strains of *Rhizobium* in pure culture and in lentil nodules using test antigens bound nonspecifically to glass slides by heat fixation. A complete agreement was found in this experiment between the ELISA and immunofluorescence tests, but the simplicity of ELISA was a real advantage.

The usefulness of indirect ELISA was then further demonstrated by Ahmad et al. (1981) examining the serological diversity of 64 slow-growing "cowpea" *Rhizobium* strains isolated from different locations in South America and West Africa. Allowing the test antigen to adsorb passively to the plate wells, they found, interestingly, that ELISA reactivity correlated with colony morphology. In general, strains that form "dry"-type colonies were more ELISA reactive than strains forming mucilage or "wet" colonies.

Although Ahmad et al. (1981) and Kishinevsky and Maoz (1983) successfully used plastic microtiter plates (Dynatech M129A and Linbro, Hamden, Connecticut) to passively adsorb the rhizobial antigens to the plate wells, the limitations of passive antigen binding in the indirect ELISA should not be overlooked. Olsen and Rice (1984) failed in their attempt to attach in this way steamed *R. meliloti* cells to Immulon 2 (Dynatech) microtiter plates. To achieve a firm attachment of test antigens to well surfaces, these authors found it necessary to evaporate rhizobial cell suspensions in plate wells at 85°C. Some difficulties with the practical use of indirect ELISA have also been reported by Tchan (1982). These were concerned primarily with the variation between batches of microtiter plates in the capacity to bind cells.

Since the passive adsorbtion of antigen to microtiter plate wells depends, among other factors, on both the nature of antigen and on the quality of plastic material (Kricka et al., 1980; Lehtonen and Viljanen, 1980), it is clearly necessary for each strain-antibody system under study to check the way in which the rhizobial antigen can be attached to solid surfaces. For example, the authors found an advantage in using preliminary heated culture antigens with some *R. trifolii* and *R. leguminosarum* strains (unpublished data).

C. Comparative Evaluation of Direct and Indirect ELISA

The performance of the direct and indirect ELISA systems was evaluated by comparing their agreement, specificity, and sensitivity using the same antigens and antisera (Kishinevsky et al., 1984). Although the two assays differ from each other in terms of the test conditions and the way in which immunoreactants are used (see Seciton II), good agreement was observed between the two tests with different antigen preparations. The results obtained showed that DAS ELISA allowed the differentiation between some cross-reacting *R. trifolii* and peanut *Rhizobium* strains, whereas the indirect ELISA indicated that they were serologically related. However, antisera adsorbed with antigens of cross-reacting strains failed completely to react in heterologous indirect ELISA reactions without loss of their serological specificity. This may be of practical importance since the indirect ELISA can be used both for defining strain relationships and as a tool for routine rhizobial identification.

V. CONCLUSIONS

There can be no doubt that the ELISA techniques have become established as rapid, reliable, and reproducible marker techniques for the identification of *Rhizobium* strains. They have many advantages over other marker techniques. Of the utmost importance is the fact that ELISA utilizes wild-type strains. Where mutant markers are used, these may be linked genetically to deleterious characteristics which may reduce symbiotic effectiveness, competitive ability, or persistency of the strains (Bromfield and Jones, 1979). In comparison with other serological techniques, ELISA has far greater sensitivity. Heated bacterial suspensions can be confidently identified at concentrations of 10^4 to 10^5 cells/ml with ELISA, whereas the lowest sensitivity level to ensure detectable bands in immunodiffusion techniques is 10^8 to 5×10^{10} cells/ml (Dudman and Brockwell, 1968; Vincent, 1970) and 1 to 10×10^9 cells/ml in the agglutination test (Vincent, 1970). In addition, the requirement for antibody is exceptionally low in ELISA; there is no dependence on the formation of immunoprecipitates and the technique can be used quantitatively. As with all techniques, there are bound to be limitations, and for ELISA, these would seem to involve only some variation in the quality of the polystyrene microtiter plates (Harding, 1982) and the inability to detect very low antigen amounts, as is found in the very small, ineffective nodules of white clover plants. The evidence for the usefulness of ELISA in many aspects of *Rhizobium* ecology (Hodgson and Waid, 1981; Asanuma et al., 1985; Asanuma and Ayanaba, 1982; Ranga Rao et al., 1982; Jones, 1983) and inoculant quality control strongly suggests that this marker technique will increase in usage for many years to come.

ACKNOWLEDGMENTS

The authors express their thanks to Drs. M. F. Clark (East Malling Research Station, U.K.) and M. Bar-Joseph (The Volcani Center, Bet Dagan, Israel), who have freely discussed details of personal experience in the use of ELISA methods, and to Dr. W. A. Rice and Mr. P. O. Olsen (Beaverlodge Research Station, Canada) for helpful advice and unpublished information.

APPENDIX

Buffers

1. Phosphate-buffered saline (PBS), pH 7.4, 0.15 *M*:

NaCl	8.0 g/liter
KCl	0.2 g/liter
KH_2PO_4	0.2 g/liter
$Na_2HPO_4 \cdot 12H_2O$	2.9 g/liter
NaN_3	0.2 g/liter

Dissolve in 1000 ml of distilled water. It is convenient to make up a 10× solution for storage and dilute as required.

2. PBS-Tween, PBS + 0.5 ml of Tween-20 per liter.
3. Carbonate buffer, pH 9.6, 0.05 *M*:

Na_2CO_3	1.59 g/liter
$NaHCO_3$	2.93 g/liter
NaN_3	0.2 g/liter

Dissolve in 1000 ml of distilled water.

4. Substrate buffer:

Diethanolamine	97 ml
NaN_3	0.2 g

Dissolve in 800 ml of distilled water and adjust pH to 9.8 with HCl. Make up volume to 1 liter with distilled water.

5. Tris buffer, pH 8.5 at 25°C:

Tris base	4.36 g/liter
Tris-HCl	2.21 g/liter

Desalting γ-Globulin Solution by Using Pharmacia PD-10 Column

1. Equilibrate PD-10 column in ±50 ml of half-strength PBS. Let it run almost dry.
2. Add 2.0 ml of samples. Let the sample run through and throw away this first volume.
3. Wash with 3.5 ml of half-strength PBS and catch up the desalted sample.

REFERENCES

Ahmad, M. H., Eaglesham, A. R. J., and Hassouna, S. (1981). *Arch. Microbiol. 130*: 281.

Asanuma, S., and Ayanaba, A. (1982). Testing of cowpea rhizobial inoculants. In *Annual Report of the International Institute of Tropical Agriculture*. IITA, Ibadan, Nigeria, pp. 82-85.

Asanuma, S., Thottappilly, G., Ayanaba, A., and Ranga Rao, V. (1985). *Can. J. Microbiol. 31*: 524.

Avrameas, S. (1969). *Immunochemistry 6*: 43.

Avrameas, S., and Ternynck, T. (1971). *Immunochemistry 8*: 1175.

Bar-Joseph, M., and Salomon, R. (1980). *J. Gen. Virol. 47*: 509.

Berger, J. A., May, S. N., Berger, L. R., and Bohlool, B. B. (1979). *Appl. Environ. Microbiol. 37*: 642.

Bergersen, F. J. (1961). *Aust. J. Biol. Sci. 14*: 349.

Bohlool, B. B., and Schmidt, E. L. (1970). *Soil Sci. 110*: 229.

Brockwell, J., Schwinghamer, E. A., and Gault, R. (1977). *Soil Biol. Biochem. 9*: 19.

Bromfield, E. S. P., and Jones, D. G. (1979). *Ann. Appl. Biol. 91*: 211.

Chase, M. W. (1967). Production of antiserum. Section A: Preparation of immunogens. In *Methods in Immunology and Immunochemistry*, Vol. 1, C. A. Williams and M. W. Chase, eds. Academic Press, Inc., New York, pp. 197-209.

Clark, M. F., and Adams, A. N. (1977). *J. Gen. Virol. 34*: 475.

Dudman, W. F. (1977). Serological methods and their application to dinitrogen-fixing organisms. In *A Treatise on Dinitrogen Fixation*, Sec. 4, *Agronomy and Ecology*, K. W. F. Hardy and A. H. Gibson, eds. John Wiley & Sons, Inc., New York, pp. 487-508.

Dudman, W. F., and Brockwell, J. (1968). *Austr. J. Agric. Res. 19*: 739.

Engvall, E. (1978). *Scand. J. Immunol. 8* (Suppl. 7): 25.

Engvall, E., and Perlmann, P. (1971). *Immunochemistry 8*: 871.

Engvall, E., and Perlmann, P. (1972). *J. Immunol. 109*: 129.

Fuhrmann, J., and Wollum, A. G., II (1985). *Appl. Environ. Microbiol. 49*(4): 1010.

Goudswaard, J., Van der Donk, J. A., Noordzij, A., Van Dam, R. H., and Vaerman, J. P. (1978). *Scand. J. Immunol. 8*: 21.

Hardarson, G., and Jones, D. G. (1979). *Ann. Appl. Biol. 92*: 229.

Harding, N. (1982). *Lab. Equip. Dig. 20*: 89.

Hodgson, A. L. M., and Waid, J. S. (1981). Use of enzyme-linked immunosorbent assay (ELISA) to identify bacteriocinogenic strains of *Rhizobium trifolii* in nodules following a mixed strain inoculation. In *Current Perspectives in Nitrogen Fixation*, A. H. Gibson and W. F. Newton, eds. Australian Academy of Science, Canberra, p. 430.

Johnston, A. W. B., and Beringer, J. E. (1975). *J. Gen. Microbiol. 87*: 343.

Jones, D. G. (1983). The ecology of *Rhizobium*. In *Temperate Legumes, Physiology, Genetics and Nodulation*, D. G. Jones and D. R. Davies, eds. Pitman Publishing Ltd., London.

Jones, D. G., and Morley, S. J. (1981). *Ann. Appl. Biol. 97*: 183.

Kishinevsky, B., and Bar-Joseph, M. (1978). *Can. J. Microbiol. 24*: 1537.

Kishinevsky, B., and Gurfel, D. (1980). *J. Appl. Bacteriol. 79*: 517.

Kishinevsky, B., and Maoz, A. (1983). *Curr. Microbiol. 9*: 45.

Kishinevsky, B., Gurfel, D., Lobel, R., and Nemas, C. (1982). *Isr. J. Bot. 31*: 119.

Kishinevsky, B., Maoz, A., Gurfel, D., and Nemas, C. (1984). A comparative evaluation of direct and indirect enzyme-linked immunosorbent assay (ELISA) in serological studies of *Rhizobium* strains. In *Symbiotic Nitrogen Fixaiton*, Vol. 1, B. S. Ghai, ed. USG Publishers and Distributors, Ludhiana, India.

Koenig, R. (1978). *J. Gen. Virol. 40*: 309.

Kricka, L. J., Carter, T. J. N., Burt, S. M., Kennedy, J. H., Holder, R. L., Halliday, M. I., Telford, M. E., and Wisdom, G. B. (1980). *Clin. Chem. 26*(6): 741.

Lehtonen, O. P., and Viljanen, M. K. (1980). *J. Immunol. Methods 34*: 61.

Maolini, R., and Masseyeff, R. (1975). *J. Immunol. Methods 8*: 223.

Martensson, A. M., and Gustafsson, J. G. (1985). *J. Gen. Microbiol. 131*: 3077.

Martensson, A. M., Gustafsson, J. G., and Ljungren, H. D. (1984). *J. Gen. Microbiol. 130*: 247.

Means, U. M., Johnson, H. W., and Date, R. A. (1964). *J. Bacteriol. 87*: 547.

Morley, S. J., and Jones, D. G. (1980). *J. Appl. Bacteriol. 49*: 103.

Nambiar, P. T. C., and Anjaiah, V. (1985). *J. Appl. Bacteriol. 58*: 187.

Nambiar, P. T. C., Strinivasa Rao, B., and Anjaiah, V. (1984). *Peanut Sci. 11*: 83.

Nassau, E., Parsons, E. R., and Johnson, G. D. (1976). *Tubercle 57*: 67.

Obaton, M. (1971). *Comptes-Re. Acad. Sci. Paris Ser. D 272*: 2630.

Olsen, P. E., and Rice, W. A. (1984). *Can. J. Microbiol. 30*(9): 1093.

Olsen, P. E., Rice, W. A., Stemke, G., and Page, W. J. (1981). Serological Identification of Canadian selected *Rhizobium meliloti* strains in commercial inoculants. In *Proceedings of the 8th North American Rhizobium Conference*, September 3-7, 1981, pp. 456-467.

Olsen, P. E., Rice, W. A., Stemke, G., and Page, W. J. (1983). *Can. J. Microbiol. 29*: 225.

Ranga Rao, V., Thottapilly, G., and Ayanaba, A. (1982). Studies on the persistence of introduced strains of *Rhizobium japonicum* in soil during fallow, and the effects on soybean growth and yield. In *Biological Nitrogen Fixation Technology for Tropical Agriculture*, papers presented at a workshop held at CIAT, March 9-13, 1981, P. H. Graham and S. C. Harris, eds. Centro Internacional de Agricultura Tropical, Cali, Colombia, pp. 309-315.

Renwick, A., and Jones, D. G. (1985). *J. Appl. Bacteriol. 58*(2): 199.

Rice, W. A., Olsen, P. E., and Page, W. J. (1984). *Can. J. Microbiol. 30*(9): 1187.

Russell, P. E., and Jones, D. G. (1975). *Soil Biol. Biochem.* 7: 15.

Saunders, M. S., and French, W. J. (1983). *Appl. Environ. Microbiol. 46*(2): 344.

Schmidt, E. L., Bankole, R. O., and Bohlool, B. B. (1968). *J. Bacteriol. 95*: 1987.

Schwinghamer, E. A., and Dudman, W. F. (1973). *J. Appl. Bacteriol. 36*: 263.

Schwinghamer, E. A., and Dudman, W. F. (1980). Methods for identifying strains of diazotrophs. In *Methods for Evaluating Biological Nitrogen Fixation*, F. J. Bergersen, ed. John Wiley & Sons, Inc., New York, pp. 337-365.

Tchan, Y. T. (1982). Application of some newer serological techniques. In *Nitrogen Fixation in Legumes*, J. M. Vincent, ed. Academic Press, Inc., New York, pp. 27-32.

Thresh, J. M., Adams, A. N., Barbara, D. J., and Clark, M. F. (1977). *Ann. Appl. Biol. 87*: 1.

Trinick, M. J. (1969). *J. Appl. Bacteriol. 32*: 181.

Van Weeman, B. K., and Schuurs, A. H. W. M. (1971). *FEBS Lett. 15*: 232.

Vincent, J. M. (1970). *A Manual for the Practical Study of the Root-Nodule Bacteria* (IBP Handbook 15). Blackwell Scientific Publications Ltd., Oxford.

Voller, A., Bidwell, D. E., and Bartlett, A. (1973). Microplate enzyme immunoassays for the immunodiagnosis of virus infections. In *Manual of Clinical Immunology*, N. R. Rose and H. Friedman, eds. American Society for Microbiology, Washington, D.C., pp. 506-512.

Voller, A., Bidwell, D. E., and Bartlett, A. (1977). *The Enzyme-Linked Immunosorbent Assay (ELISA): A Review with a Bibliography of Microplate Applications.* Flow Line Publications, Guernsey Channel Islands, pp. 1-48.

8

The Use of Intrinsic Antibiotic Resistance for Rhizobium Study

ALLAN R. J. EAGLESHAM *Boyce Thompson Institute for Plant Research, Ithaca, New York*

I. INTRODUCTION

Studies in the ecology of rhizobia* rest on the ability to recognize strains. In field inoculation trials it is often necessary to distinguish inoculum from indigenous strain(s) for the determination of success

---

*The nomenclature of Jordan (1984) and of Scholla and Elkan (1984) will be used. The terms "rhizobia" and "rhizobium" refer to bacteria in the genera *Rhizobium* and *Bradyrhizobium*; "bradyrhizobia" refers only to slow growers.

in nodule occupancy, and experiments on interstrain competitiveness often also require the enumeration of competiting strains in the soil and/or rhizosphere. Where the characterization of diversity of indigenous soil populations of rhizobia is required, simple, rapid tools are desirable for the delineation of types.

The simplest method for identifying a nodule occupant relies on an unusual appearance of the nodule. Ineffective strains sometimes form nodules that are smaller in size and/or internally white or green in color rather than pink (Vincent, 1970). Strain-specific dark coloration of nodules has been reported with several legumes (Eaglesham et al., 1982; Eaglesham, 1985) and has been used to study competitiveness among bradyrhizobia (Eaglesham et al., 1982; Rafique Uddin et al., 1984). Distinctive colony appearance of rhizobial growth from surface-sterilized nodules is another potentially useful and convenient characteristic. Within-strain variation of colony types is relatively common (Kuykendall and Elkan, 1976; Upchurch and Elkan, 1977; Meyer and Pueppke, 1980; Stowers and Eaglesham, 1984) and provides a convenient source of different colony morphologies for competitiveness work (Marques Pinto et al., 1974). Screening of bradyrhizobia in West African soils revealed the very strong predominance of a small, "dry" colony type in one location and of a large, confluent "wet" colony type in another (Sinclair and Eaglesham, 1984). The degree of absorption of the dye Congo red was found to vary greatly among strains of *Rhizobium leguminosarum* (Kneen and LaRue, 1983), and strong Congo red absorption by a strain of biovar *trifolii* was used as a marker in competitiveness studies (Bromfield and Jones, 1979). Red pigmentation of colonies (i.e., without Congo red) was used as a diagnostic tool to examine survival in soils of exotic strains effective on *Lotononis bainesii* (Diatloff, 1977).

A commonly used method of rhizobium strain identification utilizes induced resistance to high level(s) of an antibiotic(s) (as described in Chapter 9). Strains marked with antibiotic resistance can be easily and definitively recognized in cultures from soil and nodules when grown in the presence of the correct concentration of the appropriate antibiotic. Before using an antibiotically marked strain, checks should be made to ensure that its effectiveness (Schwinghamer, 1964, 1967; Schwinghamer and Dudman, 1973; Pankhurst, 1977; Hagedorn, 1979) and, perhaps even more important, its competitiveness (Bromfield and Jones, 1979) are the same as those of the wild type. In addition to possible adverse effects on symbiotic characteristics, a disadvantage of the antibiotic marker technique is that in soil work it affords no information on the indigenous rhizobium population.

Antigenic properties have been extensively used for rhizobium identification and to characterize natural populations (see Chapters 5, 6 and 7). The scope of the usefulness of the various serological techniques currently in use is largely determined by the number of antisera available and the extent of cross-reaction among the rhizobia. Antiserum production requires skilled manipulation of animals and the production of more than a few antisera is often impractical. Care must be exercised in choosing strains for antiserum production to avoid cross-reaction with competing strains.

Rhizobia exhibit a very broad range of biochemical and physiological characteristics (e.g., Graham and Parker, 1964; Jordan, 1984; Eaglesham et al., 1984); however, only a few assays are sufficiently definitive to be used alone for strain identification. Differential tolerance of antimicrobial agents among rhizobia has been known for many years (Trussell and Sarles, 1943; Fogle and Allen, 1948), and large differences in degree of tolerance to antibiotics have been demonstrated among species of rhizobia and among strains within species (Davis, 1962; Kecskés and Manninger, 1962; Graham, 1963; Schwinghamer, 1964, 1967; Chopra and Venkataraman, 1969; Elkan, 1971; Schwinghamer and Dudman, 1973; Cole and Elkan, 1979; Hagedorn, 1979). Unique "fingerprints" for 34 strains of *R. leguminosarum*, including eight strains of biovar *phaseoli*, were provided by growth responses to relatively low levels of eight antibiotics (Josey et al., 1979). Having demonstrated the usefulness of intrinsic antibiotic resistance (IAR) as a tool for delineating cultured strains of rhizobia, Beynon and Josey (1980) then used the technique to demonstrate heterogeneity of strains of *R. leguminosarum* biovar *phaseoli* in a natural soil population. These results suggested that IAR can be a powerful tool for the study of rhizobium ecology. The chief advantages of the technique are that it is broadly applicable and does not require sophisticated technology, mutant forms of rhizobia, or animal husbandry.

II. IAR AS A TOOL

The chief uses for IAR in rhizobium studies are (a) as a characteristic for comparing strains, and (b) as a marker for examination of competitive performance. Citations in which IAR in rhizobia has been studied and/or used, the method applied, and the species are listed in Table 1.

A. For Characterization of Rhizobia

Production of penicillinase (Foley and Perret, 1962) was one of a number of diagnostic features for which 97 strains of rhizobia and agrobacteria were examined (Graham and Parker, 1964). Although intraspecific differences were observed, the most striking trend was the higher frequency of occurrence of penicillinase among the slow-growing rhizobia (77% of 31 strains) than among fast growers (10% of 48 strains).

Table 1 Citations in Rhizobium Studies and Clinical Microbiology Methods Involving IAR

Method	Subject[a]							Source
	Cl.	R.f.	R.l.	R.lt.	R.m.	B.j.	B.sp.	
Agar diffusion					X			Antoun et al. (1982)
	X							Bauer et al. (1966)
							X	Broughton et al. (1975)
			X					Chanway and Holl (1986)
						X		Cole (1977)
						X		Cole and Elkan (1979)
			X		X	X	X	Davis (1962)
							X	Diatloff (1977)
						X		Elkan (1971)
			X		X	X	X	Graham (1963)
							X	Gupta et al. (1982)
		X						Heron and Peuppke (1983)
	X							Linton (1983)
			X					Mahler and Bezdicek (1978)
	X							Matsen and Barry (1974)
			X		X	X	X	Pattison and Skinner (1974)
					X			Peterson et al. (1983)
	X							Sabath and Matsen (1974)
	X							Thornsberry (1974)
Agar dilution						X	X	Ayanaba and Wong (1982)
			X					Beynon and Josey (1980)
			X				X	Bromfield et al. (1982)
			X		X	X	X	Chopra and Venkataraman (1969)

				X		Cole and Elkan (1979)
	X				X	Dakora (1985)
				X		Dobereiner et al. (1981)
					X	Eaglesham et al. (1984)
		X				Garg et al. (1985)
	X					Glynn et al. (1985)
				X		Gross et al. (1979)
X						Houang et al. (1983)
			X			Jenkins and Bottomley (1985)
	X					Josey et al. (1979)
	X		X			Kecskés and Manninger (1962)
		X			X	Kingsley and Bohlool (1983)
	X				X	Kremer and Peterson (1982)
					X	La Favre and Focht (1983)
			X			Marques Pinto et al. (1974)
	X					Materon and Hagedorn (1982)
	X					Materon and Hagedorn (1983)
					X	McLaughlin and Ahmad (1984)
	X					McLoughlin et al. (1984)
					X	Misplon and Bishop (1983)
		X			X	Pankhurst (1977)
	X		X		X	Pattison and Skinner (1974)
				X		Pilacinski and Schmidt (1981)
	X					Rai (1983)
		X			X	Rupela et al. (1982)
	X		X			Schwinghamer (1964)
	X		X			Schwinghamer (1967)
	X		X	X		Schwinghamer and Dudman (1973)
					X	Sinclair and Eaglesham (1984)
	X				X	Stein et al. (1982)

Table 1 (*Continued*)

Method	Subject[a]							Source
	Cl.	R.f.	R.l.	R.lt.	R.m.	B.j.	B.sp.	
		X						Stowers and Eaglesham (1984)
	X							Washington (1974)
	X							Washington and Barry (1974)
Broth dilution	X							Gerlach (1974)
		X						Meyer and Pueppke (1980)
	X							Tilton and Newberg (1974)
			X		X		X	Trussel and Sarles (1943)
	X							Washington and Barry (1974)

[a]Cl., clinical; R.f., *R. fredii*; R.l., *R. leguminosarum* (including biovars *trifolii* and *phaseoli*); R.lt., *R. loti*; R.m., *R. meliloti*; B.j., *B. japonicum*; B. sp., *Bradyrhizobium* sp(p).

Several studies have indicated that, in general, higher levels of resistance to antibiotics occur in slow growers than in the fast growers (Graham, 1963; Chopra and Venkataraman, 1969; Pattison and Skinner, 1974; Pankhurst, 1977; Khmelnitsky and Zlotnikov, 1982; Kremer and Peterson, 1982; Dakora, 1985). Twenty-three of 24 randomly selected strains of *Bradyrhizobium japonicum* were resistant to 200 μg/ml neomycin (Cole and Elkan, 1979), whereas a majority of 44 strains of *R. leguminosarum* were susceptible at 2.5 μg/ml (Josey et al., 1979). Pankhurst (1977) compared five slow-growing and seven fast-growing rhizobia from *Lotus* spp. (i.e., *Bradyrhizobium* sp. (*Lotus*) and *R. loti*, respectively) for their patterns of resistance to 16 antibiotics. Greater resistance to nine of the antibiotics was shown by the slow growers, and only in the case of nalidixic acid were fast growers the more resistant group. Minimum inhibitory concentrations of antibiotics in agar (i.e., by the agar-dilution method) are generally in the range 0.1 to 50 μg/ml (see the citations in Table 1). However, much higher levels of resistance among both fast and slow growers have been reported from time to time: for example, resistance to 400 μg/ml polymyxin B in *B. japonicum* (Cole and Elkan, 1979), to 540 μg/ml of streptomycin in *B.* sp. (*Vigna*) (Sinclair and Eaglesham, 1984), to 500 μg/ml kanamycin in *B.* sp. (*Vigna*) (Ayanaba and Wong, 1982), to 750 μg/ml neomycin and to 1500 μg/ml viomycin in *B.* sp. (*Lotus*) (Pankhurst, 1977), and to 2000 μg/ml chloramphenicol in *R. leguminosarum* and *R. meliloti* (Chopra and Venkataraman, 1969).

IAR has been used as a tool for the determination of the heterogeneity of rhizobia, either as assembled from different culture collections (Graham, 1963; Pankhurst, 1977; Gross et al., 1979; Kingsley and Bohlool, 1983; Meyer and Pueppke, 1980; Antoun et al., 1982) or as representatives of natural soil populations (Broughton et al., 1975; Mahler and Bezdicek, 1978; Hagedorn, 1979; Beynon and Josey, 1980; Ayanaba and Wong, 1982; Sinclair and Eaglesham, 1984; McLoughlin et al., 1984; Glynn et al., 1985).

Unique responses to eight antibiotics were reported with each of 26 strains of *R. leguminosarum* (Josey et al., 1979) and 15 out of 16 cultures of bv. *phaseoli* were distinguished on 14 antibiotics (Bromfield et al., 1982). It would, however, be a mistake to assume that different strains of rhizobia can always be distinguished by IAR. The more antibiotics that are used to compare strains, the more likely it is that differences will be discovered, but it can be difficult to be certain that IAR characteristics of two strains are identical. Cultures of "cicer" bradyrhizobia which differed markedly in their serological properties, and therefore were categorized as different strains, showed identical responses to nine antibiotics (Kingsley and Bohlool, 1983)—strong evidence that strains cannot always be "fingerprinted" by IAR. Moreover, 10 strains of *R. fredii* were found to be indistinguishable by their reactions to 10 antibiotics (Heron and Pueppke, 1983) but were clearly distinguishable by their plasmid profiles (Heron and Pueppke, 1984).

The discovery of *R. fredii* (i.e., fast-growing isolates from soybean) (Keyser et al., 1982) prompted investigation of their relative similarity to other fast growers and to *B. japonicum*. Relatively low levels of IAR were found (Stowers and Eaglesham, 1984), consistent with their taxonomic placement in the genus *Rhizobium* rather than in *Bradyrhizobium* (Scholla and Elkan, 1984).

In studies of native populations of rhizobia in which IAR was examined in conjunction with other characteristics, the IAR features correlated with serological properties and protein-gel patterns (Eaglesham et al., 1984; Jenkins and Bottomley, 1985) and with DNA-restriction and plasmid-profile patterns (Glynn et al., 1985).

B. As a Marker

Potentially the most useful function of IAR is as a tag for the determination of rhizobial competitive performance in pot and field studies. Differences in IAR can sometimes occur even in closely related strains. For example, cultures that were identical by serological criteria could be differentiated by IAR (Kingsley and Bohlool, 1983). It should be noted that the use of strains with intrinsic resistance to at least two antibiotics guards against possible complications arising from spontaneous mutations within inoculum cultures or nodules.

With model systems it is necessary to ensure that the strains of interest have different reactions to one or, preferably, two antibiotics. In a competitiveness experiment with *Medicago* spp. grown in tubes, a strain of *R. meliloti* resistant to kanamycin at 2 μg/ml was paired with each of three strains susceptible to kanamycin but resistant to streptomycin at 1.5 μg/ml (Marques Pinto et al., 1974). Thus resistant forms could be recovered quantitatively from rhizosphere mixtures, even when present in low proportion, as well as from nodules. In a similar experiment, again with *R. meliloti*, four strains were chosen which had distinct profiles of intrinsic resistance to eight antibiotics, allowing examination of the competitive interactions of all two-, three-, and four-strain mixtures (Peterson et al., 1983).

In a study of Fix^- mutants of bradyrhizobia, checks that Fix^+ nodules were due to revertants rather than to chance contaminants were achieved by confirming that their tolerance to antimicrobial agents was identical to that of the parent strain and dissimilar to those of other bradyrhizobia (Misplon and Bishop, 1983).

When IAR is to be used as a tag to assess performance of rhizobia added to soil, it is usually advisable to make preliminary assessment of the IAR characteristics of the indigenous rhizobia. The occurrence of native rhizobia with the same IAR profile as an inoculum strain does not necessarily disqualify the use of that strain provided that the native strains form few nodules on uninoculated plants (La Favre and Focht, 1983). In a study of clover nodule occupancy by four strains of *R. leguminosarum* biovar *trifolii* applied individually in a field trial, three of the four strains were distinguishable from the indigenous rhizobia by their IAR patterns (McLoughlin et al., 1984). The fourth strain had the same IAR "fingerprint" as 48% of the nodule isolates from uninoculated plots, and so yielded no useful data.

Known patterns of resistance to single levels of two or three antibiotics can be sufficient to distinguish inoculum from indigenous rhizobia (La Favre and Focht, 1983; Materon and Hagedorn, 1983). For a study of field inoculation of three legume species, each with three rhizobial strains, Kremer and Peterson (1982) determined the maximum levels of resistance of the nine strains, and of the native rhizobia of the same types, of six antibiotics. Each inoculum strain was readily differentiated from the native counterparts by its resistance or susceptibility to only two or three of the antibiotics at one specific level. Where the natural population is very diverse, however, even several antibiotics at multiple levels may not ensure differentiation of inoculum and native strains (McLoughlin et al., 1984).

Introduced bradyrhizobia in soil examined over a 12-year period showed that their IAR characteristics were only slightly less stable than their serological characteristics, which remained unchanged throughout the period (Diatloff, 1977).

III. DETERMINATION OF IAR

There are three methods for the determination of rhizobial IAR characteristics: (a) the agar (or disk)-diffusion method, using antibiotic-impregnated paper disks (e.g., Sensi-Discs, BBL, Cockeysville, Maryland; Dispenso-Discs, Difco, Detroit, Michigan) applied to rhizobium-seeded agar; (b) the agar-dilution method, using stock solutions of antibiotics added to sterilized agar media held at 45 to 50°C, then used to culture rhizobia after cooling; and (c) the broth-dilution method, similar to method b with liquid media. Each method is subject to limitations and in the interests of ensuring reproducible data, the rhizobiologist will find it beneficial to draw on the experience of clinical microbiologists who have used the techniques extensively (Table 1).

A. Agar-Diffusion Method

Although not recommended for slow-growing bacteria of clinical importance (Thornsberry, 1974), the agar-diffusion method has been used successfully with slow- as well as fast-growing rhizobia (see the citations in Table 1).

Commercially available antibiotic-impregnated paper disks are manufactured according to federally legislated quality standards and are therefore reliable. To maintain potency, stocks of disks should be stored with a desiccant in a freezer, preferably at -65°C. Supplies for day-to-day use can be transferred to a 4°C refrigerator and, to prevent condensation, should be allowed to warm to room temperature before being opened to the atmosphere. Dispensors that deposit up to 12 disks spaced equally on a 15-cm–diameter petri plate are commercially available (Difco); otherwise, the disks are placed on the seeded agar surface with sterile forceps. Synergistic and antagonistic interactions are possible between antibiotics; with this in mind, disks should separated sufficiently for each inhibition zone to be entire.

A range of disk potency is available for each antibiotic. Preliminary examination will determine the most appropriate level(s) of the chosen antibiotics. The susceptibility of a strain to an antibiotic is judged by the clear inhibition zone around the disk after appropriate incubation. The size of the clear zone is the consequence of the diffusion of the antibiotic into the agar medium and of the generation time of the rhizobium. There is a complex interaction between these parameters (Linton, 1983) and since both are influenced by a number of factors, care must be taken to ensure that the conditions used for different batches are as similar as possible. Composition of the culture medium can affect antimicrobial activity (Houang et al., 1983). The culture medium should be routinely reproducible with the same brands and reagent quality used each time, and sterilization procedures should be standardized. A constant volume of culture medium per petri plate is important and since pH is particularly critical, it should be routinely checked with a surface electrode. Rhizobia alter the pH of their culture media; therefore, incorporation of buffers [e.g., (*N*-2-hydroxyethylpiperazine-*N*'-ethanesulfonic acid) (Hepes) and [2-(*N*-morpolino)-ethanesulfonic acid] (MES)] (Cole and Elkan, 1979) may improve reproducibility.

The number of organisms applied in the lawn, and their growth stage, can affect the results of the agar-diffusion test. Strain-to-strain variation in extracellular polysaccharide production can result in cultures of equal optical density (OD) but disparate cell densities. Moreover, as a result of strain differences in generation time, cultures of equal age often have significantly different cell densities. Where reproducibility problems are encountered, the construction of standard curves of OD versus viable count for each strain of interest will help to ensure equal cell numbers in inocula, at the same growth stage, within batches and/or between batches. In the same vein, unless generation times are known, circumspection should be exercised in the interpretation of strain-to-strain variations in the sizes of inhibition zones.

As a check on between-batch reproducibility, a few strains of known response to the antibiotics should be included each time. These also serve as indicators of disk potency; alternatively, nonrhizobia can be used for the purpose (Cole and Elkan, 1979).

A modification of the agar-diffusion method involves incorporating the rhizobia in molten agar and flooding it over the previously poured and solidified culture medium (i.e., "agar overlay"). This has been found to improve reproducibility in clinical work (Washington and Barry, 1974; Matsen and Barry, 1974) and has been used with rhizobia (Peterson et al., 1983).

B. Agar-Dilution Method

As described for the agar-diffusion method, conditions and components must be standardized to ensure reproducible results. Particularly important are culture medium composition, pH and sterilization procedures, temperature of agar medium when antibiotic stock solution is added, volume of agar per petri plate, growth stage of test cultures, and size of inoculum (see Rupela et al., 1982).

A critical factor is antibiotic preparation. Many antibiotics are relatively unstable and must be stored strictly according to manufacturers' instructions. Whenever possible, freshly obtained materials should be used. Stock solutions should be used immediately, but if storage is necessary, they should be frozen, preferably at -20°C, for no longer than 14 days. Some antibiotics are not readily soluble in water and stocks are prepared in alternative solvents (e.g., nalidixic acid in 1 *M* NaOH, rifampicin in methanol). Stock solutions must be prepared with due regard to asepsis since they deteriorate if autoclaved; filter sterilization is recommended. Antibiotic stocks are added to the autoclaved agar medium held at 45 to 50°C (the same precise temperature being used each time), thoroughly mixed, poured, and cooled. The plates should be used as soon as possible after preparation, preferably after surface drying in a laminar flow cabinet. Careful hand inoculation (Sinclair and Eaglesham, 1984) or the use of a multipin inoculator (Washington, 1974; Josey et al., 1979) allows a number of strains to be tested on each plate.

An agar-overlay modification of the agar-dilution method involved relatively few rhizobia suspended in the molten agar overlay, with the sensitivity to the antibiotic expressed as percent survival (Rai, 1983). A concentration gradient technique was found to give better reproducibility in the analysis of IAR of *R. leguminosarum* biovar *phaseoli* strains than the pin inoculation/standard agar-dilution method (Bromfield et al., 1982).

C. Broth-Dilution Method

With the broth-dilution method a separate broth culture is required for every strain × antibiotic combination and probably for this reason has been little used for rhizobium work. Small volumes of broth (ca. 2 ml) are used generally, but even greater economy is possible using microdilution techniques that operate at 50-μl volumes (Gerlach, 1974). An advantage of the broth-dilution method is that turbidity measurements allow quantitative expression of data in contrast to the qualitative visual estimates with the agar-dilution method [but see the agar-overlay modification of Rai (1983)]. Where a relatively small number of strain × antibiotic combinations are to be examined, the use of side-arm flasks permits determinations of antibiotic effects on generation times. As with the other methods, there are a number of factors that can reduce within- and between-batch reproducibility, and care is again required in the preparation and execution of the technique.

D. Problems of Reproducibility

IAR has proven to be a useful tool for a number of aspects of rhizobium research, as shown in the references in Table 1. However, it would be misleading to indicate that by closely following published methods, the researcher is guaranteed reproducible IAR data. In general, reports of problems in producing dependable data seldom appear in the literature. Nevertheless, in several cases authors have described or implied batch-to-batch variability in the determination of IAR characteristics of rhizobia (Rupela et al., 1982; Scotti et al., 1982; Misplon and Bishop, 1983; McLoughlin et al., 1984; Sinclair and Eaglesham, 1984; Jenkins and Bottomley, 1985; Chanway and Holl, 1986). Even more disturbing, serious within-batch variability has also been described (Stein et al., 1982). For reasons not yet understood, some strains of rhizobia are variable in their reaction to certain antibiotics (Josey et al., 1979; Beynon and Josey, 1980), but this does not adequately explain the variability experienced by Stein and her co-workers. Lack of rigorous control of inoculum size inherent with the pin-type multi-inoculator may be a major factor in within-batch variability, and differences in incubation conditions may contribute significantly to between-batch variability (Bromfield et al., 1982). In a study of bradyrhizobia, with care taken to reproduce the growth stage of the inoculum as well as its volume, and account taken of variable incubation conditions (only in the most expensive incubators is the temperature *largely* independent of ambient), reproducibility within batches was better than 99% and between batches was 96% (Sinclair and Eaglesham, 1984). However, it does appear that the heterogeneity reported for a natural population of *R. leguminosarum* biovar *phaseoli* (Beynon and Josey, 1980) was overestimated (Stein et al., 1982). Certainly, the differentiation of strains within a population on the basis of low-level IAR is fraught with difficulties. Jenkins and Bottomley (1985) have suggested that in characterizing native populations of rhizobia (in their case, *R. meliloti*) IAR can be useful in conjunction with other methods, but should not be used exclusively.

IV. ECOLOGICAL SIGNIFICANCE OF IAR

There exists among rhizobia an enormous range of patterns of responses to antibiotics, begging the question of the ecological significance of IAR. Perhaps the most plausible rationale for rhizobial resistance to antibiotics can be made in terms of an adaptive response to the presence of antibiotic-producing competitor organisms in the soil. Chatel and Parker (1972) showed that some soils contain water-soluble compounds, believed to be of microbial origin, which were toxic to *R. leguminosarum* biovar *trifolii* but not to *Bradyrhizobium* sp. (*Lupinus*), and adverse effects on rhizobium growth in vitro of diffusable compounds from soil bacteria, fungi, and actinomycetes have been well documented (Fogle and Allen, 1948; Holland and Parker, 1966; Strzelcowa, 1968; Patel, 1974; Foo and Varma, 1976; Anusuya and Sullia, 1984). However, to what extent similar interactions occur in the soil or rhizosphere is uncertain given the vastly different conditions in culture and in soil (Trinick and Parker, 1982). Nevertheless, strong circumstantial evidence has pointed to a role of competitor soil organisms in some cases of legume crop failure resulting from poor nodulation (Damirgi and Johnson, 1966; Holland and Parker, 1966; Patel, 1974; Dobereiner et al., 1981). Problems in the establishment of clover pastures on newly cleared soils in Australia were explained as the result of the high incidence of microorganisms having bacteriocidal or bacteriostatic effects on rhizobia (Holland and Parker, 1966). Unsuccessful inoculation of soybeans in virgin cerrado soils in Brazil was reversed by using strains that were resistant to streptomycin (Scotti et al., 1982); it appears to be more than coincidental that the streptomyces population is unusually high in Brazilian cerrado soils (Coelho and Drozdowicz, 1978). Moreover, the counts of actinomycetes and of streptomycin-resistant bacteria (at 100 μg/ml) increased markedly after lime application to the cerrado (Dobereiner et al., 1981). The disturbance of virgin ecosystems therefore appears to favor development of antibiotics-producing soil microorganisms. Consistent with this, bradyrhizobia isolated from two West African soils cleared from bush within the previous 12 years showed high-level intrinsic resistance to several antibiotics (Sinclair and Eaglesham, 1984). However, in one of these soils the most sensitive bradyrhizobia coexisted with the most resistant, suggesting microsite variability. The cropping of legumes in recently cleared virgin soil may actually enhance the growth of antibiotic-producing microorganisms, as indicated by the preferential stimulation of streptomycin-resistant bacteria in the rhizosphere as compared to nonrhizosphere soil (Brown, 1961).

Resistance to several antibiotics correlated very strongly with the production of copious slime in the study of West African bradyrhizobia referred to above (Sinclair and Eaglesham, 1984). These "wet" strains, in comparison with those of the classical small, discrete, "dry" bradyrhizobium colony type, did not absorb congo-red dye and were more resistant to salt stress (Eaglesham and Ayanaba, 1984), indicating an ability to exclude a range of types of compounds. Resistance to antibiotics shown by "wet" bradyrhizobia correlated with nonaccumulation of radioactively labeled streptomycin (McLaughlin and Ahmad, 1984). In contrast, however, negative correlations were found between IAR and colony mucosity among strains of *B. japonicum* (Meyer and Pueppke, 1980; Pilacinski and Schmidt, 1981). Mutants of *R. leguminosarum* biovar *trifolii* showed two mechanisms of resistance to streptomycin. Low-level resistance (10 to 100 μg/ml) involved changes in membrane permeability and reduced uptake, whereas high-level resistance (1000 μg/ml) was due to alteration in ribosome binding (Zalazna-Kowalska, 1977).

Although the function of IAR in rhizobia does appear a priori to be an adaptive response to the presence of antibiotic-excreting organisms lending saprophytic competence, other factors may also be involved. A link between salt tolerance and resistance to antibiotics was observed also (see above) among strains of *R. leguminosarum* (Rai, 1983). This phenomenon raises the possibility that chemical fertilization of soil, particularly of those in which the cation-exchange capacity is low, may have an adverse effect on rhizobia, causing a selection for those able to exclude salts and coincidentally show resistance to antibiotics. However, the mechanisms of salt tolerance, like those of IAR, vary with rhizobium strain and so they do not necessarily correlate. For example, strains of *R. fredii* grew with 2% NaCl (Sadowsky et al., 1983), but were relatively susceptible to antibiotics (Stowers and Eaglesham, 1984). A possible further compounding factor influencing IAR is the indiscriminate use of pesticides and herbicides, which can have mutagenic as well as selective effects. Examinations of a soil population of *R. meliloti* separated by 20 years revealed significant shifts toward resistance not only to pesticides, but also to antibiotics and metals (Szende, 1982). Heavy metal pollution of soils is common, resulting from exhaust from automobiles, smelters, and power plants and also chemical fertilization. A direct link between heavy metal contamination and induced resistance to antibiotics in soil bacteria has been demonstrated (Lighthart, 1979), although it was not confirmed for *B. japonicum* in a study of three strains (Cole and Elkan, 1973).

Thus there may be a number of agents causative of resistance to antibiotics. Although not proven, there is evidence to indicate that the therapeutic use of antibiotics in animals and even in humans could lead to the induction of IAR in rhizobia. Study of soils around a mission hospital in a remote community in South Africa revealed antibiotic-resistant coliforms (Burt and Woods, 1976). The conjugative transfer of R-factor plasmids carrying multiple antibiotic resistance from *Escherichia coli* to rhizobia is readily achievable in vitro (Beringer, 1974; Johnston and Beringer, 1975; Kuykendall, 1979; Kennedy et al., 1981) and presumable can occur in vivo. The R-factor host range includes *Klebsiella, Pseudomonas, Agrobacterium*, and rhizobia, and in addition to rendering the host resistant to antibiotics can give the ability to transmit the resistance by conjugation (Falkow, 1975). Transfer of R factors among rhizobia occurs in culture (Beringer, 1974; Kuykendall, 1979; Pilacinski and Schmidt, 1981) and therefore could occur in the legume rhizosphere. R-factor transfer between rhizobia can also take place within nodules (Johnston and Beringer, 1975) and, theoretically at least, by transduction (Falkow, 1975). The presence of R factors in naturally occurring rhizobia with high-level IAR has yet to be demonstrated, although much evidence indicated that multiple antibiotic resistance among *B. japonicum* strains [also reported in other bradyrhizobia and *R. meliloti* (Khmelnitsky and Zlotnikov, 1982)] was plasmid mediated (Cole and Elkan, 1973, 1979). However, the mechanisms of resistance to antibiotics shown by *B. japonicum* differed from those shown by bacteria carrying R factors (Cole, 1977). At this time the mechanisms lending resistance to antibiotics among rhizobia are not fully understood and may be as varied as the causative agents.

ACKNOWLEDGMENTS

Thanks to Suzette Noti for typing the manuscript and to Greta Colavito for assistance in obtaining references.

REFERENCES

Anasuya, D., and Sullia, S. B. (1984). *Plant Soil* 77: 387.
Antoun, H., Bordeleau, L. M., and Prevost, D. (1982). *Plant Soil* 66: 45.
Ayanaba, A., and Wong, A. L. (1982). *Soil Biol. Biochem.* 14: 139.
Bauer, A. W., Kirby, W. M. M., Sherris, J. C., and Turck, M. (1966). *Am. J. Clin. Pathol.* 45: 493.
Beringer, J. E. (1974). *J. Gen. Microbiol.* 84: 188.
Beynon, J. L., and Josey, D. P. (1980). *J. Gen. Microbiol.* 118: 437.
Bromfield, E. S. P., and Jones, D. G. (1979). *Ann. Appl. Biol.* 91: 211.

Bromfield, E. S. P., Stein, M., and White, R. P. (1982). *Ann. Appl. Biol. 101*: 269.

Broughton, W. J., Chan, P. Y., Padmanabhan, S., and Tan, K. P. (1975). *Malay. Agric. Res. 4*: 141.

Brown, M. E. (1961). *J. Gen. Microbiol. 24*: 369.

Burt, S. J., and Woods, D. R. (1976). *Antimicrob. Agents Chemother. 10*: 567.

Chanway, C. P., and Holl, F. B. (1986). *Plant Soil 93*: 287.

Chatel, D. L., and Parker, C. A. (1972). *Soil Biol. Biochem. 4*: 289.

Chopra, C. L., and Venkataraman, G. S. (1969). *Curr. Sci. 38*: 119.

Coelho, R. R. R., and Drozdowicz, A. (1978). *Rev. Ecol. Biol. Sol 15*: 459.

Cole, M. A. (1977). Antibiotic resistance in *Rhizobium japonicum*: implications for safety and interpretation of *nif* transfer from *Rhizobium.* In *Genetic Engineering for Nitrogen Fixation*, A. Hollaender, ed. Plenum Press, New York, pp. 423-431.

Cole, M. A., and Elkan, G. H. (1973). *Antimicrob. Agents Chemother. 4*: 248.

Cole, M. A., and Elkan, G. H. (1979). *Appl. Environ. Microbiol. 37*: 867.

Dakora, F. D. (1985). *Acta Microbiol. Pol. 34*: 187.

Damirgi, S. M., and Johnson, H. W. (1966). *Agron. J. 58*: 223.

Davis, R. J. (1962). *J. Bacteriol. 84*: 187.

Diatloff, A. (1977). *Soil Biol. Biochem. 9*: 85.

Dobereiner, J., Scotti, M. R. M. M. L., Sá, N. M. H., and Vargas, M. A. T. (1981). Resistance to streptomycin of *Rhizobium* isolates from Cerrado and Amazon soils . In *Current Perspectives in Nitrogen Fixation*, A. H. Gibson and W. E. Newton, eds. Australian Academy of Science, Canberra, p. 434.

Eaglesham, A. R. J. (1985). *Trop. Agric. Trinidad 62*: 105.

Eaglesham, A. R. J., and Ayanaba, A. (1984). Tropical stress ecology of rhizobia, root nodulation and legume fixation. In *Current Developments in Biological Nitrogen Fixation*, N. S. Subba Rao, ed. Oxford & IBH Publishing Co., New Delhi, pp. 1-35.

Eaglesham, A. R. J., Ahmad, M. H., Hossouna, S., and Goldman, B. J. (1982). *Appl. Environ. Microbiol. 44*: 611.

Eaglesham, A. R. J., Stowers, M. D., Sinclair, M. J., Ayanaba, A., and Goldman, B. J. (1984). Patterns of diversity of cowpea-miscellany rhizobia from three West African locations. In *Symbiotic Nitrogen Fixation*, Vol. 1, B. S. Ghai, ed. USG Publishers and Distributors, Ludhiana, India, pp. 119-125.

Elkan, G. H. (1971). *Plant Soil, Spec. Vol.*: 85.

Falkow, S. (1975). *Infectious Multiple Drug Resistance*, Pion Ltd., London.

Fogle, C. E., and Allen, O. N. (1948). Associative growth of actinomycetes and rhizobia. In *Proceedings of the 48th General Meeting*

of the Society of American Bacteriologists, p. 53.
Foley, J. M., and Perret, C. J. (1962). *Nature 195*: 287.
Foo, E. L., and Varma, A. K. (1976). *Folia Microbiol. 21*: 315.
Garg, F. C., Beri, N., and Tauro, P. (1985). *J. Agric. Sci. Cambridge 105*: 85.
Gerlach, E. H. (1974). Microdilution I: a comparative study. In *Current Techniques for Antibiotic Susceptibility Testing*, A. Balows, ed. Charles C Thomas, Publisher, Springfield, Ill., pp. 63-76.
Glynn, P., Higgins, P., Squartini, A., and O'Gara, F. (1985). *FEMS Microbiol. Lett. 30*: 177.
Graham, P. H. (1963). *Aust. J. Biol. Sci. 16*: 557.
Graham, P. H., and Parker, C. A. (1964). *Plant Soil 20*: 383.
Gross, D. C., Vidaver, A. K., and Klucas, R. V. (1979). *J. Gen. Microbiol. 114*: 257.
Gupta, R. P., Ghai, J., Ghain, S. K., and Kalra, M. S. (1982). *Ind. J. Exp. Bot. 20*: 835.
Hagedorn, C. (1979). *Soil Sci. Soc. Am. J. 43*: 921.
Heron, D. S., and Pueppke, S. G. (1983). *Plant Physiol. 72* (Suppl.): 158.
Heron, D. S., and Pueppke, S. G. (1984). *J. Bacteriol. 160*: 1061.
Holland, A. A., and Parker, C. A. (1966). *Plant Soil 25*: 329.
Houang, E. T., Hince, C., and Howard, A. J. (1983). The effect of composition of culture media on MIC values of antibiotics. In *Antibiotics: Assessment of Antimicrobial Activity and Resistance*, A. D. Russell and L. B. Quesnel, eds. Academic Press, Inc., New York, pp. 31-48.
Jenkins, M. B., and Bottomley, P. J. (1985). *Soil Biol. Biochem. 17*: 173-179.
Johnston, A. W. B., and Beringer, J. E. (1975). *J. Gen. Microbiol. 87*: 343.
Jordan, D. C. (1984). Family III rhizobiaceae. In *Bergey's Manual of Systematic Bacteriology*, 9th ed., Vol. 1, N. R. Kreig and J. G. Holt, eds. Williams & Wilkins, Baltimore, pp. 235-251.
Josey, D. P., Beynon, J. L., Johnston, A. W. B., and Beringer, J. E. (1979). *J. Appl. Bacteriol. 46*: 343.
Kecskés, M., and Manninger, E. (1962). *Can. J. Microbiol. 8*: 157.
Kennedy, C., Dreyfus, B., and Brockwell, J. (1981). *J. Gen. Microbiol. 125*: 233.
Keyser, H. H., Bohlool, B. B., Hu, T. S., and Weber, D. F. (1982). *Science 215*: 1631.
Khmelnitsky, M. I., and Zlotnikov, K. M. (1982). *Antibiotiki Moscow 27*: 418.
Kingsley, M. T., and Bohlool, B. B. (1983). *Can. J. Microbiol. 29*: 518.
Kneen, B. E., and LaRue, T. A. (1983). *Appl. Environ. Microbiol. 45*: 340.

Kremer, R. J., and Peterson, H. L. (1982). *Appl. Environ. Microbiol. 43*: 636.

Kuykendall, L. D. (1979). *Appl. Environ. Microbiol. 37*: 862.

Kuykendall, L. D., and Elkan, G. H. (1976). *Appl. Environ. Microbiol. 32*: 511.

La Favre, J. S., and Focht, D. D. (1983). *Appl. Environ. Microbiol. 46*: 304.

Lighthart, B. (1979). *Appl. Environ. Microbiol. 37*: 859.

Linton, A. H. (1983). Theory of antibiotic inhibition zone formation, disc sensitivity methods and MIC determinations. In *Antibiotics: Assessment of Antimicrobial Activity and Resistance*, A. D. Russell and L. B. Quesnel, eds. Academic Press, Inc., New York, pp. 19-30.

Mahler, R. L., and Bezdicek, D. F. (1978). *Appl. Environ. Microbiol. 36*: 780.

Marques Pinto, C., Yao, P. Y., and Vincent, J. M. (1974). *Aust. J. Agric. Res. 25*: 317.

Materon, L. A., and Hagedorn, C. (1982). *Soil Sci. Soc. Am. J. 46*: 553.

Materon, L. A., and Hagedorn, C. (1983). *Soil Sci. Soc. Am. J. 47*: 491.

Matsen, J. M., and Barry, A. L. (1974). Susceptibility testing: diffusion test procedures. In *Manual of Clinical Microbiology*, 2nd ed., E. H. Lennette, E. H. Spaulding, and J. P. Truant, eds. American Society for Microbiology, Washington, D.C., pp. 418-427.

McLaughlin, W., and Ahmad, M. H. (1984). *FEMS Microbiol. Lett. 21*: 299.

McLoughlin, T. J., Bordeleau, L. M., and Dunican, L. K. (1984). *J. Appl. Bacteriol. 56*: 131.

Meyer, M. C., and Pueppke, S. G. (1980). *Can. J. Microbiol. 26*: 606.

Misplon, J. A., and Bishop, P. E. (1983). *Plant Soil 74*: 395.

Pankhurst, C. E. (1977). *Can. J. Microbiol. 23*: 1026.

Patel, J. J. (1974). *Plant Soil 41*: 395.

Pattison, A. C., and Skinner, F. A. (1974). *J. Appl. Bacteriol. 37*: 239.

Peterson, E. A., Sirois, J. C., Berndt, W. B., and Miller, R. W. (1983). *Can. J. Microbiol. 29*: 541.

Pilacinski, W. P., and Schmidt, E. L. (1981). *J. Bacteriol. 145*: 1025.

Rafique Uddin, M., McLaughlin, W., and Ahmad, M. H. (1984). *Plant Soil 81*: 305.

Rai, R. (1983). *J. Agric. Sci. Cambridge 100*: 81.

Rupela, O. P., Josey, D. P., Toomsan, B., Mittal, S., Dart, P. J., and Thompson, J. A. (1982). Application of inherent antibiotic resistance to ecological studies of rhizobia. In *Biological Nitrogen*

Fixation Technology for Tropical Agriculture, P. H. Graham and S. C. Harris, eds. Centro Internacional de Agricultura Tropical, Cali, Colombia, pp. 625-629.

Sabath, L. D., and Matsen, J. M. (1974). Assay of antimicrobial agents. In *Manual of Clinical Microbiology*, 2nd ed., E. H. Lennette, E. H. Spaulding, and J. P. Truant, eds. American Society of Microbiology, Washington, D.C., pp. 428-430.

Sadowsky, M. J., Keyser, H. H., and Bohlool, B. B. (1983). *Int. J. Syst. Bacteriol. 33*: 316.

Scholla, M. H., and Elkan, G. H. (1984). *Int. J. Syst. Bacteriol. 34*: 484.

Schwinghamer, E. A. (1964). *Can. J. Microbiol. 10*: 221.

Schwinghamer, E. A. (1967). *Antonie van Leeuwenhoek; J. Microbiol. Serol. 33*: 121.

Schwinghamer, E. A., and Dudman, W. F. (1973). *J. Appl. Bacteriol. 36*: 263.

Scotti, M. R. M. M. L., Sá, N. M. H., Vargas, M. A. T., and Doberеiner, J. (1982). Susceptibility of *Rhizobium* strains to antibiotics: a possible reason for legume inoculation failure in cerrado soils. In *Biological Nitrogen Fixation Technology for Tropical Agriculture*, P. H. Graham and S. C. Harris, eds. Centro Internacional de Agricultura Tropical, Cali, Colombia, pp. 195-200.

Sinclair, M. J., and Eaglesham, A. R. J. (1984). *Soil Biol. Biochem. 26*: 247.

Stein, M., Bromfield, E. S. P., and Dye, M. (1982). *Ann. Appl. Biol. 101*: 261.

Stowers, M. D., and Eaglesham, A. R. J. (1984). *Plant Soil 77*: 3.

Strzelcowa, A. (1968). *Acta Microbiol. Pol. 17*: 263.

Szende, K. (1982). *Mutat. Res. 97*: 225.

Thornsberry, C. (1974). The agar diffusion antimicrobial susceptibility test. In *Current Techniques for Antibiotic Susceptibility Testing*, A. Balows, ed. Charles C Thomas, Publisher, Springfield, Ill., pp. 6-16.

Tilton, R. C., and Newberg, L. (1974). Standardization of the microdilution susceptibility test. In *Current Techniques for Antibiotic Susceptibility Testing*, A. Balows, ed. Charles C Thomas, Publisher, Springfield, Ill., pp. 77-87.

Trinick, M. J., and Parker, C. A. (1982), *Soil Biol. Biochem. 14*: 79.

Trussell, P. C., and Sarles, W. B. (1943). *J. Bacteriol. 45*: 29.

Upchurch, R. G., and Elkan, G. H. (1977). *Can. J. Microbiol. 23*: 1118.

Vincent, J. M. (1970). *A Manual for the Practical Study of Root-Nodule Bacteria* (IBP Handbook 15). Blackwell Scientific Publications Ltd., Oxford.

Washington, J. A., II (1974). The agar-dilution technique. In *Current Techniques for Antibiotic Susceptibility Testing*, A. Balows, ed. Charles C Thomas, Publisher, Springfield, Ill., pp. 54-62.

Washington, J. A., II, and Barry, A. L. (1974). Dilution test procedures. In *Manual of Clinical Microbiology*, 2nd ed., E. H. Lennette, E. H. Spaulding, and J. P. Truant, eds. American Society for Microbiology, Washington, D.C., pp. 410-417.

Zelazna-Kowalska, I. (1977). *J. Bacteriol.* *132*: 8.

9

Isolation and Identification of Genetically Marked Strains of Nitrogen-Fixing Microsymbionts of Soybeans

L. DAVID KUYKENDALL *Agricultural Research Service, U.S. Department of Agriculture, Beltsville, Maryland*

I. INTRODUCTION

The purpose of this chapter is to provide some detailed descriptions of microbiological concepts and methods that are useful for the isolation and identification of genetically marked strains of bacteria that are nitrogen-fixing microsymbionts of legume plants. Vincent (1982) has eloquently stated the need for an adequate microbiological basis for research in nitrogen fixation in legumes as follows: "There is however a real risk when the enthusiastic nonmicrobiologist moves in to investigate a system in which a bacterium, which may be more difficult to categorize and control, plays its own major role in an intimate association with the more tractable host. Whether he be plant physiologist, biochemist, geneticist, or even a microbiologist unfamiliar with *Rhizobium*, a worker coming into this field cannot afford to take this microorganism too lightly. One of two courses is open to him. Either he takes the trouble to familiarize himself sufficiently with the nature of the microorganism he is dealing with, and the techniques needed for its purification and maintenance, or he works cooperatively with someone who has this knowledge and technical capacity." Following Vincent's advice may not be easy, but it is important. The researcher studying nitrogen fixation in legumes needs to understand and master the basic skills of obtaining a pure culture and of maintaining pure strains of bacteria. Since culture maintenance is discussed in Chapter 17, I will not discuss this particular subject in detail. Primarily, the focus herein is on microbiological procedures useful for obtaining genetically marked bacteria.

Two prerequisites that are absolutely essential for doing good research on a microorganism are as follows:

1. The cultures to be used must be ascertained to be pure and authentic.
2. A suitable culture medium for growing the organism to high cell densities must be defined.

The researcher must determine that these two conditions are met before proceeding further in a microbiological research project.

II. CULTURE PURITY

First, let us define culture purity. Certainly, for a culture to be said to be pure, there must be no contaminating organisms of distinct species or genera. A "pure culture" should also mean a lot more—a pure culture is a strain of bacteria that has genetic homogeneity; that is, assuming the absence of indigenous transposons or translocatable DNA sequences, which can be responsible for deletions and insertional inactivation of gene(s). A strain is a clone that is usually expected to have purity greater than or equal to 99.99% for any given genetic trait. Mutations in bacterial genes occur at frequencies in the range 10^{-5} to 10^{-9} per cell per generation. Most of the gene mutations that can occur in a population can be found in a 50-ml culture of bacteria at cell densities greater than or equal to 1 billion colony-forming units (CFU) per milliliter (i.e., in a total population of 50 billion bacteria). For example, suppose that we plate aliquots of a dense culture onto agar plates containing an antibiotic (e.g., kanamycin sulfate). Let us also suppose that the concentration of kanamycin sulfate added to the growth medium is one that totally inhibits the wild-type strain but allows "spontaneous" or background kanamycin-resistant mutants to grow. Then we can typically expect that following incubation at a suitable temperature for several days or 1 week, depending on the bacterial species, a few bacterial colonies will be obtained per petri dish. If, from viable count data, we know that a population of approximately 100 million (10^8) viable cells had been spread on each plate and that about 10 colonies of mutants were obtained on each plate, there is a calculated frequency of kanamycin-resistant mutants of about 1 in 10 million, or 10^{-7}. Similarly, mutant strains can be obtained for just about any possible mutation. Thus, even a pure culture has a low frequency of genetic diversity. The question is one of numbers. The frequency obtained of a given mutant from a culture should ideally reflect the frequency at which the corresponding mutation occurs.

Culture purity is not to be taken for granted. Colony morphology variants or derivatives of *Bradyrhizobium japonicum* strain USDA 110 of the slow-growing species that nodulates and fixes nitrogen in symbiosis with *Glycine max* were first described by Kuykendall and Elkan (1976). Recent studies by Mathis et al. (1986) demonstrate that substrains I-110 and L1-110 are probably from a common ancestor and that strain I-110 symbiotically fixes nitrogen, whereas strain L1-110 is deficient in symbiotic nitrogen fixation. The demonstrated presence of insertion sequence-like elements in the symbiotic region of the *B. japonicum* genome (Kaluza et al., 1985) suggests that some degree of intrinsic genetic instability may be characteristic of this organism. The substrains of USDA 110 were initially distinguished and then isolated on the basis of their very different colony morphologies formed on yeast extract-mannitol (YEM)-Hepes-MES (HM) medium (see the Appendix and Figure 1). The relatively small size of colonies formed by substrain I-110 are clearly distinguishable from the much larger colonies formed by substrain L1-110. The basis for the difference in colony size on YEM-HM medium is the ability of strain L1-110 to utilize D-mannitol since it possesses an inducible D-mannitol dehydrogenase (Kuykendall and Elkan, 1977), whereas strain I-110 does not under these conditions. Cultivation of strain USDA 110 on growth media that contain D-mannitol as a carbon source enriches for (increase the frequency of) derivatives, such as strain L1-110, that can utilize this substrate. These are not necessarily desirable circumstances for the microbiologist or other researcher interested in symbiotic nitrogen fixation, particularly since such derivatives have lost symbiotic competence. A genomic rearrangement of structural *nif* genes does not appear to account for the loss of symbiotic nitrogen fixation ability (Mathis et al., 1986).

The isolation of soybean microsymbionts from root nodules often shows occupancy of an apparently single nodule by two distinct strains (Devine et al., 1980). A nodule isolate must be presumed to require

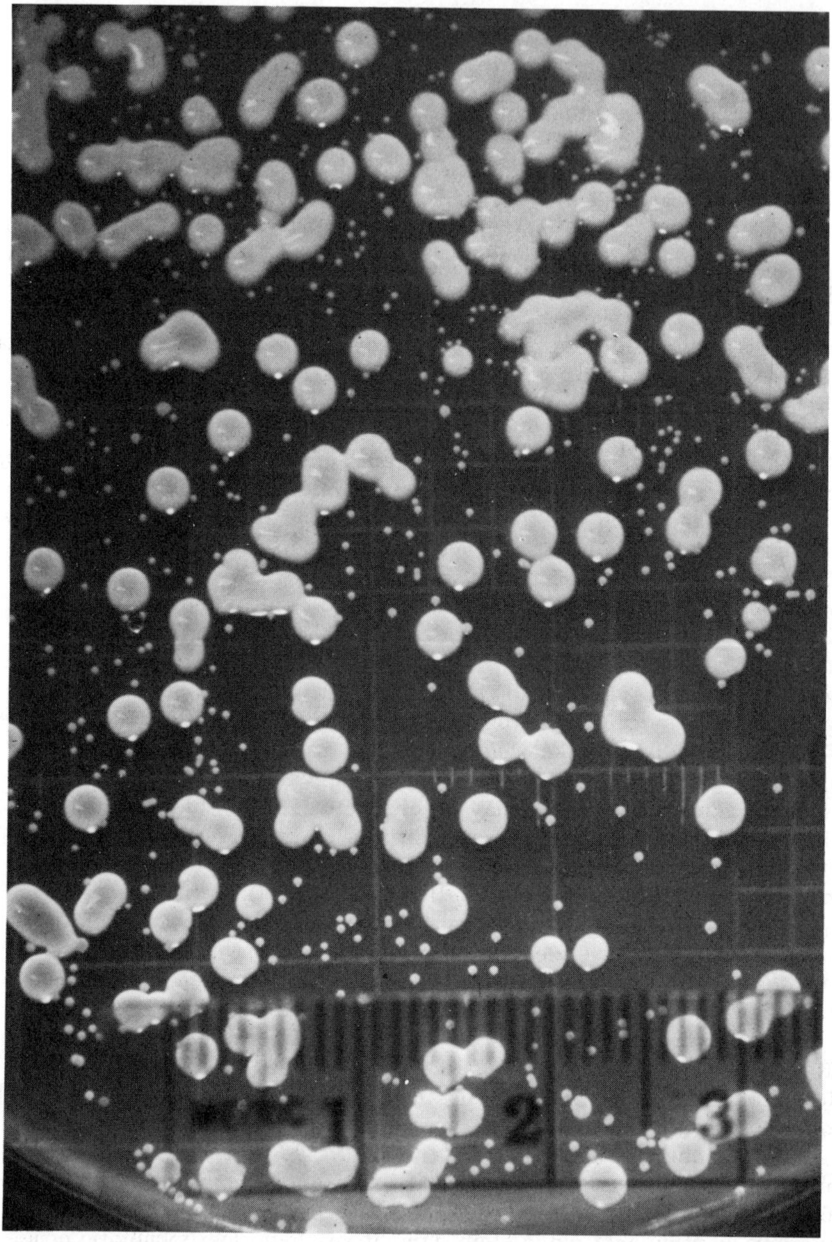

Figure 1 Colony-type variants of *Bradyrhizobium japonicum* USDA 110. The relatively small colonies were formed by substrain I-110, which symbiotically fixes nitrogen for soybeans, and the relatively large colonies were formed by substrain L1-110, which is deficient in ability to fix nitrogen symbiotically.

purification to obtain single strains. Simple streaking and restreaking may not be sufficient to obtain culture purity, since such bacteria often associate to form clumps. These clumps can be dissociated to obtain colonies arising from single cells by the use of a surfactant such as Tween (Cole, 1972). Tween-20 or Tween-40 is added to HM salts (Cole and Elkan, 1973; see the Appendix) to give a final concentration of 0.01%; this mixture is used as a diluent for preparing 10-fold serial dilutions. With thorough mixing by vortexing between dilutions, dilute cell suspensions are routinely obtained from which CFUs are derived from individual or single bacterial cells rather than from clumps of cells having as many as 20 cells. As will be described later in this chapter, the ability to detect and isolate amino acid auxotrophs of *B. japonicum* requires the use of single-cell suspensions, obtained following chemical mutagenesis, for plating onto supplemented growth medium plates prior to the replica plating to screen for auxotrophs.

It is important to have a well-defined growth medium that supports high cell densities. To illustrate, let us go back to the example of identifying kanamycin-resistant mutants. Suppose the culture used in that experiment had only contained fewer than or equal to 100 million (10^8) viable cells per milliliter. By spreading an agar growth medium plate with 0.1-ml aliquots of such a culture, thereby plating only about 10^7 cells, the researcher might not obtain any mutants resistant to kanamycin since their frequency, as we determined, was only 1 in 10 million (10^{-7}). This example serves to illustrate the importance of a good growth medium that permits high cell densities to be obtained. In this particular example, plating several plates would probably suffice to obtain the desired mutant. If the mutant occurred at a lower frequency such as 10^{-9}, then either spreading hundreds of petri plates or concentrating the cells by centrifugation prior to plating aliquots on selective media might be required to obtain such a relatively rare mutant.

B. japonicum strain I-110 grows well in HM salts + 0.1% arabinose with or without 0.1% yeast extract for enrichment (AIE medium, see the Appendix). It is important that the arabinose be sterilized separately as a 10% solution and added to sterile HM salts after autoclaving. Other *B. japonicum* strains often require either the addition of yeast extract or the addition of an alternative carbon source such as D-gluconate in order to achieve high cell densities.

To summarize the introduction: Before beginning work on the isolation, identification, and use of genetically marked strains of microsymbionts of legume plants two basics must be established:

1. The microbiological and genetic purity of the culture: has it been successfully cloned?
2. A good culture medium: does it allow for high cell densities to be obtained?

A few words on strain authenticity need to be said. A genuine, symbiotically competent soybean microsymbiont would be expected to nodulate and fix nitrogen with soybeans. A mutant or derivative strain could lack this ability but it will still retain most if not all other biochemical and/or genetic traits that are characteristic of its parent strain. Unambiguous identification of a mutant strain or strain derivative requires positive identification of biochemical and genetic traits such as antibiotic resistances. Genetically marked, well-characterized bacterial strains are of greatest use for microbiological research for precisely this reason. They are readily identifiable and consequently, so are their progeny produced through genetic manipulation. Genetic studies of bacteria require the development of many mutant strains having genetic markers useful for recombinant selection and identification. Genetically marked strains of bacteria are also useful in ecological studies. Information on the isolation and identification of mutant strains is given in the following sections.

III. ANTIBIOTIC-RESISTANT MUTANTS

A. Isolation of the Mutants

As outlined in the Introduction, antibiotic-resistant mutants of bacteria are easily obtained by the process of direct selection. As Carlton and Brown (1981) have stated: "Because of the great resolving power of direct selection (i.e., the ability to detect those infrequent mutant cells against a very great background of nonmutant cells), it is usually unnecessary to employ any enrichment techniques." In direct selection of antibiotic-resistant mutant strains of bacteria, vast populations of 100 million bacteria of a pure strain are plated onto the agar surface of petri dishes containing various concentrations of an antibiotic (e.g., streptomycin or rifampicin). Various concentrations of the selective agent are used to determine the optimum concentration for mutant selection. Ideally, an antibiotic concentration can be determined that will effectively prevent any growth by the wild-type parent strain, yet will permit the growth of resistant mutants. By using twofold increments of concentrations (e.g., 10, 20, 40, 80, 160, 320 μg/ml), the researcher is likely to determine a suitable antibiotic concentration without difficulty.

The direct isolation procedure for drug resistance mutants arising "spontaneously" (background mutants) is very simple:

1. Develop a broth culture of the bacterial strain to high cell densities of at least $\geqslant 5 \times 10^8$ CFU/ml. As inoculum for this culture, choose a well-isolated colony from a plate where colonies have arisen from single cells [use a surfactant such as Tween-20 or Tween-40 if necessary to dissociate clumps].

Also maintain a source of the parental strain by using material from the same colony to develop an agar slope culture to be stored at 4°C as soon as the culture has grown.

2. Spread 0.1-ml aliquots of the dense culture onto the surfaces (with a sterile bent 45° glass rod) of antibiotic containing growth medium. Invert the plates.
3. Incubate the plates for about 7 days for *Bradyrhizobium* or only 3 or 4 days for fast-growing *Rhizobium* species such as *R. fredii* (Scholla and Elkan, 1984).
4. Observe plates under good lighting conditions. Those petri plates giving confluent lawns of growth represent concentrations of antibiotic that are below the minimum inhibitory concentration (MIC) for that particular strain of organism. Look for plates that give distinct colonies arising from a clear background. A clear background represents complete inhibition of the parental strain.
5. Pick colonies of resistant mutants with a sterile loop and streak out on another antibiotic-supplemented plate of the same medium composition. The same antibiotic concentration as was useful for complete counterselection of wild-type bacterial cells is used for purification. This simple purification step ought to be sufficient for obtaining a pure culture of the antibiotic-resistant mutant strain of interest. Observe whether the mutant strain grows well under the particular selective condition imposed. Useful genetically marked strains of bacteria will grow under selective conditions at about the same rate that the parent strain grows under nonselective conditions (i.e., in the absence of the antibiotic).

Success using the experimental protocol outlined above for obtaining antibiotic-resistant mutants is not assured. Possible pitfalls include the following:

1. The wild-type of parental strain may already be resistant to the antibiotic that was used. If this is the case, a lawn of bacterial growth will be evident even at the highest concentration employed. Many *Bradyrhizobium* strains are naturally resistant to rifampicin at concentrations as high as 500 μg/ml.
2. Complete counterselection (no growth) of the parent strain may occur without the appearance of "spontaneous" resistant mutant colonies. The numbers of cells in the bacterial population used may not have been high enough to obtain a relatively rare mutant (one that occurs at a frequency of less than 10^{-8}).
3. Mutants with resistance to that particular drug of antibiotic may not occur at a detectable frequency. We may appropri-

ately ask ourselves the question: What is the detectable frequency in our experiment? If we spread 10 plates each with 10^8 bacterial cells on each plate, we have a good chance of obtaining a rare mutant even if the frequency with which it occurs is only 10^{-9}.

Our ability to detect even extremely rare mutant cells is what makes the technique of direct selection a highly useful procedure. In addition to use for the derivation of antibiotic-resistant mutants, direct selection may also be used to obtain mutants resistant to phages. Mutants resistant to various bacteriocidal or bacteriostatic antimetabolites can be obtained as well.

B. Identification of the Mutants

Antibiotic-resistant, genetically marked bacterial strains are readily identified on the basis of their ability to grow under selective conditions that prohibit growth of their wild-type parent strain. However, because other strains of bacteria of the same or unrelated taxa may naturally possess that particular antibiotic resistance marker, one is not necessarily assured of positive identification on the basis of one drug resistance marker.

An example of problems with positive strain identification based on a single genetic marker is as follows:

1. About one-third of the strains classified as *B. japonicum* strains in the Beltsville USDA collection are naturally resistant to high levels (⩾250 μg/ml) of rifampicin, a potent inhibitor of transcription in bacteria.
2. Rifampicin-resistant mutants, occurring at frequencies of 10^{-7} to 10^{-8}, are readily obtained from those *B. japonicum* strains that are naturally rifampicin sensitive. Naturally occurring rifampicin-sensitive strains may give a haze of growth even at 100 μg/ml rifampicin, but complete inhibition of the parent strain occurs at 250 μg/ml or 500 μg/ml.
3. Rifampicin-resistant mutant strains of *B. japonicum* are not distinguishable from naturally occurring rifampicin-resistant *Bradyrhizobium* on the sole basis of rifampicin-resistance levels. However, these two groups may be distinguished on the basis of other phenotypic characteristics, such as the ability of some strains of the latter group, but not the former, to produce rhizobitoxine (RT)-induced chlorotic symptoms *in planta*, on RT-sensitive soybean genotypes (manuscript in preparation). DNA homology group II (Hollis et al., 1981) strains that have been suggested to be designated as *Bradyrhizobium* sp. are correlated highly with rifampicin resistance

and usually with chlorosis induction. Several antibiotic resistance differences between these two groups are now being documented (O'Neill et al., manuscript in preparation).

Now that we have established that a single genetic marker may not serve to distinguish a bacterial strain from other related bacteria, let us now ask: How many genetic markers are needed? The answer may not be as simple to find as we would hope. Let us say that two good genetic markers will usually suffice for positive strain identification. Some applications, such as gene transfer experiments, require that two strains be distinguishable on the basis of at least five dissimilar genetic markers.

In the identification of genetically marked strains of bacteria, there are some possible problems that extend beyond the previously addressed question of how many markers are required for unambiguous strain identification. One such potential problem is the effect of environmental factors such as temperature, pH, and available carbon source on the phenotypic expression of a genetic determinant by a bacterial strain. Another potential problem is the effect of biotic factors such as the presence of other species and genera of microorganisms (as in ecology studies) or the presence of related organisms (as in gene transfer experiments). Another potentially major problematic factor would perhaps be limited knowledge, since we of course do not yet understand everything about bacterial genetics, let alone bacterial ecology. Discussion of all of these deficiencies and problems is clearly beyond the scope of this chapter. However, to briefly cover some possible problems with strain identification via the use of genetic markers, we can define two factors that must be kept in mind:

1. Environmental and physical factors and conditions need to be precisely defined and consistently repeated in order to obtain accurate results.
2. Biotic factors, some of which may as yet remain undefined, may influence the results of experiments. Always include an uninoculated control in plant tests done either in the greenhouse or field.

Let us define some genetic markers that can be usefully employed in positive strain identification in both genetics research and ecological studies of *Bradyrhizobium japonicum.* Table 1 shows examples of genetic markers that can readily be obtained, using direct selection technique, as I have previously described.

B. japonicum strain I-110 ARS was derived from wild-type strain I-110 by sequential mutant selections for azide, rifampicin, and streptomycin resistance markers. Similarly, *B. japonicum* strain I-110 FN

Table 1 Some Useful Genetic Markers Obtained by Sequential Direct Selections of Multiply Marked Mutants of *Bradyrhizobium japonicum* I-110

Strain designation	Selective agent (concentration)	Phenotype abbreviations
I-110 *azi-5*	Sodium azide (10 μg/ml)	Azi^R
I-110 *azi-5 rif-17*	Rifampicin SV (500 μg/ml)	Azi^R, Rif^R
I-110 *azi-5 rif-17 str-74*	Streptomycin sulfate (1000 μg/ml)	Azi^R, Rif^R, Str^R
I-110 5 *fu-1*	5-Fluorouracil (10 μg/ml)	5 Fu^R
I-110 5 *fu-1 nal-23*	Nalidixic acid (500 μg/ml)	5 Fu^R, Nal^R

Source: Kuykendall (1979).

was developed from strain I-110 using two sequential direct selections for 5-fluorouracil and nalidixic acid resistance mutants. Thus strain I-110 ARS and strain I-110 FN have five dissimilar genetic markers and are both derived from the ancestral strain I-110 following only two or three simple laboratory manipulations. R-factor transfer between genetically marked substrains of *B. japonicum* was first demonstrated (Kuykendall, 1979) using these two different genetically marked strains. Strains I-110 ARS and I-110 FN were also used in an ecological study (Kuykendall et al., 1982) that demonstrated the importance of nodule occupancy in the establishment of *B. japonicum* in subsequent crops of soybean. Thus the usefulness of genetically marked strains of *B. japonicum* in genetics experiments and ecological studies has been established and the reader is referred to the original references for detailed descriptions of these investigations. The present chapter is intended to focus on the isolation and identification of genetically marked strains of nitrogen-fixing microsymbionts. Sufficient detail is provided so that biological nitrogen fixation researchers will have the information required to obtain mutant strains for use in investigations. In the next section, the derivation of amino acid auxotrophs is described.

IV. AUXOTROPHIC MUTANTS

Beringer (1974) described the use of amino acid auxotrophs in genetic studies of *Rhizobium leguminosarum*. Kondorosi and Johnston (1982) reviewed the development of genetic maps of the chromosomes of *R. meliloti* and *R. leguminosarum* and these authors state that "it is noticeable that on these maps there are no representatives of mutations leading to certain nutritional requirements; e.g. there are no Ala^-, Asp^-, Asn^-, Hom^-, Ile^-, Lys^-, Pro^-, Thr^-, or Val^- strains." It may be that certain classes of mutants are more easily obtained than others. Tryptophan-requiring mutants are among the most readily obtained amino acid auxotrophs of *R. leguminosarum* (J. E. Beringer, personal communication). Wells and Kuykendall (1983) described the isolation and characterization of 11 tryptophan auxotrophs of *Bradyrhizobium japonicum* I-110 ARS. Table 2 summarizes some of the properties of Trp^- mutants. Eleven Trp^- mutants fell into five phenotypic classes, and one of the mutant strains, TA8, appeared to lack activity in three enzymes required for tryptophan biosynthesis: anthranilate synthetase, indole glycerol phosphate synthetase, and tryptophan synthetase.

Four of the 11 Trp^- strains were found to specifically lack anthranilate synthetase. Three tryptophan-requiring mutants had deficiencies in the last step of the tryptophan biosynthetic pathway—the tryptophan synthetase B reaction. The other three Trp^- mutants had defects in intermediate steps of the pathway. Most interestingly, evidence for a role of ability to produce indole in soybean nodulation was obtained since only those three mutants having the ability to produce this compound (the immediate precursor to tryptophan) were found to be capable of host plant nodulation. The other eight mutants, all distinguished by their ability to grow on indole-supplemented media (whereas the three tryptophan synthetase mutants cannot), were apparently only capable of nodulating the host plant following reversion to prototrophy (growth on minimal medium). The implications of these results are currently under investigation. Recently, Sadowsky et al. (1986) described the isolation and characterization of two symbiotically defective auxotrophs of *B. japonicum* strain USDA 122.

The practical isolation of auxotrophs requires a mutagenesis of the parent culture. The detection of mutants with nutritional requirement markers is then facilitated by replica plating (Lederberg and Lederberg, 1952). Besides their use in isolation of auxotrophic mutants, mutagenesis and replica plating techniques are generally used in the isolation of conditionally lethal mutants of bacteria. Most mutations are conditionally lethal, and the isolation of mutants that have lost an essential metabolic function (such as ability to synthesize a required amino acid) also requires use of an indirect method or screening procedure. Mutants that require metabolic intermediates undoubtedly occur spontaneously in large bacterial populations at frequencies such

Table 2 Some Properties of Tryptophan-Requiring Mutant Strains Derived from *Bradyrhizobium japonicum* I-110 ARS

Strain designation	Enzyme deficiencies[a]	Ability (+) or lack thereof (-) to nodulate soybeans as a Trp⁻ mutant[b]
TA1	Tryptophan synthetase	+
TA2	Tryptophan synthetase	+
TA3	Anthranilate synthetase	-
TA4	Anthranilate synthetase	-
TA5	PRT^-	-
TA6	PRT^- and low indoleglycerol phosphate synthetase (INGPS)	-
TA7	Anthranilate synthetase	-
TA8	Possible regulatory mutation-three enzymes deficient	-
TA9	Anthranilate synthetase	-
TA10	Tryptophan synthetase	+
TA11	$INGPS^-$	-

[a]By enzyme assays, except that the identification of strain TA7 as deficient in anthranilate synthetase activity was based solely on the ability of this mutant to grow on media supplemented with anthranilate. PRT, anthranilate-5-phosphoribosyl pyrophosphate phosphoribosyl transferase.

[b]Those strains identified as negative in ability to nodulate soybeans as tryptophan-requiring mutants can, however, nodulate soybeans after genetic reversion to tryptophan independence.

as those at which antibiotic resistance mutants occur. But, unlike drug resistant mutants, auxotrophs cannot be directly selected. Their isolation requires an indirect screening procedure such as replica plating. Replica plating permits the relatively facile screening of tens of thousands or even hundreds of thousands of bacterial colonies for conditional lethal phenotypes. Screening such numbers of colonies by individual colony pickings would not be feasible. The simultaneous transfer of 100 or more colonies, using sterile velveteen cloth, represented a tremendous improvement in technique. Even with replica plating technique, screening 1 million to 1 billion bacterial colonies to obtain any given conditional lethal mutant is not practical. Mutagenesis by chemical or other means is usually required and enrichment for mutants may sometimes be required. Mutagenesis increases the frequency of the appearance of mutants manyfold above the background level of "spontaneous" mutants.

Using methods of chemical mutagenesis by nitrous acid as first described by Kaudewitz (1959) and later detailed by Miller (1972), Beringer (1974), and Wells and Kuykendall (1983) obtained amino acid auxotrophs of *Rhizobium leguminosarum* and *B. japonicum*, respectively. As a chemical mutagen, nitrous acid is said to "induce primarily bidirectional transition mutations, due to its deaminating activity on cytosine and adenine. In addition, it induces deletions at a significant level by an unknown mechanism" (Carlton and Brown, 1981). Prior to mutagenesis with nitrous acid, cells are harvested by centrifugation (10,000 × *g* for 10 min to pellet the cells) from a stationary phase culture (typically, 2 to 3 days old for fast-growing *Rhizobium* species or 5 to 6 days old for slow-growing *Bradyrhizobium*). The cells are washed once in an equal volume of sterile 0.1 *M* sodium acetate buffer, pH 4.6. Fresh nitrous acid (0.05 *M* sodium nitrite in 0.1 *M* sodium acetate buffer, pH 4.6) is prepared, filter sterilized, and then used to resuspend the washed bacterial cells. A volume equal to one-fourth the original volume is used. Incubation with nitrous acid is allowed to proceed at room temperature for 30 min. Four volumes of fresh culture medium is added to slow the reaction and the cells are immediately pelleted, washed, and resuspended in fresh sterile complete growth medium. The inactivation rate is usually 10^{-3} to 10^{-4} (0.1 to 0.01% survival) for best results to be obtained. The degree of killing is assayed by doing viable counts both before and after exposure to nitrous acid.

Dilutions of suspensions of mutagenized bacterial cells (most suitable are those dilutions that give between 50 and 150 colonies per petri dish) are plated onto complete medium (full supplementation). After approximately 1-mm-diameter colonies have developed following growth on the supplemented growth medium plates, they are replica plated to a defined minimal medium and to a fresh supplemented plate as a control for effective transfer. Following incubation to allow for good growth, the plates are scored for putative auxotrophs as tentatively identified by the lack of a growth patch on minimal medium. Presumptive mutants are picked from complete or supplemented medium and then streaked on both minimal medium and supplemented medium to confirm their status as bona fide mutants having nutritional requirement markers.

Auxotrophs are useful in genetic mapping and in determining biochemical pathways. The symbiotic properties of some auxotrophic mutants of legume microsymbionts were reviewed (Kuykendall, 1981) and it was concluded that auxotrophs of *Rhizobium* that are symbiotically defective (even as an indirect consequence of a nutritional requirement) can be useful tools for studying the complex series of events in the formation of symbiotic nitrogen-fixing nodules. This conclusion was based primarily on the pioneering work of Truchet et al. (1980). The investigators showed that a leucine-requiring mutant of *R. meliloti*, while remaining in infection threads, nevertheless was able to induce nodule organogenesis. This finding suggests that the induction of nodule formation by *Rhizobium* could perhaps be similar to crown gall tumorigenesis by *Agrobacterium* wherein a segment of plasmid DNA is transferred from bacterium to plant. It could also result from a biochemical signal.

Transposon Tn5 mutagenesis can also be used for the isolation and identification of genetically marked strains of soybean microsymbionts. This technique is useful for obtaining auxotrophs as well as symbiotically defective mutants (see Rostas et al., 1984; Sadowsky et al., 1985). Transposon mutagenesis can be more useful for gene isolation using recombinant DNA techniques. The Tn5 introduction vector pGS9 developed by Selvaraj and Iyer (1983) has also been used in our laboratory at Beltsville to introduce transposon Tn5 into both fast-growing and slow-growing soybean microsymbionts *Rhizobium fredii* and *B. Japonicum*.

V. CONCLUSIONS

Detailed descriptions of microbiological methods for isolation and identification of genetically marked strains of bacteria were herein presented for the BNF researcher interested in either the genetics or ecology of the nitrogen-fixing microsymbionts of legume plants. Starting with pure cultures is important. The isolation of antibiotic resistant mutants is direct, straightforward, and simple. Chemical mutagenesis and replica plating are needed to obtain auxotrophic mutants, which are useful in studies of the biochemistry and genetics of symbiosis.

APPENDIX

Growth Media

HM Salts Recipe

Add 10 ml of each of the following seven 100× stock solutions to 930 ml of distilled H_2O to make 1 liter.

1. Hepes-MES buffer, pH 6.6 (65.0 g of Hepes (*N*-2-hydroxyethylpiperazine-*N*'-ethanesulfonic acid) 55.0 g of MES [2-(*N*-morpolino])ethanesulfonic acid]), pH adjusted to 6.6 with NaOH pellets and volume brought up to 500 ml
2. $FeCl_3 \cdot 6H_2O$ 0.67 g/liter

3. $MgSO_4 \cdot 7H_2O$ 18.0 g/liter
4. $CaCl_2 \cdot 2H_2O$ 1.30 g/liter
5. Na_2SO_4 25.0 g/liter
6. NH_4Cl 32.0 g/liter
7. Na_2HPO_4 12.5 g/liter

HM Salts + Arabinose Medium (Minimal Medium)

Add 1.0 ml of a sterile 10% solution of L(+)-arabinose for each 100 ml of HM salts, after autoclaving.

AIE Medium

Also add 1.0 ml of a sterile 10% solution of yeast extract (Difco) for each 100 ml of HM salts + arabinose, after autoclaving. *Important*: Always autoclave the carbon sources separately from basal salts. For minimal medium, use either purified agar of Noble agar.

Antibiotics (100× Stock Solutions)

Rifampicin SV (read warning label on bottle—avoid breathing dust), 50 mg/ml in 0.1 *N* NaOH
Nalidixic acid, 50 mg/ml in 0.1 *N* NaOH; may be necessary to heat gently in order to dissolve
5-Fluorouracil, 5 mg/ml in distilled water
Sodium azide, 5 mg/ml in distilled water
Streptomycin sulfate, 100 mg/ml in distilled water
Kanamycin sulfate, 10 mg/ml in distilled water

To prevent destruction by heat, prepare sterile antibiotic solutions, for addition to media, by filter sterilization.

REFERENCES

Beringer, J. E. (1974). *J. Gen. Microbiol. 84*: 188.

Carlton, B. C., and Brown, G. J. (1981). Gene mutation. In *Manual of Methods for General Bacteriology*, P. Gerhardt, R. G. E. Murray, R. N. Costilow, E. W. Nester, W. A. Wood, N. R. Krieg, and G. B. Phillips, eds. American Society for Microbiology, Washington, D.C., p. 222.

Cole, M. A. (1972). Ph.D. thesis, North Carolina State University, Raleigh, N.C.

Cole, M. A., and Elkan, G. H. (1973). *Antimicrob. Agents Chemother. 4*: 248.

Devine, T. E., Kuykendall, L. D., and Breithaupt, B. H. (1980). *Can. J. Microbiol. 26*: 179.

Hollis, A. B., Kloos, W. E., and Elkan, G. H. (1981). *J. Gen. Microbiol. 123*: 215.

Kaluza, K., Hahn, M., and Hennecke, H. (1985). *J. Bacteriol. 162*: 535.

Kaudewitz, F. (1959). *Nature, London 183*: 1829.

Kondorosi, A., and Johnston, A. W. B. (1982). The genetics of *Rhizobium*. In *Biology of the Rhizobiaceae*, K. L. Giles and A. G. Atherly, eds. Academic Press, Inc., New York, p. 191.

Kuykendall, L. D. (1979). *Appl. Environ. Microbiol. 37*: 862.

Kuykendall, L. D. (1981). Mutants of *Rhizobium* that are altered in legume interaction and nitrogen fixation. In *Biology of the Rhizobiaceae*, K. L. Giles and A. G. Atherly, eds. Academic Press, Inc., New York, p. 299.

Kuykendall, L. D., and Elkan, G. H. (1976). *Appl. Environ. Microbiol. 32*: 511.

Kuykendall, L. D., and Elkan, G. H. (1977). *J. Gen. Microbiol. 98*: 291.

Kuykendall, L. D., Devine, T. E., and Cregan, P. B. (1982). *Curr. Microbiol. 7*: 79.

Lederberg, J., and Lederberg, E. M. (1952). *J. Bacteriol. 63*: 399.

Mathis, J. N., Kuykendall, L. D., and Elkan, G. H. (1986). *Appl. Environ. Microbiol. 51*: 477.

Miller, J. H. (1972). *Experiments in Molecular Genetics*. Cold Spring Harbor Laboratory, Cold Spring Harbor, N.Y.

Rostas, K., Sista, P. R., Stanley, J., and Verma, D. P. S. (1984). *Mol. Gen. Genet. 197*: 230.

Sadowsky, M. J., Rostas, K., Sista, P. R., Bussey, H., and Verma, D. P. S. (1986). *Arch. Microbiol. 144*: 334.

Scholla, M. H., and Elkan, G. H. (1984). *Int. J. Syst. Bacteriol. 34*: 484.

Selvaraj, G., and Iyer, V. N. (1983). *J. Bacteriol. 156*: 1292.

Truchet, G., Michel, M., and Denarie, J. (1980). *Differentiation 16*: 163.

Vincent, J. M. (1982). The legume-*Rhizobium* symbiosis. In *Nitrogen Fixation in Legumes*, J. M. Vincent, ed. Academic Press, Inc., New York.

Wells, S. E., and Kuykendall, L. D. (1983). *J. Bacteriol. 156*: 1356.

10

Measurement of Nitrogen Fixation by Direct Means

JOE EUGENE LEPO and SUZANNE M. FERRENBACH* *The University of Mississippi, University, Mississippi*

**Present affiliation*: University of Tennessee Health Science Center, Memphis, Tennessee

I. INTRODUCTION

In this chapter we describe methods that may be used to quantitate increases in the total nitrogen content of a system or to detect characteristic changes in the nitrogen-containing components of a system. The scope of the chapter is limited to direct measurement of the nitrogen per se as opposed to methods that measure activities correlated with nitrogen fixation, such as acetylene reduction or hydrogen evolution.

The measurement of nitrogen fixation by direct means has been reviewed by Bergersen (1980). Although a majority of these methods are standard in many labs, refinements and modifications continue. It is our purpose to describe the principles behind the methodology, to include procedures for established "old standards" with selected alternative modifications, and to describe any appropriate, more recently developed approaches to the direct measurement of biological nitrogen fixation.

Biological fixation of atmospheric nitrogen should be detectable as increases in the total nitrogen content of a system. This approach to quantifying nitrogen fixation has been used successfully in historic studies such as that of Helriegel and Wilfarth (1888). Difficulties arising from direct analytical methods have been reviewed elsewhere (see Wilson, 1940, pp. 95-98; Martin and Skyring, 1962; Parker, 1961) and will not be treated in depth here. Some of these problems include analytical difficulties encountered because the amount of nitrogen added to the system is small relative to the preexisting pool; errors resulting from losses to the system via denitrification or by leaching, thus diminishing the net gain; the possible existence of nitrogen-containing compounds that resist analysis; variation in the nitrogen content of biological material; and simple analytical error. On the other hand, direct measurement of nitrogen accumulation is effectively applicable to simple systems such as the cell-free extracts which accumulate NH_3 that is not assimilated into other products (see Koch et al., 1967; Maryan and Vorley, 1979; and Section III).

Although certain radio- or mass-isotopic methods constitute "direct" estimations of nitrogen fixation, these techniques are treated in Chapter 11 and thus are not discussed here.

II. ESTIMATION OF N_2 FIXATION BY ANALYSIS OF TOTAL NITROGEN

Measurement of the increase in total fixed nitrogen in an organism or cell-free extract provides a direct estimate of N_2 fixation. However, in systems initially high in nitrogen, such as legume root nodules, the small proportion of additional fixed nitrogen may be obscured by the experimental error of the method. Methods such as these are adequate for cultures of microorganisms capable of high N_2-fixation rates.

Commonly used analyses for total nitrogen are based on one of two digestion procedures:

1. *Oxidative method.* Material to be analyzed is oxidized in the presence of copper oxide to produce N_2, the volume of which is determined. This method is now usually restricted to commercially produced apparatus such as the Coleman nitrogen analyzer. These methods vary with the design of the apparatus and are usually described by the manufacturer.
2. *Kjeldahl method.* A more commonly employed method (Bremner and Tabatabi, 1972; Hiller et al., 1948; McKenzie and Wallace, 1953; Nelson and Sommers, 1973), wherein organic and mineral nitrogen is reduced to NH_3 in hot, concentrated sulfuric acid in the presence of a catalyst. The NH_3 is recovered by distillation or diffusion and estimated colorimetrically, by titration or by using an ammonia electrode.

We will describe the most frequently utilized method together with some variations that may be useful. Other descriptions of these procedures have been provided by Bergersen (1980), Umbreit and Bond (1936), and Parker (1961). Bremner (1965) has described the nitrogen analysis of soils.

A. Kjeldahl Digestion to Ammonia

All materials selected for analysis should have similar nitrogen content (2 to 10 mg of nitrogen). Plant material should be thoroughly dried (12 hr in an 80°C oven) and ground or milled to a homogeneous mixture. Samples of 200 to 300 mg are weighed for digestion. Adaptation of these methods to a microscale of 1 to 14 μg of nitrogen may be made provided that suitable precautions are undertaken to ensure replicability and accuracy of the analyses; however, the use of samples as large as practicable will minimize errors (Bergersen, 1980).

Digestion Reagent

Catalyst (K_2SO_4:$CuSO_4$:Se; 100:10:1) may be added directly to 50- or 100-ml Kjeldahl flasks (1.1 g each) together with sufficient concentrated sulfuric acid to allow digestion of the biological sample (2 to 5 ml for 200 to 300 mg of dried plant material; as low as 1 ml for 10 mg of dried bacteria). A few Carborundum boiling chips should also be added; the flasks can be predigested overnight in a 130°C sand bath.

Alternatively, a digestion reagent of the following composition is highly recommended by Bergersen (1980). Potassium sulfate (100 g) to elevate the boiling point and powdered metallic selenium (1 g) are dissolved in 1 liter of concentrated sulfuric acid by heating at 250 to 300°C for 3 to 4 hr. Usually, one can expect complete recovery of tryptophan and nicotinic acid nitrogen if the digestion is continued for 1 to 2 hr after clearing. Because the reagents may become contaminated with ammonia from laboratory air, a blank digestion should always be done.

Digestion Procedure

Kjeldahl flasks of 50- or 100-ml capacity that have the volume marked can be used as digestion vessels. The digestion furnace should be set up within a vented fume hood, with a rack supporting the neck of the flasks at 45° from the vertical. The flask surfaces above the level of the digestion liquid should not be in contact with the heating elements. Loss of ammonia during the process can be minimized by limiting the refluxing of the acid to the lower third of the vessel neck. During the digestion, flasks should be rotated and, if necessary, cooled and washed down with distilled water to ensure complete processing of material carried up the flask during the early stages of digestion. Boiling chips or glass beads should be used.

Digestion temperature is crucial because high temperatures promote efficient digestion, but temperatures above 330 to 350°C can lead to losses of NH_3. We use 280°C as an approximate starting temperature; 6 hr at this temperature is usually sufficient to digest 200 mg of plant material to clearness; samples may retain a slight amber color but should appear transparent. The material being analyzed will dictate the volume of the digestion reagent used and the digestion time. For plant material, 2 to 5 ml of reagent is usually sufficient for samples of 200 to 300 mg dry weight. More digestion reagent may be required for samples high in carbohydrate (e.g., bacterial cultures with residual medium components). Bacterial cell powders of 10 to 50 mg dry weight, which have been washed and then dried, are usually cleared with 1 ml of digestion reagent. Plant material may be problematic if it contains high amounts of nitrate. Carbon present during the digestion may not be sufficient to reduce the nitrate to ammonia; thus Bergersen (1980) suggests a modification of the digestion procedure to effect conversion of nitrate to nitrosalicylic acid, which digests readily to produce ammonia. The weighed sample in the digestion vessel is thoroughly mixed with 1 to 3 ml of salicyclic acid (5% w/v) in concentrated sulfuric acid for at least 20 min. Sodium thiosulfate (0.3 to 1 g) is then added and the mixture gently heated until fuming. After cooling, the digestion reagent is added and digestion is conducted in the usual fashion. Alternatively, nitrate remaining after digestion may be recovered by boiling the digest with Devarda's alloy as described in Section IV.B.

After digestion, distilled water is used to bring the contents of the flask up to the marked volume and an appropriate method for analysis of ammonia is applied as described below in Sections II.B and III.B.

Bergersen (1980) suggests that ordinary Erlenmeyer flasks (nominal capacity 25 to 100 ml) may substitute for Kjeldahl flasks; these are heated on a thermostatically controlled electric hot plate (Figure 1). After evaporation of water from the sample, the necks of the flasks are closed with small glass funnels, the tips of which have been sealed and drawn to a point (Figure 1).

For microscale digestions, standardized Pyrex test tubes can serve as digestion vessels with glass marbles used to seal the test tube mouths (Figure 2). We have used the method suggested by Bergersen (1980) that employs metal heating blocks set on top of electric hot plates that are thermostatically controlled. Suitable steel or anodized-aluminum blocks can be purchased with a variety of hole sizes to fit the test tubes. The block temperature can be checked with a mercury thermometer; the heating plate control setting that will result in 280 to 330°C

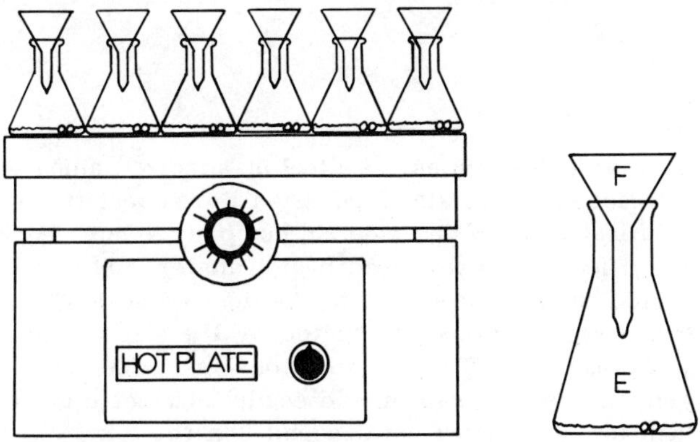

Figure 1 Kjeldahl digestion in Erlenmeyer flasks. A thermostatically controlled electric hot plate placed inside a fume hood serves as a Kjeldahl furance; the temperature is set as described in Section II.A, "Digestion Procedure." Pyrex Erlenmeyer flasks (E) of 100-ml capacity should be used for plant material of 200 to 300 mg; smaller flasks with 25- or 50-ml capacities are suitable for micro- or semimicro digestions. Small funnels (F) with the tips sealed serve as test tube closures; boiling chips or glass beads should be used. (Adapted from Bergersen, 1980. Courtesy of John Wiley & Sons Ltd., Chichester, West Sussex, England.)

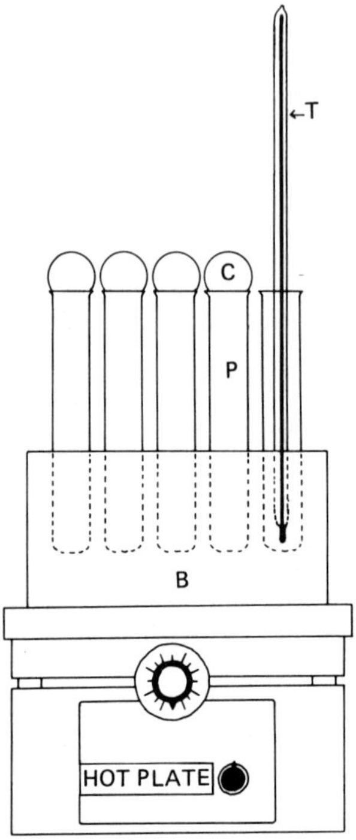

Figure 2 Kjeldahl digestion in test tubes. A steel or anodized aluminum block (B) drilled to accommodate standard-sized Pyrex test tubes (P) rests on a thermostatically controlled electric hot plate. The depth of the holes should be sufficient to allow good heat transfer. The Pyrex test tubes are selected to be appropriate to the desired scale of digestion; the number of tubes per block is limited by the size of the hot plate. A mercury thermometer (T) imersed in one test tube containing digestion reagent is used to monitor and empirically set the temperature of the block to 320 to 330°C as described in the text. Refluxing should rise no farther than half the distance from the top of the block to the tops of the tubes. Closures (C) should be either glass marbles or funnels as described in Figure 1. (Adapted from Bergersen, 1980.)

can be empirically determined; this setting is generally quite replicable for a given hot plate/block combination. Alternatively, an electric contact thermometer may be employed to regulate the hot plate temperature, although we find this an unnecessary complication.

B. Distillation of Ammonia from Digests and Estimation by Titration

Ammonia may be recovered from digested material by distillation using commercially available stills (Figure 3). Their use has been described by Humphries (1956), Bremner and Edwards (1965), and more recently by Bergersen (1980). The ammonia is trapped in 2% boric acid containing 5 ml/liter 0.1% methyl red in 95% ethanol and 20 ml/liter 0.1% bromocresol green in 95% ethanol. It may be necessary to increase the indicator concentrations with some dye batches. For micro-Kjeldahl distillations, the boric acid concentration should be reduced to about 0.2%.

Steam for the distillation is supplied at a controlled rate from a steam generator as shown in Figure 3; the distilled water used in the generator should be acidified with about 1 ml/liter H_2SO_4 to trap any NH_3 that may be present, and the level of the water should be maintained at more than half its initial volume. Before use, the distillation apparatus should be cleaned and steamed out for 10 to 30 min. The rate of steam flow (controlled by adjusting boiling rate in the steam generator) and the flow rate of the cooling water in the condenser should be regulated so that the distillate emerging from the condenser is close to room temperature. An aliquot of digested sample is added with the steam flowing and rinsed in with distilled water. When steam is passing freely through the sample, the tip of the condenser is lowered to immerse it in the boric acid/indicator solution. For Kjeldahl distillation samples in the range 0.05 to 5.0 mg of nitrogen, 20 ml of the boric acid solution is sufficient. The sample should then be rendered alkaline to added phenolphthalein by the slow addition of 40% NaOH; this alkali should be freshly prepared and protected from atmospheric CO_2. After about 5 min the condenser tip is raised above the boric acid solution and 30 to 35 ml of distillate is collected. Distillation is terminated by closing the steam line (venting the generator to the atmosphere); as steam trapped in the system condenses, a vacuum forms by which the remaining sample of digest is sucked out of the distillation chamber and may be discharged through the waste port. The system is then cleaned and readied for the next sample by flushing distilled water through the sample inlet port and draining through the waste port.

If the samples contain ^{15}N, additional precautions must be taken to avoid possible contamination from trace amounts of highly labeled NH_3 remaining in the still. Traces of ammonia adhering to the glass

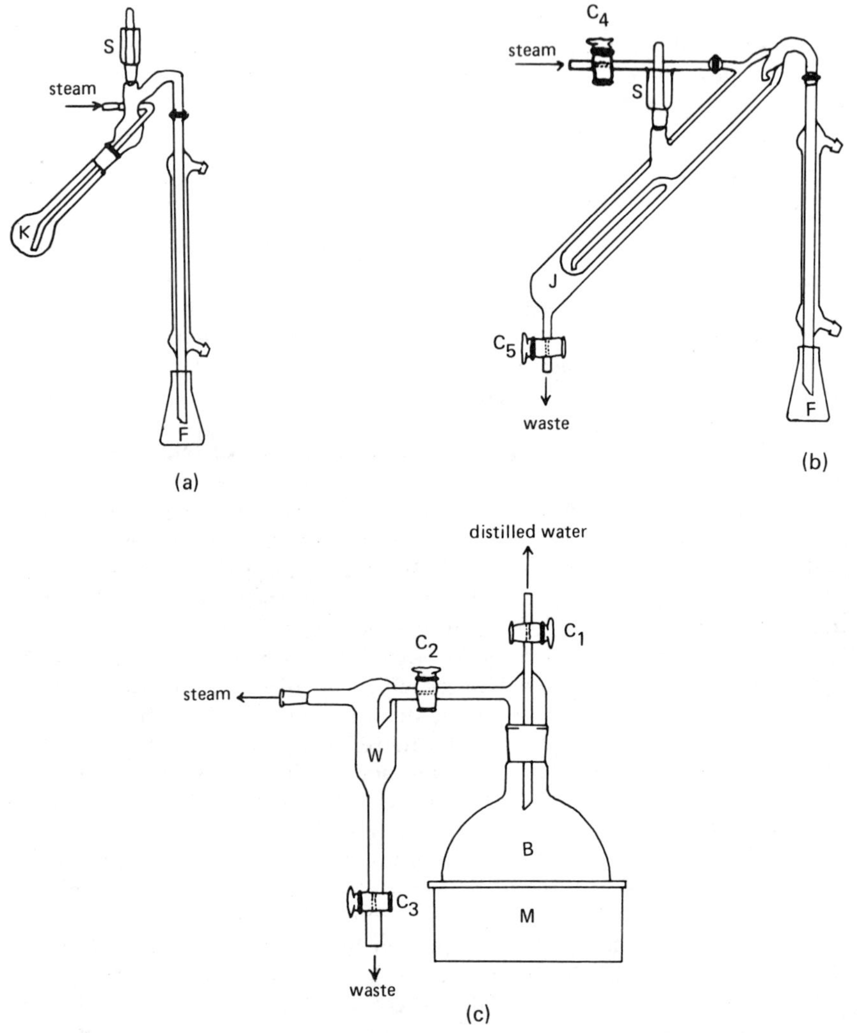

Figure 3 Stills for the recovery of ammonia: (a) semimicro still similar to the assembly of Bremner and Edwards (1965), which couples directly to Kjeldahl flasks (K) of 50- or 100-ml capacity; (b) Markham still; (c) steam generator. B is a 2- to 3-liter distillation flask resting on an electric heating mantle (M). Acidified distilled water is supplied through stopcock C_1, which also opens to the atmosphere to vent steam when necessary. Digested sample and NaOH are introduced into the stills through the sample port (S) and distillate is collected in boric acid/indicator solution in Erlenmeyer flasks (F). When the Bremner

surfaces of the still can be removed by the distillation of about 15 ml of ethanol between samples (Bremner and Edwards, 1965). However, Bergersen (1980) claims that cross-contamination does not occur when using the Markham still (Figure 3), provided that at least 100 ml of distillate is collected for each sample.

Distillates having 1 to 10 mg of nitrogen as ammonium are titrated with standardized, approximately N/28 HCl (31.8 ml of concentrated HCl diluted to 10 liters with distilled water) to a faint pink end point (pH 4.9). If microgram quantities of ammonium N are expected in the distillates, the titrating acid should be 10 times more dilute. Microtitrations are facilitated by use of commercially available micrometer syringes (AGLA micrometer syringe, Burroughs Wellcome & Co., London, England). A 1-ml volume of N/28 HCl is equivalent to 0.5 mg of nitrogen. The titrating acid is standardized with standard borax, prepared as follows. A saturated solution of borax is prepared at 55°C and allowed to recrystallize two times. The crystals are successively washed with cold water, dry ethanol, and diethyl ether. They are then allowed to equilibrate for 24 hr in a desiccator over water saturated with sucrose and NaCl; the crystals should be stored in a tightly stoppered bottle inside a desiccator (Humphries, 1956). For use in standardizing the HCl, approximately 0.3 g of the recrystallized borax is accurately weighed, dissolved in water, and diluted to 50 ml with water. One drop of 0.1% ethanolic methyl red is added to 10 ml of this solution, which is then titrated with the HCl solution to be standardized.

$$\text{Normality of acid} = \frac{\text{wt. borax} \times 200}{190.72 \times \text{ml acid}}$$

III. ESTIMATION OF AMMONIA

Whereas intact organisms fix N_2 and convert it rapidly to amino acids and proteins, cell-free extracts from N_2-fixing organisms accumulate ammonia. Crude extracts of microorganisms might retain glutamine synthetase sufficient to convert ammonia to glutamine, thereby intro-

---

and Edwards still (a) is used C_1 is closed and C_2 is open; when heat to B is removed, steam condenses, producing a vacuum in B which draws the sample residue from the Kjeldahl flask (K) into the waste chamber (W), from which it may be expelled through C_3. The same result is achieved with the Markham still (b) by closing the stopcock C_4 (making sure to open C_1); steam condensing in the jacket (J) creates a partial vacuum drawing wastes into J, from which they are discharged through C_5. (Redrawn from Bergersen, 1980. Courtesy of John Wiley & Sons Ltd., Chichester, West Sussex, England.)

ducing an appreciable error into estimates of N_2 fixation based on ammonia production. In contrast, purified preparations of nitrogenase accumulate ammonia quantitatively, its increase thus serving as an excellent index of N_2 fixation.

Mortensen (1961) demonstrated that quantitative determination of ammonia is a valid index of N_2 fixation. He utilized the Conway microdiffusion technique followed by titration of the ammonia. Others (Dilworth et al., 1965; Burris, 1972; Bulen et al., 1965) have utilized microdiffusion and nesslerization to quantitate production of ammonia. The distillation and titration of ammonia is described in Sections II.B and IV.B as part of procedures for total nitrogen and nonprotein nitrogen analyses, respectively. Microdiffusion is described immediately below. The use of colorimetry, ammonia-specific electrodes, the AutoAnalyzer, and enzymatically based assays is described in Section III.B.

A. Microdiffusion

Ammonia produced by N_2 fixation can be recovered for analysis by the microdiffusion method of Conway (1950); this is more conveniently performed as described by Burris (1972) in small Erlenmeyer flasks or in 15-ml serum bottles (see Figure 4) rather than the special Conway diffusion vessels. The end of a glass rod is rounded by heating and then etched by using HF or by griding with a sanding belt. The rod is inserted in a one-hole rubber stopper to fit the mouth of the serum bottle. The end of the glass rod is dipped into 10% H_2SO_4 and blotted to a paper towel to remove the excess acid.

Samples (e.g., Kjeldahl digests) containing from 1 to 14 μg of nitrogen as ammonia can be placed inside the bottles and made alkaline as described in the legend to Figure 4. The stopper with the acid-dipped rod is seated into the serum bottle while being careful not to touch the rod to the bottle. Diffusion may be accelerated by placing a group of bottles on a rotator to agitate the contents gently for 3 hr at room temperature. If a rotator is not used, overnight diffusion at room temperature will allow quantitative recovery of ammonia as ammonium ion on the acid-coated glass rod; it can then be estimated using one of several of the methods described in Section III.B. For some colorimetric assays, the rod may be removed and the tip soaked in 1 ml of distilled water in a small test tube to which the reagents for ammonia assay may subsequently be added.

N_2-fixation reactions can also be run in the serum bottles capped with serum bottle stoppers (Burris, 1972); the reaction is terminated by injection of 1 ml of saturated K_2CO_3. Immediately affter adding the K_2CO_3 to the reaction bottle, the acid-dipped rod is inserted to initiate microdiffusion as described above. A control should always be included in which argon replaces the N_2 atmosphere in the reaction vessel. A nitrogen-fixation assay for use with purified nitrogenase

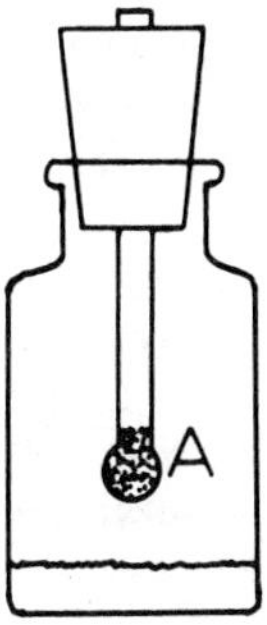

Figure 4 Apparatus for recovery of ammonia by microdiffusion. Ammonia diffuses overnight from the alkaline solution in the serum bottle to 10% H_2SO_4 held on the etched portion (A) of the central glass rod. The capacity of the serum bottles is dependent on the type and scale of the application. In vitro nitrogenase activity can be assayed in 7-ml serum bottles after the method Maryan and Vorley (1979; see Section III.B, "Direct Assay for Nitrogenase"); reaction components are easily added, and headspace samples taken through serum stoppers with hypodermic syringes; if components of the nitrogenase reaction (e.g., buffer systems) interfere with colorimetric ammonia quantitation, microdiffusion is conveniently done in these same serum bottles after stopping the nitrogenase reaction by injection of trichloroacetic acid. Larger serum bottles or Erlenmeyer flasks can also be used for diffusion of ammonia from larger-scale Kjeldahl digests. An aliquot containing 1 to 14 μg of nitrogen is compatible with colorimetric ammonia assays (see Section III.B). The sample is titrated to pH 4 to 5 with NaOH, using one drop of neutral red as an indicator. When the solution is at room temperature, 1 ml of 20% NaOH is added and the bottle quickly sealed with the rubber stopper bearing the acid-dipped glass rod. (Redrawn from Burris, 1972.)

components is described in Section III.B; this method may not require microdiffusion provided that the reaction uses components that do not interfere with the ammonia assay selected.

B. Methods for Measurement of Ammonia

Nesslerization

As described below, nesslerization produces a color intensity that follows Beer's law to about 45 μg of NH_3-N; beyond that amount turbidity may be a problem. Nessler's reagent is prepared as follows (Umbreit et al., 1964). Dissolve 4 g of KI and 4 g of HgI in 25 ml of water.

Select light-colored pieces of gum ghatti and grind 1.75 g in a mortar; add the gum ghatti to 750 ml of water and reflux until it is dissolved. Add the KI-HgI solution to the gum ghatti solution, dilute to 1 liter, and filter.

After microdiffusion is complete (described above), the stopper with the glass rod is carefully removed from the serum bottle and the rod used as a stirring rod to mix 2 ml of the Nessler's reagent, 2 ml of H_2O, and 3 ml of 2 *N* NaOH in a small beaker; this transfers $(NH_4)_2SO_4$ from the glass rod to the Nessler's solution. The color must be allowed to develop for 15 min at room temperature before reading its absorbance at 490 nm against a reagent blank.

Indophenol Method

This procedure allows determination of ammonia by the catalyzed indophenol reaction, which produces a stable blue color. The method described below is that of Chaney and Marbach (1962), in which sodium nitroprusside serves as a catalyst to speed the reaction and increase to a maximum the conversion of ammonia to indophenol.

Reagents: The reagents may conveniently be combined in the following manner:

Reagent	Dilute form (g/liter)	Concentrated form	
		g/liter	Molarity
Solution 1			
Phenol	10.0	50.0	0.5
Sodium nitroprusside	0.05	0.25	0.001
Solution 2			
Sodium hydroxide	5.0	25.0	0.625
Sodium hypochlorite	0.42	2.1	0.03

Each of the two mixtures above is stable for 60 days or more if kept cool and in amber bottles protected from light. Usually, solutions 1 and 2 are prepared in the "concentrated form" described above and diluted 1:5 before use. Sodium hypochlorite in the form of commercial bleach can be used, and other chemicals are of reagent grade.

Procedure: Five milliliters of solution 1, followed by 5 ml of solution 2 are added to 1-ml aliquots of the sample to be assayed for NH_3. Color development requires 60 min at 23°C; this time can be shortened to 35 and 20 min at 30 and 40°C, respectively.

Alternatively, one may use 1 ml each of the concentrated reagents and develop the color at 50 to 60°C. Under these conditions, color development is complete in about 3 min and the solution can be diluted to a convenient volume for spectrophotometric determination at 625 nm. The color remains stable for about 24 hr.

The reaction generates a linear response from below 0.05 m*M* to beyond 1.0 m*M* NH_3. Using the dilute reagents, a 0.5 m*M* standard should produce an absorbance of around 1.0 in a 1-cm light path against a reagent blank. The absorbance can be referred to a calibration curve to enable calculation of the concentration of NH_3.

Direct assay of nitrogenase by colorimetric determination of ammonia: Maryan and Vorley (1979) describe an improvement on the indophenol method for ammonia determinations that allows omission of the microdiffusion steps that are normally required in the presence of certain buffers or assay components. This method can be applied as a direct assay of nitrogenase activity.

Nitrogenase assays to measure NH_3 formation are set up as described by Eady et al. (1972) under N_2 (1 atm) in 7-ml glass serum bottles fitted with serum stoppers. The assays contain, in a total volume of 1.0 ml, adenosine triphosphate (ATP; 5 m*M*), $MgCl_2$ (12.5 m*M*), $Na_2S_2O_4$ (30 m*M*), creatine phosphate (12 m*M*), creatine kinase (100 μg), and one of the following buffers at a concentration of 25 m*M*, pH 7.5. Buffers that can be used in this procedure include *N*-2-hydroxyethylpiperazine-*N*'-2-ethanesulfonic acid (Hepes), *N*,*N*-bis(2-hydroxyethyl)-2-aminoethanesulfonic acid (BES), and *N*,*N*-bis(2-hydroxyethyl)-glycine (Bicine). The stock solution of $Na_2S_2O_4$ (300 m*M*) is also made up in a 25 m*M* solution of the same buffer. Some buffering systems severely inhibit color formation by the indophenol reaction with NH_3; such unsuitable buffers include tris(hydroxymethyl)aminomethane (Tris), *N*-tris(hydroxymethyl)methyl-2-aminoethanesulfonic acid (TES), and *N*-tris(hydroxymethyl)methylglycine; *N*-(2-hydroxy-1,1-bis[hydroxymethyl]ethyl)glycine (Tricine). The severe inhibition effected by Tris not only precludes its use in the assay but also as a buffer in which the nitrogenase proteins are prepared.

The assays are initiated by injection into the serum bottle of a 1:1 mixture of the nitrogenase proteins (0.05 ml) in 25 m*M* buffer and the reaction is terminated by the injection of 0.1 ml of trichloroacetic acid (TCA; 30% w/v). Following injection of 0.5 ml of water, a 0.5-ml gas sample may be taken for H_2 analysis by gas chromatography using a thermal conductivity detector or by a hydrogen electrode (Hanus et al., 1978). This allows for the estimation of hydrogen evolution concomitant with N_2 reduction. The addition of water ensures that the gas sample is taken at 1 atm of pressure and increases the volume of the assay mixture to 1.6 ml, a more convenient volume for centrifugation.

It is crucial that the dithionite ions be allowed to decompose completely, either by standing for 1 hr or by gentle sparging with air for 2 min. Failure to do this will result in irreversible inhibition of color development. Because of the acid conditions following the termination of the reaction with TCA, the sparging treatment does not result in the loss of NH_3. The contents of the assay bottles are transferred to glass tubes and centrifuged for 10 min at 10,000 × *g*.

Ammonia determination is by the procedure of Chaney and Marbach (1962) described earlier in this section, except that ammonium chloride standards from 0.02 to 1.0 m*M* are prepared in the appropriate buffer in order to generate the calibration curve.

Ammonia Electrode

NH_3-sensing electrodes are limited in their sensitivity and accuracy relative to the colorimetric procedure described above. However, if sufficient sample size is available, convenient and rapid determinations of ammonium resulting from Kjeldahl digestion can be made. Ammonium ion in the digests must be converted to NH_3 by the addition of NaOH immediately before measurement with the electrode. We have employed the method described by Bremner and Tabatabai (1972). An Orion model 95-10 ammonia electrode (Orion Research Inc., Cambridge, Massachusetts) connected to a suitable pH-millivolt meter (in the mV mode) is checked for response and calibrated in terms of the millivolts per decade of ammonia nitrogen as described in the instruction manual. Such an electrode responds proportionately to the logarithm of a wide range of NH_3 concentrations (10^{-6} to 10^{0} *M*).

A standard curve of electrode potentials plotted against the logarithm of NH_3-N should be generated using several concentrations of NH_4Cl (e.g., 1, 3, 10, 30 mg of nitrogen) in Kjeldahl flasks to each of which are added 4 ml of concentrated H_2SO_4 and sufficient water to bring the volume up to the nominal volume of the flask; these standards should also be used to demonstrate the proper electrode response within the range of ammonium concentrations to be analyzed.

The assays are done one at a time; a 5-ml aliquot from an ammonium standard or from the Kjeldahl-digested plant material (see Section II.B) is added to 91 ml of distilled water in a 150-ml beaker. While this is being magnetically stirred, 4 ml of 10 *N* NaOH is added and the electrode potential is recorded after 30 sec (3 min or more may be required for the reading to stabilize). About 1 cm of the ammonia electrode should be submerged; the electrode should make a 20° angle with the vertical to avoid air entrapment on the membrane. The ammonia nitrogen concentration can then be read from the reference curve.

Titration of Ammonia

Ammonia may be estimated by titration against standardized HCl. This method is often used following the distillation of ammonia from Kjeldahl digests. Thus the procedure is detailed in Section II.B.

Use of Autoanalyzers to Estimate Ammonia

If large numbers of total nitrogen determinations are required, the use of automatic analyzers offers considerably advantage. The Technicon AutoAnalyzer (Technicon Industrial Systems, Tarrytown, New York) has been the standard apparatus for automation of many analytical processes, including ammonia analysis. A typical flow sequence for this instrument is shown in Figure 5. A newer technology, flow injection analysis (FIA) (Ruzicka and Hansen, 1975), that is now available commercially (from Tecator AB, Hogänas, Sweden, or Tecator Inc., Herndon, Virginia) differs in that the sample is injected after the pump into the stream of mixed reagents. This configuration minimizes sample carryover and cross-contamination and offers several advantages over the AutoAnalyzer system, including shorter sampling times (approximately 30 sec), high sample throughput (up to 200 samples per hour), minimum sample injection (30 μl), and conservation of reagents (typically from 100 to 300 μl per sample). The Tecator apparatus (FIAstar Model 5020) is an integrated microprocessor-controlled FIA system with sample carriers having 40 or 100 positions. Dedicated manifolds and application protocols are available for ammonia assay utilizing a number of detection systems, including phenol-hypochlorite methods (see below), nesslerization (Krug et al., 1979; Bergamin et al., 1980), and gas-sensing electrodes (Meyerhoff and Fraticelli, 1981). The system is applicable to analyses of nitrate, nitrite, amino acids, and total nitrogen; interested readers should contact Tecator for further information on these analytical techniques.

The procedure below is that described by Williams and Twine (1967) for use with the AutoAnalyzer.

Reagents

1. Sodium phenate. Dissolve 500 g of phenol in 90 ml of distilled water; mix with 2.5 liters of 20% w/v NaOH.
2. Sodium hydroxide, 0.52%.
3. Sodium hypochlorite. Dilute 50 ml of commercial sodium hypochlorite (12% NaOCl) to 200 ml with distilled water.
4. Diluent. Mix 3 ml of Teepol detergent with 10 liter of distilled water.

Procedure: Prepare ammonium standards by dissolving 9.433 g of $(NH_4)_2SO_4$ in 1 liter of distilled water; 1 ml = 2.000 mg N. Dilute to give a range of standards containing 0 to 150 μg/ml N and 5 ml of

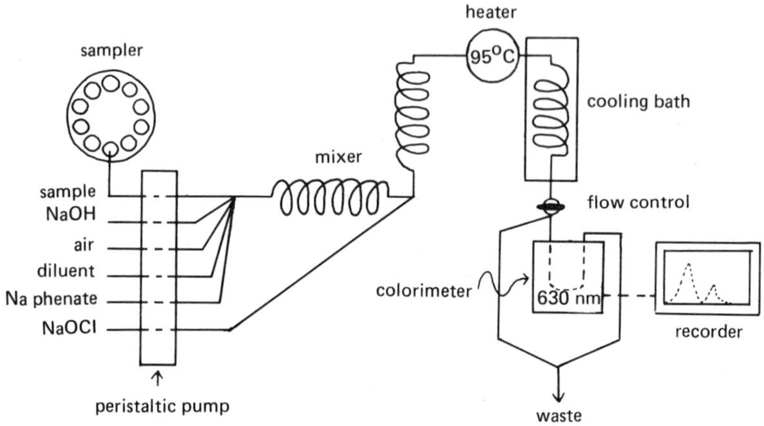

Figure 5 Flow diagram for AutoAnalyzer. Samples are "sipped" from sample cups and reagents are drawn from reservoirs by peristaltic pumps. The reagents (with the exception of NaOCl) are added to the sample and the flow passes through a mixer, after which the NaOCl is added and the reaction components are mixed and heated. After cooling, a portion of the mixture is directed through a flow cuvette in a colorimeter, where the absorbance at 630 nm is read and recorded by a strip-chart recorder. Although flow rates may be adjusted for optimal performance, those suggested (in ml/min) by Williams and Twine (1967) are: sample, 0.42; NaOH, 2.90; air, 3.40; diluent, 3.90; Na phenate, 0.80; NaOCl, 0.60; effluent from the colorimeter (assuming a 16-mm tubular flow cuvette), 1.60. (Adapted from Bergersen, 1980.)

digestion mixture per 100 ml. Samples from digestions containing 5 ml of digestion reagent are diluted to 100 ml with distilled water. These should be mixed and let stand overnight to allow solids to settle. Appropriate aliquots are then decanted into the sample cups.

A flow diagram is shown in Figure 5. Forty samples per hour may be processed; it is not necessary to use a wash between samples. Reagent blanks and standards are interspersed with replicated samples. Heating at 95°C accelerates color development at 630 nm; the colorimeter is coupled to a strip-chart recorder on which absorbances are sequentially recorded. If replicates differ by more than 5 to 10%, analyses should be repeated. Results are not influenced by Ca, Mg, or Se (50 ppm) or by Mn, Fe, or Al (40 ppm); these concentrations are greater than those likely to be found in digests of biological materials. However, results are affected by small variations in H_2SO_4 concentration in the digestion reagent; thus careful measurement of the digestion reagent is necessary to ensure consistency.

Measurement of Ammonia by Enzymatic Means

By virtue of the specificity conferred by the enzymatic methods, the microdiffusion or distillation step is not necessary to separate the ammonia from other components in a mixture. Moreover, these methods can be quite sensitive, especially when coupled with radiotracer-labeled substrates. However, care should be taken to ensure that inhibitors, enzymes, or enzyme substrates in materials to be analyzed are either eliminated or neutralized; and if necessary, that suitable control reactions be included to adjust the final ammonia determinations.

Below are described two enzymatic assays for ammonia both of which are based on commercially available glutamate dehydrogenase. The first of these is procedurally the simpler and provides rapid determination of ammonia down to 10 nmol. The second requires standard incubation times of 2 hr and is procedurally more complex; however, it offers the specific determination of ammonia in biologically derived mixtures down to 50 pmol.

Method of Levitzki (1970): This method employs the oxidation of nicotinamide adenine dinucleotide (NADH) by the reductive amination of α-ketoglutarate with ammonia that is catalyzed by L-glutamate dehydrogenase (GDH; E.C. 1.4.1.3, type II). The GDH, purified from bovine liver, can be obtained from Sigma Chemical Co. (St. Louis, Missouri). The substrates α-ketoglutarate and NADH are supplied in saturating quantities. The K_m value of GDH for ammonia is rather high (around 57 m*M*); thus ammonia concentrations will be rate limiting for the initial reaction velocity.

The rate of the reaction catalyzed by GDH, α-ketoglutarate + NH_3 + NADH → L(+)-glutamic acid + NAD is thus proportional to the concentration of ammonia from 20 to 200 nmol. The reaction progress is followed spectrophotometrically at 340 nm over a period of 3 to 5 min, during which it is linear with time.

Reaction mixture components are added to 1.5 ml of quartz cuvettes in the following amounts:

Component	Concentration	Volume
NADH	2.4 m*M*	100 μl
α-Ketoglutarate	1 *M*, pH 7.6	50 μl
K-PO_4	0.4 *M*, pH 7.6	300 μl
H_2O	—	250 μl
NH_3 standard or sample	10-200 nmol	100 μl
GDH	0.5 mg/ml	200 μl
	Total volume	1.0 ml

Ideally, the cuvette chamber temperature should be maintained at 25°C, but this is not crucial if reasonable care is exercised (i.e., if temperature fluctuations within the chamber are small and if a standard curve is prepared for each set of assays done on a particular day). After preincubation for 5 min, the reaction is initiated by addition of the GDH and the rate of NADH oxidation is followed at 340 nm on a 0.5-optical density (OD) full-range scale. The background absorbance under these conditions is around 1.65. A control reaction should also be run without ammonia; this rate should be subtracted from all other determinations to yield a net rate of change in absorbance. A calibration curve can thereby be generated that plots the net rate of absorbance change at 340 nm against the concentration of ammonia. Reproducibility of the assay is ±5%.

This assay method offers specificity and relative freedom from interference. However, the GDH is subject to inhibition by ATP and GTP in the materials being assayed. This problem can be overcome by the addition of sufficient adenosine 5'-diphosphate (ADP; a stimulator of GDH activity) to the reaction mixture to overcome inhibition by the other nucleotides. A separate calibration curve should be produced for each source of biological material that might contain the inhibiting nucleotides and sufficient ADP added to yield linear rates over a 3- to 5-min period.

Method of Kalb et al. (1978): The following procedure enhances the sensitivity of the GDH methodologies through use of radiochemically labeled α-ketoglutarate. Thus ammonia can be measured down to 50 pmol and accurate determinations in the range 0.1 to 10 nmol are possible. GDH and other enzymes are available commercially (Sigma Chemical Co., St. Louis, Missouri). The assay is based on the reaction shown in our discussion of Levitski's method; α-[1-^{14}C]ketoglutarate is reductively aminated, yielding labeled glutamate.

When the substrates α-ketoglutarate and NAD(P)H are supplied in excess, quantitative amounts of glutamate are produced. The remaining radioactive α-ketoglutarate is specifically decarboxylated by treatment with H_2O_2 and the volatile $^{14}CO_2$ is removed by heating. Because GDH catalyzes a reversible reaction, the presence of NAD(P) and an excess of nonradioactive glutamate can lead to erroneously high values for ammonia. Samples containing these interfering substances are treated as "complex systems." "Simple systems" are those in which all the components in the sample to be analyzed are known and inhibition or exchange reactions are not expected. A complex system is one containing unknown amounts of different compounds which could interfere with the assay; complex systems include cell extracts.

The assay can be modified for use with any system in which NH_3 is a product or a reactant. For example, the method might be adapted to estimate the nanomolar quantities of ammonium that accumulate during rapid stopped-flow experiments with nitrogenase. It is also possible to determine glutamine in bacterial cell extracts and other biological materials if ammonia is assayed before and after its hydrolysis from glutamine. Kalb et al. (1978) suggest dividing the samples containing glutamine; one-half is assayed for ammonia. To the other half, 1.0 *M* phosphate buffer (pH 6.5) is added to a final concentration of 0.1 *M* and heated to 100°C for 90 min. This procedure quantitatively removes the α-amino group from glutamine; ammonia determined on this half of the sample reflects the free ammonium plus that liberated from glutamine. Thus glutamine can be estimated as the difference between the amount of nitrogen in the halves of the sample.

1. *Procedure for simple mixtures.* The reaction mixture contains 300 m*M* 2(*N*-morpholino)ethanesulfonic acid (MES) buffer (pH 7.0), 2.5 m*M* ADP (to stabilize the GDH), 0.3 m*M* α-ketoglutarate, 0.2 mg/ml GDH (8 to 9 units), and 2 m*M* NADH in distilled water. The NADH is added from a stock concentrate dissolved in 100 m*M* carbonate buffer (pH 10.6). The concentrated NADH is heated at 100°C for 5 min before being used to destroy any NAD present. The reaction mixture is incubated at room temperature for 1 hr to allow for the utilization of endogenous ammonium present in the stock solutions.

During this incubation, samples to be analyzed are dispensed into scintillation minivials (15 mm × 45 mm; Sargent-Welch Scientific Co., Skokie, Illinois). Ammonium standards ranging from 0.1 to 10.0 nmol are delivered to a set of minivials. The final sample volumes are adjusted to 0.1 ml with double-distilled water; a blank with no added ammonium is included.

After the 1-hr incubation of the reaction mixture, approximately 20 nmol of α-[1-^{14}C]ketoglutarate (1.124 μCi) is added per milliliter of reaction mixture, the solution is mixed, and 50 μl is then delivered to each scintillation vial. Total radioactivity of 50 μl of reaction mixture is determined separately. The vials are incubated at room temperature for 2 hr to allow the enzymatic reaction to go to completion. The reaction is stopped by adding 100 μl of acetone to each, and the samples are heated for 10 min at 100°C to inactivate the catalase present in commercial GDH preparations. To release radioactivity from the excess α-[1-^{14}C]ketoglutarate, 50 μl of fresh 4 *M* H_2O_2 is added and the vials are heated for 10 min at 100°C. To remove any remaining $^{14}CO_2$ and to decrease any chemiluminescence from residual H_2O_2, 100 μl of HCl is added to each vial and the vials are again heated at 100°C for 20 min. Three milliliters of scintillation cocktail suitable for aqueous samples (e.g., PCS cocktail from Amersham Corp., Arlington Heights, Illinois) is added to each vial and radioactivity is determined by liquid scintillation spectrometry. After subtracting the blank (no added ammonium) value from the others, ammonium concentration can be calculated by reference to the original specific radioactivity of the α-ketoglutarate or from a standard curve.

2. *Procedure for complex systems.* Cell extracts or other complex mixtures will contain varying amounts of compounds such as nicotinamide adenine dinucleotide phosphate (NADP), glutamate, or α-ketoglutarate which interfere with the accuracy of these determinations. α-Ketoglutarate is removed from samples by treatment with H_2O_2; NAD(P) is removed with pentachloroacetone. To prevent decarboxylation of the α-[1-^{14}C]ketoglutarate, a large excess of catalase is included in the reaction mixture. NADP formed during the course of analysis is removed by the addition of an excess of glucose-6-phosphate and glucose-6-phosphate dehydrogenase.

The reaction mixture contains 300 m*M* MES buffer (pH 7.0), 2.5 m*M* ADP, 0.3 m*M* α-ketoglutarate, 1 m*M* NADP, 0.5 m*M* dithiothreitol, 10 m*M* mg-acetate, 30 m*M* glucose-6-phosphate, 0.2 mg/ml GDH, 1.25 mg/ml catalase, and 0.1 mg/ml glucose-6-phosphate dehydrogenase (20 units) in double-distilled water. To allow for utilization of endogenous ammonium and for conversion of NADP to its reduced form NADPH), the mixture is incubated at room temperature for 1 hr. Approximately 20 nmol of the radiolabeled α-ketoglutarate is added per milliliter of the reaction mixture.

Samples and standards are prepared as described above for simple systems. Pentachloroacetate (4.4 m*M*, 10 μl) is added to each scintillation vial 30 min before the addition of the reaction mixture; 5 μl of fresh 220 m*M* H_2O_2 is added to each vial 10 min before the addition of the reaction mixture. After addition of 50 μl of the reaction mixture to each minivial, the treatment of all samples is identical to that described for the simple system.

IV. ANALYSIS OF NONPROTEIN NITROGEN

Schemes for the estimation of nonprotein nitrogen are usually based on procedures that remove protein and subsequently fractionate the remaining nonprotein nitrogenous materials for analysis by various means. In this section we outline some strategies for these fractionations and analyses.

A. Extraction

Nonprotein nitrogenous compounds can be recovered from the supernatant fluids produced from grinding plant materials in acid strong enough to precipitate all proteins. Cell-free extracts of bacteria and other biological materials can also be treated in this fashion.

Aprison et al. (1954) used 3 *N* HCl to extract early products of nitrogen fixation from root nodule material. A similar procedure is suggested by Bergersen (1980): 1 g (fresh weight) of nodules or similar tissue is crushed with a glass rod in 5 ml of 2 *N* HCl, the material is centrifuged, and the extraction of the pellet repeated twice with 2 ml of 3 *N* HCl. The supernatant fluids are combined, digested, and analyzed for total nitrogen as described in Section II.A. Centrifugally separated tissue fractions can be similarly extracted by adding equal volumes of 6 *N* HCl to suspensions of particles, or centrifugally clarified buffer extracts. After standing at room temperature for 15 min, the mixture is clarified by centrifugation and the pellets reextracted with 3 *N* HCl; combined supernatant fractions are analyzed as before.

Bacterial suspensions may be extracted in a similar fashion, but usually 10% w/v trichloroacetic acid is superior to 3 *N* HCl. The removal of soluble proteins from buffer extracts is best accomplished by making the solution 0.1 *N* in H_2SO_4 and 1 to 2% w/v in sodium tungstate, which effectively precipitates proteins and all but the smallest polypeptides (Bergersen, 1980). Bacteria, tissues, and other biological material may also be extracted with boiling 80% ethanol. Such extracts are clarified by filtration or centrifugation, the ethanol removed in a rotary evaporator, and the sample redissolved in an appropriate buffer. This is crucial when analysis of ^{15}N is to be used because residual ethanol will generate spurious peaks in the mass spectrometer.

B. Ammonia, Amide, and Nitrite plus Nitrate Nitrogen

A useful method for analysis of nonprotein nitrogen described by Bergersen (1980) involves the successive removal of ammonia, amide, and nitrate plus nitrite nitrogen. A micro-Kjeldahl distillation apparatus such as that illustrated in Figure 3a is required.

Reagents

1. Sodium borate, pH 10. A saturated solution of borax ($NaB_4O_7 \cdot 10H_2O$) is adjusted to pH 10 with NaOH and kept tightly stoppered to exclude CO_2.
2. NaOH, 40% w/v.
3. Devarda's alloy (*Merck Index*, Vol. 10, 2905; obtainable from Aldrich Chemical Co., Milwaukee, Wisconsin) sieved to remove fine powder but passing 85 mesh.
4. Boric acid, 2% w/v.

Procedure

The sample size should be adjusted to contain a minimum of 100 μg of nitrogen each. The sample is introduced into the distillation flask after being adjusted to approximately pH 4. An equal volume of the borate solution is added and free NH_3 is distilled into 5 to 10 ml of 2% boric acid. When distillation is complete, a fresh boric acid flask is placed in position, 5 ml of 40% NaOH added to the sample, and the amide nitrogen is distilled into 5 to 10 ml of 2% boric acid. When this distillation is complete, the steam flow is stopped and another boric acid flask is placed in position. The Kjeldahl flask is removed, 0.2 g of Devarda's alloy added, and the flask quickly reconnected to the still. Steam flow is resumed, slowly at first to avoid splashing due to the reaction, and the nitrite and nitrate nitrogen is collected as distilled ammonia. Ammonia in the distillate is estimated as described in Section III.B. After these distillations, most of the nitrogen remaining in the sample is probably amino and nucleotide nitrogen; this may be estimated by difference or analyzed in a separate aliquot of extract.

Pace et al. (1982) contend that from 10 to 60% of the nitrate present in plant tissue extracts is reduced during Kjeldahl digestion, even in the absence of reducing agents. This would result in underestimation of nitrate by the foregoing method. To overcome this problem, it is possible to use a pretreatment protocol that they have developed for plant material high in nitrates. Material to be analyzed is pretreated with 30% hydrogen peroxide in dilute H_2SO_4 to partially oxidize organic compounds; the free nitrates are volatilized as nitric acid by treatment with concentrated sulfuric acid at 95°C. Such treatment prevents recovery of nitrate nitrogen in the initial digest.

C. Amino Nitrogen

α-Amino nitrogen can be quantified after removal of ammonia and amide nitrogen from extracts as described above using a ninhydrin method in which the colored reaction product is decomposed in acid to produce ammonia (Kennedy, 1965). Whereas the ninhydrin reaction is commonly used in the analysis of resolved amino acids, it is not valid for the estimation of mixtures of amino acids in unresolved extracts (see Section IV.D).

Procedure

Residues of extracts from which the ammonia and amide nitrogen have been distilled are neutralized with H_2SO_4 and evaporated to dryness in a rotary evaporator. They are then dissolved in hot 80% ethanol to remove excess salts, transferred to a fresh flask, and again evaporated to dryness. This residue, which should contain 50 to 100 μg of amino nitrogen, is dissolved in 1 ml of water; 1 ml of pH 5.4 acetate buffer and 2.0 ml of ninhydrin are added. The mixture is heated for 10 min at 100°C; 4.0 ml of 2 *N* HCl is added and heating continued for 10 min. The ninhydrin-ammonia complex is thus destroyed and the released ninhydrin is subsequently decomposed by treatment with 0.25 ml of hydrogen peroxide for 5 min at 100°C. Ammonia is recovered by steam distillation and estimated by titration or by colorimetric analysis of a measured aliquot of distillate (see Sections II and III). It may also be used for isotopic analysis (see Chapter 9).

D. Amino Acid Analysis

Because the methods chosen for amino acid analysis will depend on the equipment available and the material subjected to analysis, a detailed treatment of these methods is beyond the scope of this chapter. Thus we will present an outline of some established methods, the principles involved, and their application to nitrogen-fixation studies.

Paper and Thin-Layer Chromatography

Paper and thin-layer chromatographic resolution of amino acids from mixtures is in principle effected by interactions of the solutes between the mobile and the stationary phases of a chromatographic system. If the stationary phase is a component of the developing solvent, the interaction may be defined as a liquid-liquid partitioning of the solute components (partition); if the stationary phase is solid, the mobile, liquid phase competes with the solid adsorbant for solute components (adsorption). In "normal-phase" chromatography, the stationary phase is polar and the mobile phase is nonpolar; thus the more hydrophobic amino acids migrate more rapidly along the chromatogram. Identification of particular amino acids can be made based on characteristic R_f values obtained with specific solvent systems, supporting media (e.g., paper, silica gel), and developing conditions. Paper chromatographic methods suitable for the resolution of amino acids have been described by Hellmann (1956); information concerning preparation of solvents, reagents, paper selection, and R_f values is obtainable from publications by Smith (1969) and Zweig and Whitaker (1971). Paper electrophoresis alone, or a combination of electrophoretic and chromatographic techniques, often enable resolution of complex mixtures containing amino acids (Dreyer and Bynum, 1967; Smith, 1969; Zweig and Whitaker, 1971). Following detection with appropriate reagents, estimation of amino acids on paper or thin-layer chromatograms can be made by direct photometric scanning of the colored spots. More commonly, however, the spots are excised from the chromatogram, eluted, and amino acids in the eluate estimated colorimetrically.

Enhanced resolution of complex mixtures containing amino acids can be effected by employing thin-layer chromatography on silica gel, alumina, or cellulose. The methods may be adapted in scale from micro to semipreparative, and they are generally applicable to nitrogen-fixation studies. Brenner and Neiderwieser (1967), Pataki (1968), and Stahl and Ashwood (1969) have published thorough descriptions of these methods. A bibliography of paper and thin-layer chromatographic applications is also available (Macek et al., 1976).

Resolved amino acids, eluted from paper or thin-layer chromatograms, can be estimated using an adaptation of the ninhydrin method of Moore and Stein (1948). Ninhydrin reagent solutions are commercially available (Sigma Chemical Co., St. Louis, Missouri). Ninhydrin reagent and samples are kept in a boiling-water bath for 20 min; diluent (*n*-propanol:water, 1:1) is added to the solution and mixed by vortexing. After 15 min, and within 1 hr, the absorbance at 570 nm is compared with absorbance of leucine standards. Proline and hydroxyproline are read at 440 nm. The color formed by the reaction with leucine is proportionate to amino acid in the range 0.04 to 0.8 μmol per assay. Estimation of other amino acids can be made by reference to the leucine standard curve with appropriate adjustments for the amount of diluent and the "leucine equivalence factor." These arithmetic manipulations are described in detial in the original publication (Moore and Stein, 1948).

If ninhydrin is used, the amino nitrogen may be recovered from paper or thin-layer chromatograms as described above, and used for isotopic analysis (Kennedy, 1965) provided that the spot contains sufficient nitrogen. Alternatively, the resolved amino acid may be subjected to Kjeldahl digestion as described in Section II.A, and the total nitrogen analyzed. The application of thin-layer chromatographic and electrophoretic methods in the study of ^{13}N-labeled products of nitrogen fixation have been described by Wolk et al. (1976) and Meeks et al. (1977).

Ion-Exchange Chromatography

The separation of amino acids by ion-exchange chromatography is based principally on charge characteristics, although size and hydrophobicity play lesser but nontheless important roles in the process. Since the original description of ion-exchange chromatographic resolution of amino acid mixtures (Moore and Stein, 1951), the method has become the standard on which most systems of automated amino acid analysis is based. Reference works by Spackman et al. (1958), Jacobs (1966), Spackman (1967), and Hare (1977) provide detail sufficient for application to nitrogen-fixation studies.

High-Performance Liquid Chromatography

A discussion of application of high-performance liquid chromatography (HPLC) to analyses of protein, together with several reference works, is presented in the introduction to Section V. Whereas these methods have been developed primarily for use with protein hydrolysates, they are readily adaptable for nitrogen-fixation studies that require amino acid analyses. In general, HPLC methodologies that are based on reversed-phase or ion-exchange principles may be used to resolve amino acid mixtures and to quantitate the components. In reversed-phase chromatography, the stationary phase is nonpolar and the mobile phase is polar; the nonpolar or hydrophobic interactions determine the migration velocity along the stationary phase and polar solutes are eluted earlier from the column than are nonpolar solutes. This process is the opposite of that operating in most thin-layer chromatographic systems in which the stationary phase is polar.

A method described by Janes and Gilligan (1983) involves precolumn derivitization of the amino acids with *o*-phthaldialdehyde before HPLC reversed-phase chromatography is accomplished using a methanol gradient. The derivitized amino acids are fluorometrically detectable in the column effluent. The method offers a sensitivity limit below 100 fmol, a reproducibility of 1.5%, and the ability to resolve complex peptide hydrolysates or amino acid mixtures in less than 18 min. Thus these methods are attractive because of their high resolution, sensitivity, and speed. However, because most of these methods focus on the analysis of nanomole and subnanomole quantities, they are unsuitable for stable isotope analysis. However, preparative apparatus with effluent-splitting and fraction-collecting capability are available and easily adapted for these purposes.

E. Ureides

Certain legumes export almost all of the fixed nitrogen from their nodules as ureides; for example, 70 to 90% of the fixed nitrogen from soybean nodules is transported as allantoin and allantoic acid (for reviews see Reynolds et al., 1982; Robertson and Farnden, 1980; and Emerich et al., 1983). Since the use of ureide analysis as a means of estimating biological nitrogen fixation potential is covered in Chapter 11, only cursory consideration of two methods are made below.

Reynolds et al. (1982) use the following extraction procedure for the assay of ureide concentration in plant stems and other parts. About 1 g of the plant material is ground under liquid N_2 and then resuspended into 10 volumes of hot 50 m*M* potassium phosphate buffer, pH 7.5, using a Ten-Brock homogenizer. The suspension is then centrifuged at 15,000 × *g* for 15 min and the supernatant fluid is used in the determination of allantoin and allantoic acid by the method of Young and Conway (1942).

An alternative procedure developed by Herridge (1977) involves preparation of ethanolic extracts of fresh plant tissues which are then subjected to thin-layer chromatography using a butanol-acetic acid-water solvent. The ureide spots are developed with Ehrlich's reagent (4-dimethylaminobenzoaldehyde; Aldrich Chemical Co., Milwaukee, Wisconsin) and the yellow color estimated with an integrating scanning densitometer (e.g., Chromoscan, Joyce Loebl, Middlesex, England). Standards run on the same chromatogram are used to generate a calibration curve for the densitometer scans; estimation of allantoin and allantoic acid is possible in the range 0 to 1.0 μg (see also Herridge and Pate, 1977; this procedure is detailed by Bergersen, 1980).

V. MEASUREMENT OF PROTEIN

Because biological activity often parallels protein concentration in a living system or cell-free extract, the determination of protein is fundamental to establishing meaningful relative rates of many physiological processes. Biological nitrogen-fixation rates are often related to the amount of protein as a means for comparison with other systems; furthermore, the proportion of fixed nitrogen incorporated into protein (relative to nonprotein nitrogen) is frequently of interest. Below are described several standard methods for estimation of proteins in tissues, bacterial suspensions, and cell-free extracts, together with the principles on which they are based as well as their potential artifacts and limitations.

Methods involving purification and characterization of specific proteins will not be discussed here. The reader is instead referred to standard works on ion-exchange chromatography of proteins (e.g., Fasold et al., 1961; Kirkland, 1971; Jakoby, 1971), gel filtration chromatography (e.g., Determann, 1968; Kirkland, 1971; Jakoby, 1971), preparative techniques of electrophoresis (e.g., Jakoby, 1971, 1984), and affinity chromatography (Jakoby and Wilcheck, 1974; Jakoby, 1984). The 1984 *Methods in Enzymology*, which deals with the purification of enzymes (Jakoby, 1984, Vol. 104C), and the recent *Modern Methods in Protein Chemistry* (Tschesche, 1983), are particularly useful in that they present a collection of reviewed methodologies, including two-dimensional electrophoresis, a number of other electrophoretic techniques, isoelectric focusing, immunoelectrophoretic methods, and high-performance liquid chromatographic (HPLC) methods applicable to assay and study of proteins and peptides.

In the past several years it has become apparent that HPLC is revolutionizing the analysis, characterization, and isolation of peptides and proteins. Many different column packings and solvent systems have been utilized, but the literature has been dominated by reversed-phase HPLC; these methodologies are recently reviewed by Hearn (1982, 1984). Furthermore, HPLC versions of size exclusion (Unger, 1984) and ion exchange (Regnier, 1984) -chromatographic methods are now attracting more attention. Coverage of these applications of HPLC is not possible here; however, those interested will find detailed descriptions in the references supplied in this paragraph. In addition to the two volumes metioned in the preceding paragraph (Tschesche, 1983; Jakoby, 1984), application of HPLC to protein and peptide chemistry has been the subject of recent symposia, the proceedings of which include details of many such applications (Hearn et al., 1983; Lottspeich et al., 1981).

A. Protein Nitrogen

The trichloroacetic acid (TCA) and tungstic acid methods (described in Section IV.A) are efficient means of precipitating protein so that nonprotein nitrogen can be separated as a supernatant fluid. Precipitation of soluble proteins with 0.1 *N* H_2SO_4 and 1 to 2% sodium tungstate may also be used. Total nitrogen content of the precipitated protein can then be estimated after Kjeldahl digestion by one of the several means discussed above.

B. Colorimetric Protein Assays

Three popular colorimetric assays for protein are discussed below.

1. The Bradford dye-binding assay (Bradford, 1976) is based on the binding of Coomassie Brilliant Blue G-250 to protein, which is accompanied by a shift in absorption maximum from 465 to 595 nm. This method combines sensitivity (as low as 1 μg in the microassay version) with simplicity and freedom from interference by a number of common components of biological preparations.
2. The assay of Lowry et al. (1951) is based on the Folin-Ciocalteau phenol reaction with the aromatic amino acids and thus varies with the content of these amino acids in a particular protein. It is, however, a widely used and highly sensitive method (1 to 200 μg of protein).
3. The biuret assay (Goa, 1953) is based on a characteristic reaction between copper sulfate in strongly alkaline solution and a minimally requisite tripeptide which results in a violet color. Although it is less sensitive (0.01 to 2 mg of protein) than preceding two assays, it is less prone to interference than the Lowry method, offers broad-spectrum consistency among proteins (not a characteristic of the Bradford or the Lowry methods), and is useful in determination of whole-cell protein for bacteria or for protein content of plant material.

Bradford Method

The practical usefulness of Coomassie Brilliant Blue G-250 dye binding as a protein assay was first demonstrated by Bradford (1976). Dye-albumin complexes exhibit a constant extinction coefficient over a 10-fold range of albumin concentration (Spector, 1978); thus Beer's law is applicable, and accurate protein determinations can be made over a broader range, although the response may not be linear. The Bradford dye-binding assay offers reduced chemical interference relative to other colorimetric procedures such as the Lowry or the biuret method.

Both the Lowry and the dye-binding methods show considerable variation from protein to protein, whereas the biuret and Kjeldahl methods show broad-spectrum consistency. The primary attraction of the dye-binding method lies in the simplicity and rapidity with which is can be performed; assays can be read within 5 min following addition of only one reagent.

Assay kits based on the Bradford dye-binding principle can be purchased from commercial sources (e.g., Bio-Rad Laboratories, Richmond, California). The primary advantage is one of convenience: in most instances, the protein reagent need only be diluted and filtered; however, we find the original Bradford formula and procedure adequately facile. However, the detailed assay protocols and the updated listings of the effect of interfering substances that are included with these kits may be useful to some workers.

Protein reagent: Dissolve 100 mg of Coomassie Brilliant Blue G-250 (Sigma Chemical Co., St. Louis, Missouri) in 50 ml of 95% ethanol. Add 100 ml of 85% (w/v) phosphoric acid; add distilled water to bring the resulting solution to 1 liter.

Procedure: Pipette protein solution (up to 0.1 ml) containing 10 to 100 μg of protein into 12 mm × 100 mm test tubes; adjust the volume 10 0.1 ml with the appropriate buffer. Protein standards (0.1 to 1.0 mg/ml bovine serum albumin) are prepared using the same buffer and 0.1 ml of each is added to a separate test tube. Add 5 ml of the protein reagent and mix by vortexing or gentle inversion (avoid foaming). Read the absorbance at 595 nm after 2 min and before 1 hr in 3-ml cuvettes against a blank containing 0.1 ml of the buffer and 5 ml of protein reagent. A standard curve is generated by plotting the weight of the standard protein against the absorbance. The unknown protein concentrations can be derived by reference to this curve.

Procedure for the Bradford microassay: Protein solution containing 1 to 10 μg in a volume of up to 0.1 ml is pipetted into 12 mm × 100 mm test tubes. The final sample volume adjusted to 0.1 ml with the appropriate buffer and 1 ml of protein reagent added and mixed as above. Protein standards are prepared and treated as above. Absorbance is read at 595 nm as in the standard method except that 1-ml cuvettes are used.

Comments: The protein standards for the calibration curve should be treated in a manner identical to that of the unknown samples. That is, if the unknown samples are to be digested by the NaOH procedures described in the section "Digestion of Whole Bacterial Cells" below, the standards for the calibration curve should be treated likewise.

The protein-dye complex in solution has a tendency to bind to quartz cuvettes resulting in a blue-colored cuvette; the error introduced is small (about 1%) and the binding is not a problem with glass or plastic cuvettes. However, the color can interfere with other uses of the cuvettes. The stain may be removed by soaking the cuvettes in 0.1 *N* HCl for a few hours or by rinsing the cuvettes with concentrated glassware detergent followed by water and then acetone.

The Lowry Method (Lowry et al., 1951)

Reagents

1. Lowry A:

Na_2CO_3	20.0 g
NaOH	4.0 g
NaK tartrate	0.2 g

   Use distilled water to bring to a total volume of 1 liter.
2. Lowry B:

$CuSO_4 \cdot 5H_2O$	5.0 g

   Use distilled water to bring to a total volume of 1 liter.
3. Lowry C (prepared fresh daily):

Lowry A	50 parts
Lowry B	1 part
4. Folin-Ciocalteau phenol reagent. This reagent can be purchased from several commercial sources (Sigma Chemical Co., St. Louis, Missouri; Fisher Scientific Co., New York) in concentrated (2 *N*) form and must be diluted to 1 *N*.

Procedure: The samples should contain between 10 and 200 μg of protein in 1 ml. Add 5 ml of reagent 3 and incubate at room temperature for at least 10 min. Add and mix 0.5 ml of reagent 4; let stand for at least 30 min. Read absorbance at 660 nm against a reagent blank containing no added protein. The assay can be adapted to measure 1 to 100 μg of protein in about 0.2 ml. Use 1.0 ml of reagent 3 and 0.2 ml of reagent 4. Read the absorbance at 750 nm for samples below 25 μg of protein and 500 nm for higher protein concentrations. Standards are usually prepared from Cohn crystalline fraction V of bovine serum albumin (BSA). However, the Lowry and the Bradford Assays are more valid if calibrated with weighed amounts of the protein being measured because different proteins will not be equivalent in color production.

Proteins precipitated by the tungstate method, as described in Section IV.A will usually dissolve easily in reagent 3, but TCA- or HCl-precipitated proteins may require heating to facilitate dissolving; in these cases, the standards should be treated in the same manner.

Modified Lowry Protein Assay for Dilute Solutions (Cabib and Polacheck, 1984)

Solutions containing less than 25 to 50 μg/ml of protein must be concentrated before assays can be run with the standard Lowry method.

In the procedure described below, trichloroacetic acid (TCA) is used to quantitatively precipitate soluble protein in the presence of yeast-soluble RNA as a carrier. The precipitated protein is dissolved in NaOH and measured by a modification of the Lowry procedure. The method offers the further advantage of greatly reducing or eliminating interference from substances that are not precipitated with the protein.

A plot of absorbance against protein yields a straight line as opposed to the convex curve generated in this range with the direct Lowry. Some substances that ordinarily interfere with the Lowry procedure (e.g., 1% sodium dodecyl sulfate; 0.05 *M* Tris-chloride) do not affect results of this modified procedure. However, if Tris is present, it is necessary to wash the TCA precipitate once with 1 ml of cold 10% TCA; and the presence of 0.1% Triton X-100 in the sample reduces absorbance values to about 55% of that usually obtained.

Reagents

1. Yeast-soluble RNA, 5 mg/ml. Each batch should be checked and those producing a blank higher than 0.15, as measured against water, are rejected (Calbiochem-Behring product has been satisfactory).
2. Trichloroacetic acid, 100 g per 100 ml of solution.
3. NaOH, 0.1 *M*
4. Lowry reagent

Na carbonate	3% w/v in 0.1 *M* NaOH
Na tartrate	4% w/v
$CuSO_4 \cdot 5H_2O$	2% w/v

    Mix 9.6 ml of sodium carbonate-sodium hydroxide reagent with 0.2 ml each of sodium tartrate and copper sulfate.
5. Folin reagent, 2 *N* (Sigma Chemical Co., St. Louis, Missouri).
6. Bovine serum albumin, 1 mg/ml.

Procedure: Mix 1 ml of sample, containing between 5 and 25 μg of protein in a 10 mm × 75 mm Pyrex test tube (DuPont, No. 00100), with 25 μl of the RNA followed by 0.11 ml of TCA. The tube contents are mixed by vortexing and allowed to stand in ice for 45 min. The tubes are then centrifuged for 10 min at 27,000 × *g* (Sorval RC5 with SS34 rotor or equivalent) and the supernatant fluid is aspirated and discarded. The pellet is suspended in 0.5 ml of 0.1 *M* NaOH by vortexing. If complete dissolution is not apparent, the tubes are placed in a boiling water bath for 5 min and then cooled in tap water. After addition of 0.5 ml of Lowry reagent, the tubes are incubated for 10 min at 37°C. Folin reagent, 50 μl, is added and immediately vortexed. After an additional 15 min at 37°, the absorbance at 750 nm is measured. A series of standards having from 5 to 25 μg of bovine serum albumin is processed and measured at the same time.

The Biuret Method (Goa, 1953)

Biuret reaction: Tripeptides and proteins that possess the structure below give the violet color reaction of biuret with copper sulfate in alkaline solution. This absorbance at 330 nm can be read against a standard curve prepared with known concentraitons of bovine serum albumin or other suitable protein.

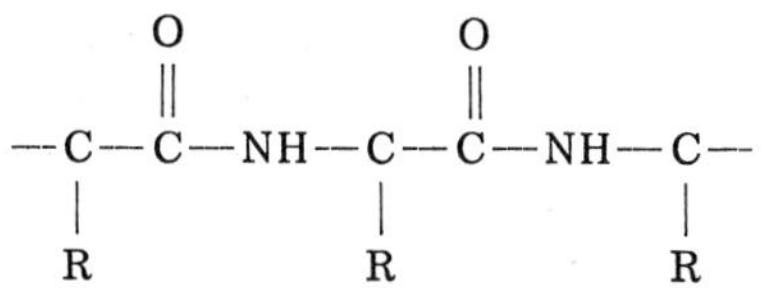

Dipeptides and amino acids (with the exception of histidine, serine, and threonine) do not give this reaction.

Reagents

1. NaOH, 3% w/v.
2. Trichloroacetic acid (TCA), 11% or 26% w/v.
3. Benedict's reagent. Dissolve 100 g of Na_2CO_3 and 173 g of sodium citrate in about 500 ml of distilled water; heat gently and stir; do not boil. Filter the solution while warm. Separately, dissolve 17.3 g of $CuSO_4 \cdot H_2O$ in about 100 ml of distilled water. Mix the two solutions in a 1-liter volumetric flask, cool, and add distilled water to 1 liter. Store in well-sealed brown glass bottles.

Procedure A (0.1 to 2 mg of protein): Centrifuge (5000 rpm, 15 min) the bacterial suspension in a glass centrifuge tube; resuspend the pellet in 1 ml of water. Add 4 ml of 11% TCA and incubate 10 min at room temperature. Centrifuge again and discard the supernatant fluid. Repeat the TCA wash steps at least once to remove nonprotein nitrogenous compounds that react with Benedict's solution. Disperse the pelleted cells in 4 ml of 3% NaOH, warming if necessary until clear. Add 0.2 ml of Benedict's solution; after 15 min, read the absorbance at 330 nm against a reagent blank.

Procedure B (10 to 400 μg of protein). Suspend the centrifuged bacteria in 0.2 ml of distilled water and add 0.1 ml of 26% TCA. Then proceed as in procedure A but use these smaller volumes of reagents. Finally, disperse the cells in 0.5 ml of 3% NaOH and add 25 μl of Benedict's reagent. Record the absorbance at 330 nm using microcuvettes.

For both procedures, protein is estimated in the sample by comparison of absorbance of samples with those from a standard curve; protein standards (usually, bovine serum albumin at appropriate concentrations) should be treated in the same manner as the sample proteins. Precipitates that form in the reaction tubes may interfere with absorbance determinations; these can be removed by centrifugation without affecting the color production.

Digestion of Whole Bacterial Cells

The digestion of whole bacterial cells in heated NaOH solutions prior to assay for protein by the biuret (Goa, 1953) or the Lowry (Lowry et al., 1951) methods is a routine procedure in many laboratories. These procedures are typified by that of Stickland (1951) for use with a biuret reaction. More recently, the Stickland digestion and modifications of it have been employed prior to estimation of whole-cell protein with the Bradford (1976) dye-binding assay.

The Stickland digestion requires treatment at 100° for 5 min in the presence of 3% NaOH. Such treatment lyses whole bacterial cells and allows reagents of the protein assay access to partially hydrolyzed protein. This method presents little problem for assays conducted by the biuret or the Lowry methods. Because these methods detect, respectively, peptides of three or more amino acids and the aromatic amino acids, they are not noticeably affected by digestion to rather short polypeptides fragments. The dye-binding assays, however, demand polypeptide chains of greater length for efficient binding of the Coomassie Brilliant Blue dye.

Thus we have found that such digestion, if not conducted under rather stringently defined conditions, can result in large errors in the estimation of protein concentration of whole cells by the Bradford method; the biuret and Lowry methods are not significantly affected. Heating suspensions of *Rhizobium* for 5 min at 100°C in 3% NaOH was optimal; longer or shorter treatment periods resulted in an apparent decrease in whole-cell protein. Standard curves based on protein standards that have been subjected to the same digestion process as the whole-cell suspensions must be utilized. A linear response of the dye-binding assay is not to be expected, but accurate determinations can be made from such curves.

REFERENCES

Aprison, M. H., Magee, W. E., and Burris, R. H. (1954). *J. Biol. Chem. 208*: 29.

Bergamin, H. F., Reis, B. F., Jacintho, A. O., and Zagatto, E. A. G. (1980). *Anal. Chim. Acta. 117*: 81.

Bergersen, F. J. (1980). In *Methods for Evaluating Biological Nitrogen Fixation*, F. J. Bergersen, ed. John Wiley & Sons, Inc., New York, pp. 65-110.

Bradford, M. M. (1976). *Anal. Biochem. 72*: 248.

Bremner, J. M. (1965). In *Methods of Soil Analysis*, C. A. Black, ed. American Society of Agronomy, Madison, Wis., Part 2, Chaps. 83-86, pp. 1149, 1179, 1238, 1256.

Bremner, J. M., and Edwards, A. P. (1965). *Soil Sci. Soc. Am. Proc. 29*: 504.

Bremner, J. M., and Tabatabai, M. A. (1972). *Commun. Soil Sci. Plant Anal. 3*: 159.

Brenner, M., and Niederwieser, A. (1967). *Methods Enzymol. 11*: 39.

Bulen, W. A., Burns, R. C., and LeComte, J. R. (1965). *Proc. Natl. Acad. Sci. USA 53*: 532.

Burris, R. H. (1972). *Methods Enzymol. 24B*: 415.

Cabib, E., and Polacheck, I. (1984). *Methods Enzymol. 104C*: 415.

Chaney, A. L., and Marback, E. P. (1962). *Clin. Chem. 8*: 130.

Conway, E. J. (1950). *Microdiffusion Analysis and Volumetric Error*, 3rd ed. Crosby, Lockwood, and Sons, London.

Determann, H. (1968). *Gel Chromatography*. Springer-Verlag, New York.

Dilworth, M. J., Subramanian, D., Munson, T. O., and Burris, R. H. (1965). *Biochim. Biophys. Acta 99*: 486.

Dreyer, W. J., and Bynum, E. (1967). *Methods Enzymol. 11*: 32.

Eady, R. R., Smith, B. E., Cook, K. A., and Postgate, J. R. (1972). *Biochem. J. 128*: 655.

Emerich, D. W., Lepo, J. E., and Evans, H. J. (1983). In *Nitrogen Fixation*, Vol. 3, W. J. Broughton, ed. Oxford University Press, Oxford, pp. 213-244.

Fasold, H., Crundlack, G., and Turba, F. (1961). In *Chromatography*, E. Hettmann, ed. Van Nostrand Reinhold Company, New York; Chapman & Hall Ltd., London, p. 378.

Goa, J. (1953). *Scand. J. Clin. Lab. Invest. 5*: 218.

Hanus, F. J., Carter, K. R., and Evans, H. J. (1978). *Methods Enzymol. 64*: 731.

Hare, P. E. (1977). *Methods Enzymol. 47*: 3.

Hearn, M. T. W. (1982). *Adv. Chromatogr. 20*: 1.

Hearn. M. T. W. (1984). *Methods Enzymol. 104*: 190.

Hearn, M. T. W., Regnier, F. E., and Wehr, C. T., eds. (1983). *High-Performance Liquid Chromatography of Proteins and Peptides*, Academic Press, Inc., New York.

Hellmann, J. (1956). In *Modern Methods of Plant Analysis*, Vol. 1, K. Paech and M. V. Tracey, eds. Springer-Verlag, Berlin, p. 126.

Helriegel, H., and Wilfarth, H. (1888). *Untersuchungen über die Stickstoffnarhrung der Gramineen und Leguminosen*. Beilageheft Ztschr. Ver Rubenzucker-Industrie Deutsch. Reichs.

Herridge, D. F. (1977). Carbon and nitrogen of two annual legumes, Ph.D. thesis, University of Western Australia, p. 16.

Herridge, D. F., and Pate, J. S. (1977). *Plant Physiol. 60*: 759.

Hiller, A., Plazin, J., and Van Slyke, D. D. (1948). *J. Biol. Chem. 176*: 1401.
Humphries, E. C. (1956). In *Modern Methods of Plant Analysis*, Vol. 1, K. Paech and M. V. Tracey, eds. Springer-Verlag, Berlin, p. 479.
Jacobs, S. (1966). In *Methods of Biochemical Analysis*, Vol. 14, D. Glick, eds. Wiley-Interscience, New York, p. 177.
Jakoby, W. B., ed. (1971). *Methods Enzymol. 22.*
Jakoby, W. B., ed. (1984). *Methods Enzymol. 104C.*
Jakoby, W. B., and Wilcheck, M., eds. (1974). *Methods Enzymol. 34.*
Jones, B. N., and Gilligan, J. P. (1983). *J. Chromatogr. 266*: 371.
Kalb, V. G., Donohue, J. T., Corrigan, M. G., and Bernlohr, R. W. (1978). *Anal. Biochem. 90*: 47.
Kennedy, I. R. (1965). *Anal. Biochem. 11*: 105.
Kirkland, J. J., ed. (1971). *Modern Practice of Liquid Chromatography*, Wiley-Interscience, New York.
Koch, B., Evans, H. J., and Russell, S. (1967). *Plant Physiol. 42*: 466.
Krug, F. J., Ruzicka, J., and Hansen, E. H. (1979). *Analyst 104*: 47.
Levitzki, A. (1970). *Anal. Biochem. 33*: 335.
Lottspeich, F., Henschen, A., and Hupe, K.-P. (1981). *High-Performance Liquid Chromatography in Protein and Peptide Chemistry*. Walter de Gruyter & Company, Mouton Publishers, West Berlin, 388pp.
Lowry, O., Rosebrough, N., Farr, A., and Randall, R. (1951). *J. Biol. Chem. 64*: 265.
Macek, K., Hais, I. M., Kopecký, J., Schwarz, V., Gasparič, J., Churáček, J. (1976). *J. Chromatog. 5* (Suppl.): 744 pp.
Martin, A. E., and Skyring, G. W. (1962). *Commonw. Br. Pastures Field Crops Technical Bulletin 46.*
Maryan, P. S., and Vorley, W. T. (1979). *Lab. Pract. 28*: 251.
McKenzie, H. A., and Wallace, H. S. (1953). *Aust. J. Chem. 7*: 55.
Meeks, J. C., Wolk, C. P., Thomas, J., Lockau, W., Shaffer, P. W., Austin, S. M. Chien, W.-S., and Galonsky, A. (1977). *J. Biol. Chem. 252*: 7894.
Meyerhoff, M. E., and Fracticelli, Y. M. (1981). *Anal. Lett. 14*: 415.
Moore, S., and Stein, W. H. (1948). *J. Biol. Chem. 176*: 367.
Moore, S., and Stein, W. H. (1951). *J. Biol. Chem. 192*: 663.
Mortensen, L. E. (1961). *Anal. Biochem. 2*: 216.
Nelson, D. W., and Sommers, L. E. (1973). *Agron. J. 65*: 109.
Pace, G. M., MacKown, C. T., and Volk, R. J. (1982). *Plant Phsiol. 69*: 32.
Parker, C. A. (1961). *Aust. J. Exp. Biol. 39*: 515.

Pataki, G. (1969). *Techniques of Thin Layer Chromatography in Amino Acid and Peptide Chemistry*, 2nd ed. Ann Arbor Science Publishers, Ann Arbor, Mich.

Regnier, F. E. (1984). *Methods Enzymol. 104*: 107.

Reynolds, P. H. S., Blevins, D. G., Boland, M. J., Schubert, K. R., and Randall, D. D. (1982). *Physiol. Plant. 55*: 255.

Robertson, J. G., and Farnden, K. J. F. (1980). In *Biochemistry of Plants*, Vol. 5, P. K. Stumpf and E. E. Conn, eds. Academic Press, Inc., New York, p. 65.

Ruzicka, J., and Hansen, E. H. (1975). *Anal. Chim. Acta 78*: 145.

Smith, I. (1969). *Chromotographic and Electrophoretic Techniques*, Vol. 1 and 2, 3rd ed. Heinemann, London.

Spackman, D. H. (1967). *Methods Enzymol. 11*: 3.

Spackman, D. H., Stein, W. H., and Moore, S. (1958). *Anal. Chem. 30*: 1190.

Spector, T. (1978). *Anal. Biochem. 83*: 142.

Stahl, E., and Ashwood, P. (1969). *Thin Layer Chromatography*, 2nd ed. Springer-Verlag, Berlin.

Stickland, L. J. (1951). *J. Gen. Microbiol. 5*: 698.

Tschesche, H., ed. (1983). *Modern Methods in Protein Chemistry*, Walter de Gruyter & Company, Mouton Publishers, West Berlin.

Umbreit, W. W., and Bond, V. S. (1936). *Ind. Eng. Chem. Anal. Ed. 8*: 276.

Umbreit, W. W., Burris, R. H., and Stauffer, S. F. (1964). *Manometric Techniques*. Burgess Publishing Co., Minneapolis, Minn., pp. 266-267.

Unger, K. (1984). *Methods Enzymol. 104*: 154.

Williams, C. H., and Twine, J. R. (1967). *CSIRO Aust. Div. Plant Ind. Tech. Paper 24.*

Wilson, P. W. (1940). *The Biochemistry of Symbolic Nitrogen Fixation*. University of Wisconsin Press, Madison, Wis., 302pp.

Wolk, C. P., Thomas, J., Shaffer, P. W., Austin, S. M., and Galonsky, A. (1976). *J. Biol. Chem. 251*: 5027.

Young, E. G., and Conway, C. F. (1942). *J. Biol. Chem. 142*: 839.

Zweig, G., and Whitaker, J. R. (1971). *Paper Chromatography and Electrophoresis*, Vols. 1 and 2. Academic Press, Inc., New York.

11

Measurement of Biological Nitrogen Fixation by ^{15}N Techniques

DENNIS D. FOCHT *University of California–Riverside, Riverside, California*

MARK POTH *Forest Service, U.S. Department of Agriculture, Riverside, California*

I. INTRODUCTION

Biological nitrogen fixation (BNF) is but one reaction of the nitrogen triangle (Figure 1), albeit the most significant one in terms of the input of biologically useful nitrogen to crops and soils. The opposing modes of exchange between the atmosphere and the terrestrial or aquatic environment as inputs (BNF) or as losses (denitrification) are not readily measured because N_2 comprises 78.1% (v/v) of the atmospheric gases. The other source of plant nitrogen obtained from mineralization of the large organic nitrogen pool, is comparatively small, transient, and highly variable. Therefore, nitrogen isotopes are the only means by which one can quantitatively distinguish between nitrogen sources in biological nitrogen-fixing systems.

There are eight naturally occurring isotopes of nitrogen yet only two—aside from ^{14}N--are sufficiently long-lived to be of any use in nitrogen tracer studies. ^{13}N has a half-life of 10 min, which limits its applicability to short-term (several hours) measurements. ^{13}N is produced from a cyclotron by the $^{13}C(p,n)^{13}N$ nuclear reaction with amorphous carbon (Austin et al., 1975) or by the $^{16}O(p,\alpha)^{13}N$ nuclear reaction with H_2O (Tiedje et al., 1979). ^{13}N decay is by B^+ (1.19 MeV) and γ(0.5 MeV) emission. ^{13}N isotopes have limited usage as tracers in terms of cost, radiation exposure to researchers, and restricted proximity to a cyclotron (Sprent, 1979). Clearly, ^{13}N is inapplicable for field studies. ^{15}N, which represents 0.366% of the natural abundance of nitrogen, is a stable isotope that presents no radiation hazards, can be used in field studies, and although not cheap ($100 per gram of nitrogen), represents less capital costs than ^{13}N. The natural abundance level of ^{15}N is the factor that limits its sensitivity. According to calculations made by Tiedje (1981), the minimum

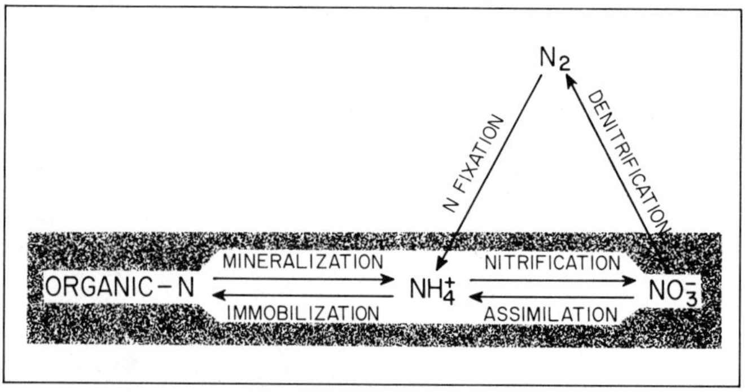

Figure 1 The nitrogen triangle (the nitrogen cycle in simplest form as redrawn from Focht and Martin, 1979) and associated branch point with organic nitrogen.

detectable amount of ^{13}N is 3×10^{-20} mol, while it is only 7×10^{-9} moles for ^{15}N with the older mass spectrometers. Newer, more sensitive mass spectrometers would extend this sensitivity about 100-fold.

II. ANALYTICAL METHODS

A. Instrumentation

^{15}N content is normally determined by converting the nitrogen in the sample to dinitrogen, measuring the ratio of mass 28 to mass 29 dinitrogen, and employing the equations shown in Section II.C to claculate atom % ^{15}N. Mass spectrometry is the oldest, most reliable, and most widely used method of determining ^{15}N. A relatively recent alternative method is ultraviolet (UV) optical emission spectroscopy (Fiedler and Proksch, 1972). Another technique holding promise for the future is infrared spectroscopy using tunable diode lasers (Whalen and Yoshinari, 1985).

When $^{14}N_2$, $^{14}N^{15}N_2$, and $^{15}N_2$ are excited by microwave radiation they emit light in direct proportion to their concentrations in the UV region at 297.7, 298.3 and 298.9 nm, respectively (Fiedler and Proksch, 1972). This phenomenon can be used to measure atom % ^{15}N. An optical emission spectrograph consists of the following: (a) a microwave generator to excite the dinitrogen to be analyzed which is contained in a quartz or specialty glass tube, (b) a prism or grating system to separate the emitted light into the desired wavelengths, and (c) a photomultiplier system to measure the light produced. With optical emission spectroscopy, it is possible to measure ^{15}N content directly in amino acids without conversion to $^{15}N_2$ (Muhammad and Kumazawa, 1975). By contrast, only very expensive and sophisticated mass spectrometers, which can measure the molecular ion of a pure compound accurately to four decimal points, and which have a computer to distinguish the isotope contribution of ^{15}N from the natural abundance of ^{2}H, ^{13}C, ^{17}O, ^{18}O, and ^{37}S heavy isotopes in the compound, can do this. The optical emission spectrograph is advantageous in requiring only a small sample (<10 μg) and allowing repeated measurements. However, as Bergersen (1980) points out, the coefficient of variation of about 3% for optical emission determinations is very much higher than mass spectrometry. Because of this, sample atom % ^{15}N values must have a greater absolute difference to distinguish statistically between samples. To conservatively differentiate a sample from natural abundance, a minimum of 0.05 atom % excess ^{15}N enrichment is suggested with less expensive mass spectrometers. By comparison, the best mass spectrometers can distinguish 0.0003 atom % excess ^{15}N.

Improvements and innovations in mass spectrometer design have been rapid, and a number of new types have become available. Presently available designs include time-of-flight instruments, ion-cyclotron resonance instruments, quadrupole magnetic designs, double and triple collector magnetic sector instruments, and tandem mass spectrometer systems (Howe et al., 1981; McLafferty, 1983; Rose and Johnstone, 1982). The reader is referred to these references for specifics of the various designs.

Once a sample has been introduced into a mass spectrometer it is ionized, typically by electron impact, via electrons produced from a heated filament under high vacuum (usually 10^{-7} torr). The charged particles produced from the ion source, which have a characteristic mass-to-charge (m/Z) ratio, are then separated. This separation is most often accomplished magnetically. Finally, as the ions strike the collector and are electrically neutralized, a voltage is produced in proportion to the abundance of the ion. The voltage is recorded for the m/z values of interest. For atom % ^{15}N, the ratio of m/z 28 to m/z 29 is determined. Single-collector instruments accomplish this by rapid switching between the measurement of the two ions of interest. Double-collector instruments measure both simultaneously. The simultaneous measurement of the two m/z species affords a more precise determination (Hauck, 1982) and such instruments are preferred for delta determinations (Shearer et al., 1974), which are defined in Section IV.A. Single-collector quadrupole instruments (Focht and Stolzy, 1978; Focht et al., 1980; St. John and Hollacher, 1976) are the least precise, yet the most sensitive and versatile for the analysis of gaseous components since separation can be achieved with an interfaced gas chromatograph (GC). Quadrupole mass spectrometers can be routinely set to monitor several different ions and are useful for rapid GC-Mass spectroscopic (MS) analysis of N_2, N_2O, and deuterium isotopes of H_2. The advent of high-performance liquid chromatography-mass spectrometer interfaces (HPLC-MS) opens new possibilities for direct ^{15}N determinations on uriedes and metabolites (Garteiz and Vestal, 1985).

Older mass spectrometers usually require sample sizes greater than 500 μg of nitrogen Some of the newer instruments are more sensitive: the commercially run Nuclide RMS 3-60 (Isotope Services, Inc., Los Alamos, New Mexico), for example, can analyze 20-μg nitrogen samples of ammonium salts. However, as sample size decreases, error introduced by background contamination increases and greater care must be taken during sample analysis.

B. Sample Preparation for Mass Spectrometry

The sample must be converted to N_2 to facilitate ^{15}N analysis. This can be done in several ways: (a) combustion of the sample in oxygen with a metallic catalyst (usually Cu) to N_2 (the Dumas procedure); (b) acid digestion of the sample to ammonium (Kjeldahl procedure) and subsequent conversion of the ammonium to N_2 (Rittenberg procedure), or (c) the ter Mullen procedure in which heat, hydrogen, and a metal catalyst are used to convert the sample to ammonium, which is collected and oxidized to N_2 (Jones, 1973). Because the Kjeldahl is the most common of the three methods used for the determination of total nitrogen in plant and soil samples, and is less expensive in terms of equipment cost, it is the preferred method and the only one that we will discuss in further detail. The reader is referred to Jones (1973) and Bremner and Mulvaney (1982) for further details of the other procedures.

Preparing Materials for Kjeldahl Digestion

The plant or soil material to be analyzed should be dried and ground to a uniform consistency so that a representative sample can be obtained. Plant samples of 200 to 300 mg and soil samples of 500 to 1000 mg with a total nitrogen content of 1 to 10 mg nitrogen should be weighed in the digestion vessel. If the recovery of ^{15}N-labeled nitrate is required, that is, if labeled nitrate is expected to represent a significant fraction of the total label in the sample, special precautions should be taken to avoid nitrate loss. To include nitrate, the sample should be pretreated for at least 20 min, within the digestion vessel, with 1 to 3 ml of a 5% (w/v) of salicylic acid in concentrated sulfuric acid: this results in the formation of phenol disulfonic acid, which couples with nitrate to form a diazo compound. Powdered sodium thiosulfate pentahydrate (approximately 0.5 g) is added and the mixture is carefully heated until fuming (Bergersen, 1980). Once the flask has cooled, the digest procedure below is performed. Alternatively, nitrate can be reduced with potassium permanganate and iron to ammonium for subsequent recovery (Bremner and Mulvaney, 1982). A simpler alternative is to use Devarda's alloy during the distillation procedure (see *Total Nitrogen Determination and Separation of* NH_3 in Section II.A).

Digest Reagents and Apparatus

The Kjeldahl procedure hydrolyzes organic nitrogen to ammonium. The semimicro procedure should contain enough dry sample to represent roughly 1 mg of nitrogen in the Kjeldahl flask. Add 1 g of Kjeldahl catalyst mixture and 3 ml of concentrated H_2SO_4. The catalyst consists of the following: 10 g of K_2SO_4, 10 g of $CuSO_4{\cdot}5H_2O$, and 1 g of Se, which has been ground up with a mortar and pestle and stored at ambient temperature in a light-shielded bottle. The flask is then heated at about 250 to 300°C for 3 to 4 hr with occasional rotation to ensure that any material on the sides of the falsk is digested. After digestion, the sample is cooled and 15 ml of distilled water is added *slowly* down the side of the flask. Some plant material tends to form resistant organic chars during digestion, especially schlerophyllous shrubs. This can be overcome by increasing the digestion time or alternatively using the modified digest method of Parkinson and Allen (1975), which incorporates hydrogen peroxide into the digest solution to aid in organic matter oxidation.

The apparatus for Kjeldahl digestion must be able to heat the sample to approximately 300°C, yet have enough heat control to remain below 350°C in order to avoid NH_3 loss and allow the heated acid to reflux down the lower third of the tube or flask. Kjeldahl digestion flasks are typically inclined at a 45° angle and must be rotated to ensure complete digestion of the char material that can form on the sides of the flask. Commercially available Kjeldahl digestion racks normally allow only six digestions to be run per apparatus, and two of these setups can effectively occupy a laboratory hood. Glass beads can be used in Erlenmeyer flasks (100 ml) to prevent bumping during initial stages of digestion (Bergersen, 1980). Flasks are closed with small glass funnels containing sealed tips once the boiling has stopped. Temperature control of the hot plate can be automated (Bergersen, 1980), but air currents in the hood may cool some flasks in outside or edge positions and inhibit complete digestion of the sample. Digestions can also be done in glass tubes set into holes in a steel or aluminum block which is heated. The commercial systems (Tecator Group, Hoganag, Sweden; Technicon Industrial systems, Tarrytown, New York) have 40-tube capacity as well as temperature programming; this allows the digestion to be started at a low temperature, automatically raised to the set digestion temperature, and shut off when finished. The commercially available systems are somewhat expensive. Variations of this design have been fabricated in the laboratory (Bergersen, 1980).

Total Nitrogen Determination and Separation of NH_3

There are two basic strategies in determining total nitrogen and atom % ^{15}N of samples from Kjeldahl digests. The first strategy involves total nitrogen determination of NH_4^+ by conversion to NH_3, followed by distillation into standardized acid and subsequent titration, of a subsample, with standardized base. The remaining NH_4^+ is then converted to N_2 by the Rittenberg procedure and analyzed for % atom ^{15}N by mass spectrometry. The alternative strategy involves the removal of a small portion of the Kjeldahl digest for either specific ion electrode or an automated colorimetric procedure for determination of NH_3, while the remainder of the digest is reserved for ^{15}N analysis. Inasmuch as this procedure requires the same distillation and conversion process as the other strategy, there is nothing to be gained in sample processing. However, analytical time is greatly reduced by substituting the titration step with an automated colorimetric analysis for NH_4^+.

The procedure used for the automated NH_4^+ determination will depend on the equipment available. The specific ion electrode procedure is described in detail by Keeney and Nelson (1982). The setup of automated systems for Kjeldahl analysis is specific to each system. Bergersen (1980) describes the use of a continuous-flow analysis system (Technicon AutoAnalyzer II), although recent advances using microbore technology allow a faster analysis rate (Alpkem, Clackamas, Oregon; Lachat, Mequon, Wisconsin).

The procedure that involves the most time is the steam distillation of NH_3. There are a number of commercially available stills which can be obtained through most general scientific supply companies. The steam generator should first be washed in dilute sulfuric acid and rinsed thoroughly to remove any traces of ammonia. It is also a good procedure to steam the entire system for 10 to 30 min before use if the apparatus has not been in immediate prior use. The end of the condenser is immersed in 5 ml of boric acid-indicator solution contained in a 50-ml Erlenmeyer flask. This solution is made by dissolving 80 g of boric acid in 3800 ml of water with heating and mixing, cooling the solution, and adding 100 ml of 95% methanol containing 0.099 g of bromcresol green and 0.066 g of methyl red. Sodium hydroxide (0.1 *N*) is carefully added as needed (to 4 liters) to achieve a reddish-purple tint (Bremner, 1965a). The Kjeldahl flask is connected to the generator and 15 ml of 12 *M* NaOH is added from the alkali reservoir of the apparatus. The reservoir is closed and the steam is turned on. It normally takes about 5 to 10 min to distill over all NH_3 in a volume of about 15 to 20 ml. It is best to standardize the time for all distillations and make any corrections necessary against a standardized solution of 1 mg of NH_4^+-N per milliliter. In cases where glass joints (usually older units) have been frozen to preclude operation of the alkali well or on units which do not have an alkali well, the alkali can be poured down the neck of the tilted digestion flask. The flask is immediately connected to the distillation apparatus (do not mix the layers). The collected distillate is titrated with 0.005 *M* H_2SO_4 using a 10-ml burette that is graduated at 0.01-ml intervals. The end-point color change is from green to pink. One milliliter of standardized acid is equivalent to 0.14 mg of nitrogen, which is present as NH_4^+.

The volume of distillate collected, regardless of whether it is titrated or not, is too great for ^{15}N analysis and so must be reduced to <2 ml. If the distillate is not titrated, it should be acidified with 2 ml of 0.04 *M* sulfuric acid before it is concentrated by heat and evaporation. Sulfuric acid has been recommended over HCl since there can be a significant loss of nitrogen due to NH_4Cl sublimation if the sample is taken to dryness (Hauck, 1982; Reeder, 1984). A hot plate is sufficient to give a reasonable concentration rate as long as the temperature does not promote vigorous boiling. Hauck (1982) describes an apparatus that can be used to pass NH_3 free air over the samples as they are concentrated. We have found that the diffusion of NH_3 into the flasks during evaporation is negligible since NH_3 diffusion into the flask cannot overcome the mass flow of water vapor out of the flask. Hence, as long as the sample is removed from the hot plate and sealed before it dries, there is no need to purge the flasks with NH_3-free air. Nevertheless, controls and blanks are recommended.

An alternative method for the separation of NH_3 from the digest is the diffusion technique (Conway, 1947). The procedure involves the addition of NaOH to the digest to give an alkaline pH within a closed chamber. The chamber can be a Conway dish or a flask (Bergersen, 1980) or any other suitable container. Included in the chamber, in a test tube or on an etched glass rod or a square of filter paper, is sufficient acid solution to absorb the NH_3. The resulting small volume of NH_3 in acid is suitable for ^{15}N analysis by mass spectrometry.

C. ^{15}N Determination by Mass Spectrometry

Procedure

The $^{15}N_4^+$ sample, which has been concentrated in a small volume (<2 ml) of acid, must be converted to $^{15}N_2$ by the Rittenberg procedure (Rittenberg, 1948) and then admitted to the mass spectrometer, with the apparatus and protocol delineated in Figure 2 for the determination of atom % ^{15}N. This is normally accomplished with alkaline NaHBrO, although LiHBrO has been reported to react more rapidly and give better yields (Porter and O'Deen, 1977).

Alkaline sodium hypobromite is prepared by dissolving 50 g of NaOH in about 400 ml of water in a 1-liter flask. Precipitate is filtered on a glass fiber filter. The flask is placed in a plastic pan of ice and the pan is set on a magnetic stir plate. A magnetic stir bar is placed into the flask and the solution mixed as 16 ml of bromine is dripped in over about a 30-min period. The temperature should not be allowed to rise above 5°C. The reagent can be stored for several weeks if tightly stoppered and refrigerated. One milliliter of this solution is enough to convert 2 mg of NH_3 to N_2.

Once the sample (about 2 ml) and the hypobromite solution (1 ml) have been put in separate arms of the Rittenberg tube (Figure 2), the air must be pumped out of the tube. The apparatus is designed to perform the following steps: (a) evacuation of the Rittenberg tube while it is isolated from the MS, (b) the reaction of the hypobromite with the sample to produce N_2 and freezing of the reacted mixture in liquid N_2 to minimize water vapor production, and (c) the admission of the sample to the mass spectrometer at a controlled rate through a liquid nitrogen trap to remove various trace impurities, such as O_2 and Br. Care should be taken to assure that the myriad of connections that make up the apparatus are secure at the very low pressures of operation. To test for leaks, set the MS to monitor m/z of 40 and use argon to check the integrity of connections.

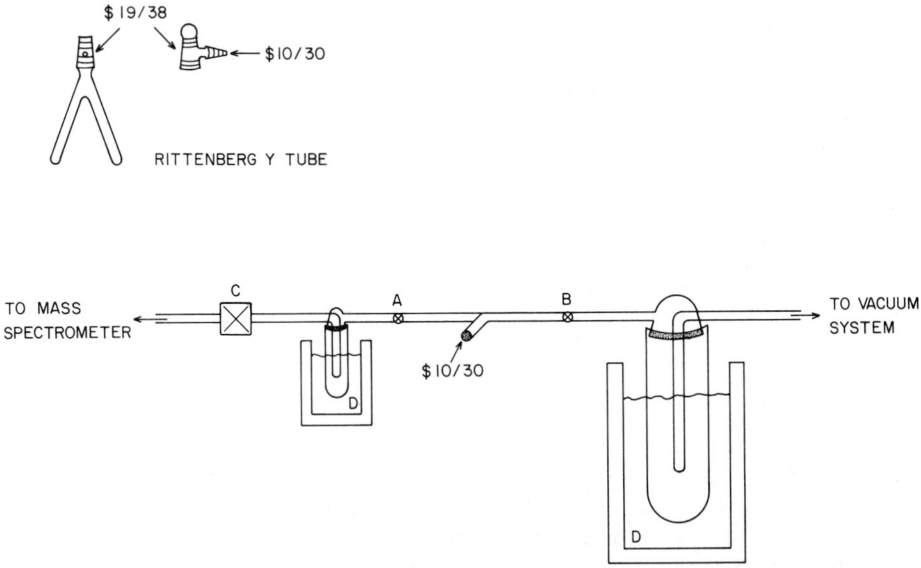

Figure 2 Apparatus and protocol for atom % ^{15}N determination by mass spectrometry. A and B are high-vacuum stopcocks with double O-rings, C is an adjustable leak valve, and D represents liquid nitrogen traps. The sample is placed in one arm of the Rittenberg Y tube, and 2 ml of hypobromite solution is placed in the other arm. The tube is capped and attached at the port between stopcocks A and B. Stopcock B is opened and the Rittenberg tube evacuated to less than 0.001 torr. The stopcock is closed, and the tube is rotated to mix the sample and the hypobromite. The mixture in the Rittenberg tube is frozen by immersing the tube in liquid N_2 while in place (generally about 30 sec). Stopcock A is opened, the N_2 enters the mass spectrometer and the ratio of m/z 28 to 29 recorded. Stopcock A is closed, the Rittenberg tube removed, and the process is repeated for the next sample.

The rate that samples can be run will be determined by how quickly the pumps used can evacuate the Rittenberg. Bergersen (1980) and Hauck (1982) recommend an oil diffusion pump backed by a mechanical pump for rates of up to 50 analyses in a typical working day. We use a high-efficiency Alcatel mechanical pump which is normally used to pump out the solid probe inlet on the Finnigan 4000 quadrupole mass spectrometer. This effectively evacuates the Rittenberg tube to less than 0.001 torr in 30 sec and enables us to run 20 analyses per hour. When running at these rates, water tends to collect in the MS inlet liquid N_2 trap. This is especially a problem if the trap at the MS inlet is made from 1/8-in. stainless steel tubing (Bremner, 1965b; Hauck, 1982). The use of a small glass trap (Figure 2) allows a visual check of ice formation which can be cleared before a sample is lost because of an ice blockage.

Calculations

Regardless of the enrichment of $^{15}N_2$ that is used for direct measurement of BNF or of $^{15}N_4$ or $^{15}NO_3^-$ for isotope dilution measurements, it is rarely necessary to determine the mass 30 isotope, so that only the ratio of 28/29 masses need be considered. The atom % ^{15}N (A) is represented by

$$A = \frac{100a}{a + b} \quad (1)$$

where a and b are the respective atoms of ^{15}N and ^{14}N which comprise the mixture. When two monoatomic species a and b form a dimer, the di-atomic species combine in the proportion of $(a + b)^2$, so that masses 30, 29, and 28 are represented by a^2, 2ab, and b^2, respectively. Thus the ratio R of 28/29 = b/2a; substituting this into equation (1) gives

$$A = \frac{100}{1 + 2R} \quad (2)$$

This equation is always valid when there is a randomly distributed isotopic composition, which is always assumed with the binomial theorem. It is unsuitable, for example, in determining the $^{15}N_2$ content of a gaseous mixture that contains two different sources not at equilibrium: All three masses would have to be measured and calculated from

$$A = \frac{\text{mass } 30 + \frac{1}{2} \text{ mass } 29}{\text{mass } 30 + \text{mass } 28 + \text{mass } 30} \times 100 \quad (3)$$

This is rarely a problem in BNF studies since the two nonequilibrium gases of N_2, if well mixed, give random incorporation of ^{15}N into the organic nitrogen pool. The problem of nonuniformity is generally confined to denitrification studies, where highly enriched $^{15}NO_3^-$ sources dimerize to give $^{15}N_2$ which are not isotopically at equilibrium with the large ambient mixture that is measured (see Hauck and Melsted, 1956).

Some investigators (Bremner, 1965b; Rennie et al., 1978; Bergersen, 1980) recommend employing a correction factor for contamination by air that may occur when the sample is introduced to the mass spectrometer. This is done by monitoring the ion current for m/z 32 at the same time that m/z 28 and 29 are measured. There are many problems with this correction factor, not the least of which is the variability in the 28/32 ratio, which is not due exclusively to N_2 and O_2. Assuming that all of the other possible gases (NO_x, CO, O_2) which could be generated during the Rittenberg conversion do not pass through the N_2 trap, the air leak on the other side of the trap causes O_2 to react with trace amounts of diffusion pump oil, which cannot be entirely removed from the ion source; this forms CO, which contributes to the m/z 28 and 29 peak. If air leakage is a problem, it should be corrected before analyses are performed.

Another potential source of error involves the use of mixed units in ^{15}N calculations. Typically, units of atom % ^{15}N have been used for ^{15}N concentrations and the total amount of nitrogen given in mass units (i.e., milligrams, grams, or kilograms) (Bremner, 1965a; Hauck and Bremner, 1976; Rennie et al., 1978; Hauck, 1982; Danso et al., 1983). This incorrectly mixes the units since atom % ^{15}N is a mol % (i.e., it is based on the number of ^{15}N atoms present, not a mass % based on the mass of the ^{15}N present) (Bremner, 1965b). Although the relationship between atom % and mass % appears linear, examination of the residuals demonstrates the nonlinear nature of the relationship (Figure 3).

Any particular study using ^{15}N could involve a series of calculations in which the isotope dilution equation is used (see Section IV.B). It could be used to measure the amounts of ^{15}N in the soil ammonium and nitrate pools by employing the standard addition technique (Hauck, 1982) as well as in calculations similar to the example. Since the error with mixed units may vary depending on the particular determination involved, the overall combined errors could either "cancel out" and be negligible or be additive and thwart attempted ^{15}N balances. To eliminate potential mixed unit errors, atom % ^{15}N should be used consistently with molar amounts of nitrogen (which is what is measured colorimetrically and by titration).

III. $^{15}N_2$ USAGE AS A DIRECT TRACER

The first successful use of ^{15}N as a tracer in BNF was demonstrated by Burris and Miller (1941) with *Azotobacter vinelandii*. Burris (1942) later found that the addition of $^{15}N_2$ to the gas headspace of closed culture vessels led to its incorporation into glutamate in a 90-min period; this provided the first definitive proof that BNF leads to the formation of amino acids. The use of $^{15}N_2$ is unquestionably the most accurate method for measuring N_2 fixaiton and should always be used for calibrating the acetylene reduction method. Unfortunately, $^{15}N_2$ experiments must be done in clsoed systems. Although small incubation chambers with intact plant-soil systems have worked satisfactorily, $^{15}N_2$ studies have been mainly useful in biochemical studies involving short-term measurements of bacterial suspension, root nodules, or excised roots.

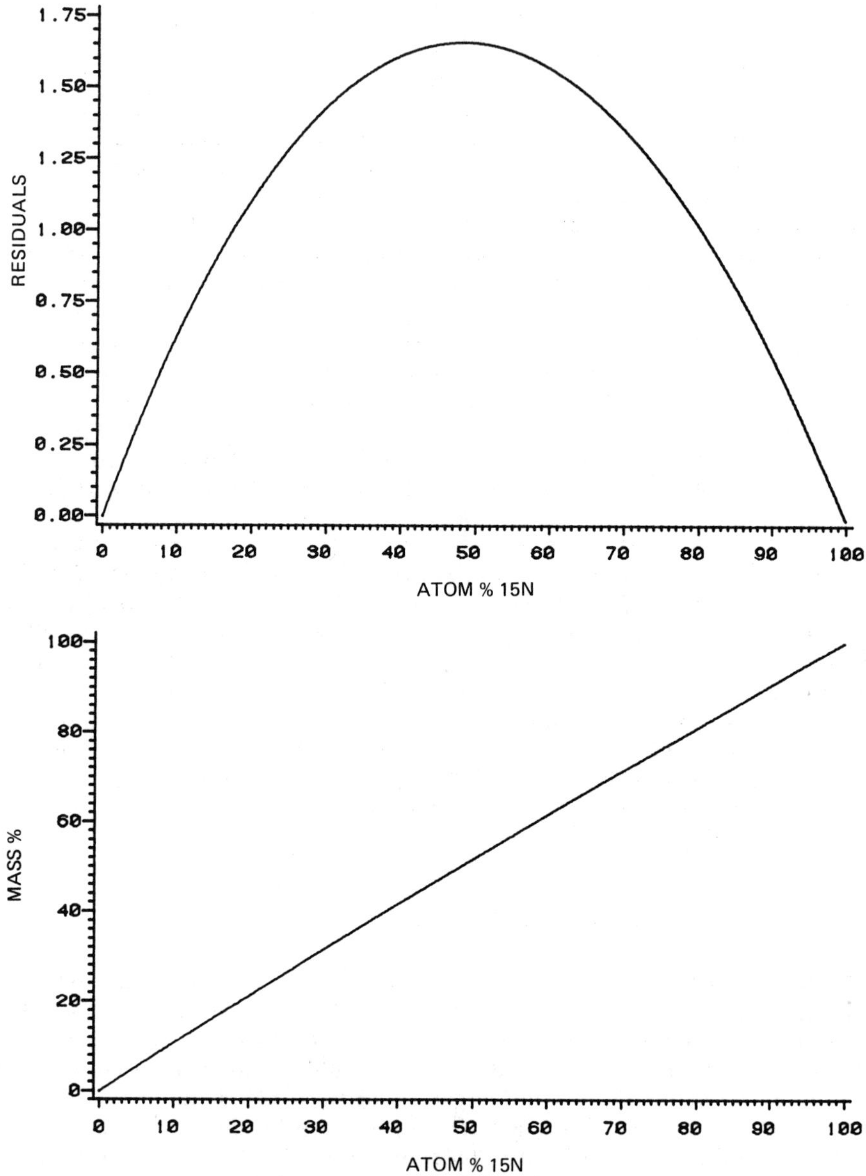

Figure 3 The lower plot shows the relationship between the atom % ^{15}N and mass % ^{15}N, while the upper residuals (mass % - atom %) plot clearly shows that the relationship is nonlinear.

A. Preparation of $^{15}N_2$

$^{15}N_2$ can be purchased from commercial suppliers in small stainless steel tanks (lecture bottles) or in break-seal glass tubes. Alternatively, $^{15}N_2$ can be made in the laboratory by conversion of $^{15}NH_3$ to $^{15}N_2$ by the Rittenberg procedure (discussed in Section II.C) or by heating over a CuO column at 600°C (Bergersen, 1980). The choice largely depends on cost versus convenience. Commercially made $^{15}N_2$ is more expensive, and the researcher has a limited choice in ordering the minimal quantity needed for the experiment. Lecture bottles of $^{15}N_2$ are used in the same manner as dealing with other gases, but break-seal glass tubes require a few manipulations involving the transfer of the gas to the desired vessel. The procedure recommended in Figure 4 is probably the easiest manner of storing gas from break-seal containers at a constant ambient pressure. This is a modification of Silvester's (1983) procedure. A solution of 0.01 *M* K_2SO_4 is used to absorb any traces of $^{15}NH_3$ and to reduce solubility of ambient N_2.

Preparation of $^{15}N_2$ by conversion from $^{15}NH_4^+$ is more tedious but is also much cheaper, not only in terms of comparison to commercially produced ^{15}N, but in terms of cheaper cost per atom ^{15}N for lower fractional enrichments. Thus if a 30% $^{15}N_2$ composition is suitable for an experiment, considerable saving in cost could be realized by purchasing the lower enrichment. The decision of cost versus convenience is one that must be decided by the investigator.

The CuO reduction procedure used by Bergersen (1980) results in complete stoichiometric conversion of NH_3 to N_2 with no discernible impurities of NO_x, unlike the Rittenberg conversion. It is, however, a slow procedure (1 to 2 hr) and involves far more elaborate construction and complexity. We have preferred the Rittenberg conversion because of its simplicity and speed. The same methodology is employed in Seciton II.C, whereby impurities are removed in a liquid N_2 trap. If large volumes of $^{15}N_2$ gas are desired (>250 ml), several tubes can be used simultaneously in place in a gas manifold. The procedure should involve at least enough NH_4^+ to produce a pressure of 1 atm in the tube as calculated from the gas volume and from an approximation of 100% conversion. For example, a tube of 63 ml total volume would contain 60 ml after allowing for the 3-ml volume of reagents: therefore, 60 ml × (mmol N_2)/(22.4 ml N_2) × (30 mg ^{15}N)/(mmol N_2 = 80 mg of nitrogen, 359 mg of $(NH_4)_2SO_4$. After the Rittenberg conversion has been performed in each tube, the tubes are connected to the gas manifold with the valves still shut. The manifold, trap, and reservoir are pumped down, and when completed, the valve to the vacuum line is turned off. The valves to the Rittenberg tubes are opened *slowly* to avoid carryover of reagent into the manifold. (Alternatively, the reagent can be frozen in liquid N_2.) When the pressure has equilibrated as noted by the manometer stability, the valve to the

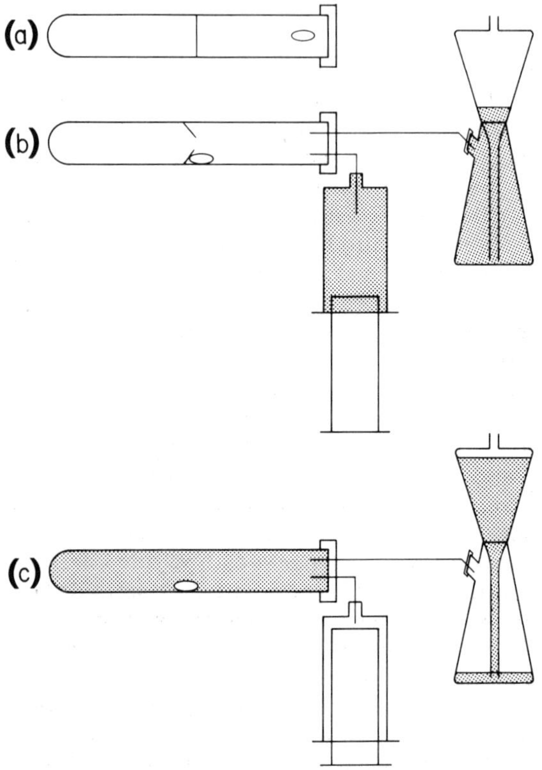

Figure 4 Procedure for transfer of gas from a break seal container to a Burris bottle (open to the atmosphere at the top and filled with water or other appropriate solution in the bottom). A magnetic stirring bar is placed in the open end of the break-seal vial, which is then capped with a serum stopper. (a) The vial is pumped out through a syringe needle injected into the serum cap. The needle is removed, and the stirring bar brought against the glass wall to break the partition and liberate the gas into the chamber. The vial is connected to the Burris bottle between syringe needles and butyl rubber high-vacuum tubing. (b) Water or other appropriate solution is injected by a syringe through another needle in the serum cap into the vial to completely displace all the gas, which is transferred to the Burris bottle. (c) The Burris bottle is disconnected from the syringe needle. Gas samples removed from the lower chamber of the Burris bottle will be displaced by an equivalent amount of water from the top and will remain at ambient atmospheric pressure.

reservoir is turned off to isolate the $^{15}N_2$ gas. If the reservoir is much larger than the gas volume in the manifold and Rittenberg tubes, the $^{15}N_2$ remaining behind is probably not worth recovering. If this is not the case, another reservoir can be attached, pumped down separately by isolating the manifold and Rittenberg tubes from the reservoir and vacuum pump, and the process repeated. The N_2 content of the reservoir can be determined directly from the manometer reading. The reservoir can be brought to 1 atm by admitting water or K_2SO_4-H_2SO_4 solution, or by a desired gas mixture (e.g., Ar, Ar-O_2).

For smaller volumes of $^{15}N_2$, as normally used for high enrichments, a Burris bottle (Figure 4) can be used in place of the reservoir. The Burris bottle, filled entirely with solution in the lower chamber and a few milliliters excess in the top chamber, is connected with the top port to the manifold assembly with the Rittenberg tube isolated by the closed valve. The manifold and the top chamber of the Burris bottle are pumped down. When the liquid rises in the top chamber so that 5 to 10 ml of gas space is created in the bottom chamber, the vacuum valve is closed, the valve between the Burris bottle and manifold port is closed, and the Burris bottle is removed from the manifold. The valve between the Rittenberg tube and the manifold is opened. The bottom port of the Burris bottle is then connected to the manifold port, and the valve is turned slowly to let the gas pass into the bottom of the Burris bottle. (Rapid introduction of gas will cause it to bubble up through the tube to the top chamber.) When the system has equilibrated, the manifold valve to the Burris bottle is closed and the Burris bottle is removed. The top chamber is returned to atmospheric pressure by removing the serum cap or by inserting a syringe needle. Water from the top chamber will flow back down again, and the $^{15}N_2$ gas in the bottom will now be at ambient pressure. If desired, another Burris bottle can be installed to scavenge the remaining $^{15}N_2$.

B. Gas Manifolds

The preparation and mixing of gas mixtures requires the use of a gas manifold complete with vacuum-tight valves and connections, vacuum pump, collection vessels, and manometer (or pressure gauge). The use of inexpensive vacuum pumps which maintain a vacuum of about 60 cum (0.08 atm) necessitates at least two flushings of the manifold. More expensive high-vacuum pumps, such as those used in back of diffusion pumps on mass spectrometers, will maintain a vacuum of 10^{-3} torr (1.3×10^{-6} atm), and do not necessitate flushing of the manifold. Because of the low costs for Pyrex glass valves, glass manifolds have been most widely used. The major difficulty, however, has been the frequent plugging of the stopcocks by the high-vacuum grease that is used. This generally necessitates a thorough disassembly and cleaning of the entire apparatus at least once a year or more often, depending on usage. Not infrequently, one or more ports may become blocked, and if the investigator is unaware of this during flushing and evacuation of all flasks on the manifold, spurious results may occur. It is thus necessary to isolate and check each port prior to every experiment. The use of stainless steel manifolds (Peterson et al., 1982) when coupled with nonlubricating high-vacuum Teflon (such as those used in Figure 2) or stainless steel valves (San Diego Valve and Fitting, San Diego, California) eliminates the problems associated with glass manifolds, albeit not without cost. Because of the greater strength and rigidity of steel, the size and gas volume can be greatly reduced, which minimizes waste of ^{15}N isotopes and facilitates more rapid evacuation and flushing of the manifold.

When anaerobic N_2 fixation is to be studied, argon is the least expensive and most widely used gas to replace N_2 and O_2. Since most commercial-grade gas mixtures contain about 0.5 to 1% air impurities, an O_2 scrubber must be placed between the gas tank and the manifold. Aqueous solutions of O_2 absorbing compounds are not recommended because of frequent problems that arise with displacement of the fluid into the evacuated manifold upon introduciton of the gas mixture. Solid alkaline pyrogallol should also be avoided since it produces CO, which not only interferes with nitrogenase activity but also with mass spectral analysis (Bergersen, 1980). Metallic catalysts of either copper filings (400°C) or manganese oxide supported by fireclay (100°C) will effectively remove oxygen (Jones, 1973). Most gas manifolds contain a heating element around a column that is packed with the copper catalyst (Bergersen, 1980; Peterson et al., 1982). Although O_2 removal can be accomplished at room temperatures with commercially available Oxysorb cylinders, more efficient removal occurs at higher temperatures; generally 130° is used. Some investigators have preferred to have an extra O_2 scrubber in reserve so that experiments will not be held up. Regeneration of the catalyst is accomplished by passing H_2 throught the cylinder at about 1 to 5 ml/min at 130°C for 1 hr. This can be done directly on the manifold or by removing the column and placing it into a gas chromatographic oven where the column would go and running H_2 gas through it. A restrictor valve on the exit port should be used to prevent mass flow of H_2 through large channels between the catalyst. When the O_2 scrubber is recharged, it should be thoroughly purged of H_2 by continuous evacuation overnight. In studies involving nitrogenase reductase, removal of H_2 is absolutely essential, and the manifold should be checked for H_2 by gas chromatography (LaFavre and Focht, 1983) prior to the start of the experiment.

C. Experimental Design

Before proceding with any $^{15}N_2$ experiment, it is necessary to have some idea of the dynamics in the system to be studied in order to determine how and how much $^{15}N_2$ should be added. First, the atmosphere regardless of its composition should have at least 30% N_2 in order to fully saturate nitrogenase, the K_m of which is 0.06 atm (Bergersen, 1980). Second, if the rate of N_2 fixation is very high, a lower enrichment of $^{15}N_2$ can be used at lower cost and greater convenience. In certain cases, it may be feasible to simply inject 2.5 to 25 ml of highly enriched (100%) $^{15}N_2$ gas with a syringe into a 250-ml flask containing ambient gas and therefore eliminate the use of gas manifold and special gas mixtures that replace N_2 with Ar. One can determine how much $^{15}N_2$ should be used by the universal mixing equation, which describes any system whereby two sources of different ^{15}N composition mix to form one as follows:

$$P_1X + P_2Y = P_3(X + Y) \tag{4a}$$

where P_1 and P_2 are the two different percentages (or fractional) enrichment of the ^{15}N sources, X and Y are the respective corresponding amounts of nitrogen (in atoms or moles) in these sources, and P_3 is the percentage (or fractional) ^{15}N content of the mixture. If we let Y represent the moles of nitrogen in the N_2 fixing system (e.g., nitrogen molar content of nodules or bacteria) and P_2 equal the ambient $^{15}N_2$ content (0.366), and use 0.376 for P_3 as the least detectable and significant enrichment above background, equation (4a) simplifies to

$$X = \frac{0.010Y}{P_1 - 0.376} \tag{4b}$$

where X is the amount of nitrogen fixed (moles) and P_1 is the % ^{15}N in the tracer. If we consider a very high N_2-fixing system such as pigeon peas (*Cajanus cajan*), an upper limit of ARA of about 18 μmol per plant per hour is observed from 200 mg dry mass of nodules (LaFavre and Focht, 1983; Hernandez and Focht, 1985). With a 3.5% nitrogen content of nodules, Y = 500 μmol of nitrogen [i.e., 200 (10^3) × 0.035 × 1/4]. Assuming 4 μmol of C_2H_4 per micromole of N_2 (2 μmol nitrogen), X = 9 μmol of nitrogen. Thus P_1 = 0.932% nitrogen, and it would be possible to use normal atmospheric gas as the diluent over the incubation time of the experiment. However, this would not be recommended if comparisons were to be made with inhibitors that would give lower rates of BNF.

With slower N_2-fixing systems, ambient N_2 must be diluted with argon. (Helium can also be used, but it is more expensive and diffuses more rapidly through selas than argon.) This is most readily accomplished by flushing out the gas manifold and the incubation vessels (containing the culture, nodules, or soil plant system) and replacing the atmosphere with a mixture of Ar, O_2, and $^{15}N_2$ at volume ratios of 5:2:3. Equaiton (4b) can be rearranged in order to decide the appropriate incubation time t necessary to observe the necessary 0.01% excess ^{15}N enrichment to give

$$t = \frac{0.01Y}{r(P_1 - 0.376)} \tag{4c}$$

where r is the rate of N_2 fixation. (The units for Y and r must be the same--either in moles or mass nitrogen.) Using the data from De-Polli et al., 1977, Table 3, row 4), Y = 2890 μg of nitrogen, P_1 = 96, and r = 8 μg of nitrogen per week, it would take 6.2 hr before any significant enrichment would be observed from 96% excess $^{15}N_2$. By comparison, the N_2-fixing example given previously with *Cajanus cajan* would require 22 sec with P_1 = 96% to obtain an excess 0.01% ^{15}N. Inasmuch as De Polli et al (1977) conducted their experiment for 29 days, whereupon the excess percent enrichment of ^{15}N in the plant exceeded 1%, they may have been able to utilize a lower enrichment of $^{15}N_2$. However, their use of the high enrichment facilitated detection of ^{15}N translocated to other parts of the plant, which had lower enrichments than the roots. Thus if one is going to invest considerable time in the design and construction of the incubation chamber and in the execution of the experiment, the extra cost in using the highest enrichment possible is generally worth the price to ensure success.

D. Sealed Plant-Soil Incubators

Long-term $^{15}N_2$ studies with intact plant-soil systems can be performed, but constant monitoring of the gaseous headspace is essential. Slight changes in atom % ^{15}N may occur from uptake, mixing with ambient N_2 originally present in the soil and plant, or from diffusion of ambient N_2 through chamber seals. The greatest problems, however, are in maintaining the ambient levels of CO_2 and O_2. Stability of CO_2 concentrations during light and dark periods can be controlled by an automatic CO_2 sensor that activates a valve to a CO_2 tank or to a soda-lime column as the gaseous mixture is continually circulated by a pump (Eskew et al., 1981). The increase in O_2 partial pressure as a result of photosynthesis is manually controlled by venting the overpressure in the chamber when it reaches 1.2 atm, and the atmosphere is restored to its original content by closing the vent valve and opening a valve to a 5-liter reservoir containing an argon-$^{15}N_2$ mixture (4:1) until the entire gaseous composition is equilibrated with a circulating pump (Eskew et al., 1981).

If carefully monitored, the sealed plant-soil incubation systems can be very useful in providing information on the dynamics of BNF and the subsequent translocation of ^{15}N to shoots and leaves. This intact system is particularly necessary for asymbiotic or associative N_2 fixation since it is much slower than symbiotic N_2 fixation, the latter of which can be determined on detached nodules, excised roots, or the entire detached plant—with small legumes—in a 30 to 90-min incubation period. Studies by De-Polli et al. (1977) with *Digitaria decumbens* and *Paspalum notatum* and by Eskew et al. (1981) with rice (*Oryza sativa*) showed significant ^{15}N enrichment in roots but very little to none in shoots or leaves after 3 days. At 12 to 14 days, significant ^{15}N enrichment was observed in all parts of plants in both studies, although the below-ground portion was highest.

Incubation periods of sufficient duration (with intact soil-plant systems) must be undertaken, not only because of the slower rates of BNF and the subsequent translocation of ^{15}N amino acids to the aerial portions of the plant, but also to allow adequate time for equilibration of $^{15}N_2$ headspace gas and the ambient N_2 in the soil-plant system. When the air-filled pore content drops below 15% of the total pore space volume in soil, gases move solely by liquid diffusion (Wesseling and van Wijk, 1957; Fluhler, 1973), which is at least 10^5 times slower than diffusion through air. Letey et al. (1980) determined that N_2 produced by denitrification in soil would take 3 to 7 days, depending on the air-filled pore content, before evolving into the gaseous headspace. Thus it is not unreasonable to assume the reverse in $^{15}N_2$-fixing systems. Even though the amount of dissolved ambient N_2 is probably too small to effect significant changes in the $^{15}N_2$ content of the headspace, it may supply the major localized source of N_2 in the root system during the initial stages of incubation. Thus total estimates of BNF at the end of the incubation period may be slightly conservative, and any interpretation of a lag period for induction of nitrogenase during this period should be exercised judiciously.

IV. ^{15}N USAGE AS AN INDIRECT TRACER

Because $^{15}N_2$ usage is prohibitive in agronomic research, BNF is measured directly from plant ^{15}N enrichment by using the difference between ambient $^{15}N_2$ enrichment (0.3663 atom %) and that added to the soil. The major limitation with these methods is the uncertainty and variability of the ^{15}N enrichment of the mineral nitrogen pool from which the plant draws upon. Nevertheless, experiments on the scale of field plots can be performed at a reasonable cost.

It is worthwhile to recall a caveat raised by LaRue and Patterson (1981) in assessing agronomic studies involving BNF. ^{15}N studies on BNF, by the nature of their cost and design, invariably are conducted with far more care and attention than a farmer would denote to a large field, and therefore frequently represent the upper limit of N_2 fixation. Researchers should therefore keep in mind that their extensive technological input into a relatively small area of land may be unrealistic in evaluating BNF inputs in crop production. When considered in this light, however, ^{15}N tracers can be valuable tools for assessing comparative differences in BNF among environmental parameters and among inoculant strains. Unlike the acetylene reduction assays, which are point measurements, ^{15}N studies are integrative and therefore subjected to less temporal variability.

A. $\delta\, ^{15}N$

It has been observed since the original study of Cheng et al. (1964) that most soils are naturally enriched in ^{15}N. Although the mechanism that causes this enrichment is not clearly understood, it is general knowledge that biological fractionation of all isotopes does in fact occur because of the kinetic differences in velocities of the two different masses (i.e., the inverse-square-root proportion of mass to velocity). Thus the product will be enriched in the light isotope, while the remaining substrate is enriched in the heavy isotope. The fractionation of $^{28}N_2/^{29}N_2$ during BNF is very small: a ratio of 1.004 has generally been observed (Hoering and Ford, 1960; Delwiche and Steyn, 1970). On the other end, the fractionation of nitrate during denitrification is comparatively much larger, being on the order of 1.03 (Wellman et al., 1968; Delwiche and Steyn, 1970). Thus the comparatively small enrichment of $^{14}NH_4^+$ to soil as a result of BNF (about 0.4%) is overshadowed by the much larger enrichment of residual $^{15}NO_3^-$ (about 3%). Recent studies in fact reveal that soils under legumes have lower $\delta\, ^{15}N$ values—with some slightly negative—than soils not harboring legumes (Virginia and Delwiche, 1982; Shearer et al., 1983; Marriotti et al., 1983). If a very small correction factor is considered for dinitrogen fixation and if the excess atom % ^{15}N natural abundance of the soil is known, it is theoretically very simple to calculate BNF from equation (4a) once the ^{15}N content in the plant has been determined.

Despite the apparent straightforward nature of the $\delta\, ^{15}N$ concept, the method in reality is fraught with at least one major difficulty concerning analysis. The excess natural abundance of ^{15}N in soil is very low, and the upper limit is equal to a $\delta\, ^{15}N$ value of 17 (Cheng et al., 1964). The $\delta\, ^{15}N$ unit is expressed in parts per thousand difference from the ambient value as such:

$$15N = \frac{\text{atom \% } ^{15}N_{\text{sample}} - \text{atom \% } ^{15}N_{\text{ambient}}}{\text{atom \% } ^{15}N_{\text{ambient}}} \times 1000 \qquad (5)$$

Taking 0.3663% as the ambient value, one $\delta\, ^{15}N$ unit is equal to 0.00037 excess atom % ^{15}N, which for many years was considered to be the standard deviation for measuring atmospheric $^{15}N_2$ (Junk and Svec, 1958). Inasmuch as Bergersen (1980) and we (Seciton II.C) have considered 0.01% as a conservative figure to be used for significance in $^{15}N_2$ studies, and whereas the highest excess natural abundance ($\delta^{15}N = 17$) reported in soil is 0.005%, it is clear that $\delta^{15}N$ measurements require the use of highly stable and specialized dual collector mass spectrometers, which must be operated with great skill and care.

Since the establishment of 0.3663% ± 0.0003 ^{15}N as the value for atmospheric N_2 (Junk and Svec, 1958), technological advances in mass spectrometry have made it possible to measure the ambient value with greater precision, and accuracy. Shearer et al. (1974, 1978) measured ^{15}N from ambient ammonium sulfate with a standard deviation of about ±0.4 $\delta^{15}N$ and consistently observed the standard to be 1.47 $\delta^{15}N$ units higher than ambient. More recent investigations from Marriotti's laboratory (Marriotti, 1983, 1984; Marriotti et al., 1983) put the standard deviation at ±0.026 $\delta^{15}N$. The greater precision obtained by Marriotti is due to the use of a dual-inlet, dual-collector mass spectrometers of the VG series (VG Isoges Ltd., Middlewick, Cheshire, England). The greater precision obtained with a dual-inlet, dual-collector system becomes very obvious since the sample signals (from the 28 and 29 masses) are simultaneously recorded with the same two signals from the standard sample. Instead of two separate errors in measuring the sample and the ambient values, equation (5) simplifies to a single measurement to give

$$\delta^{15}N = \left(\frac{A_s}{A_a} - 1\right) \times 1000 \tag{6a}$$

where A_s and A_a are the respective atom ^{15}N contents of sample and ambient. The actual measurement from equation (2) is therefore

$$\delta^{15}N = \left(\frac{1 + 2R_a}{1 + 2R_s} - 1\right) \times 1000 \tag{6b}$$

where R_a and R_s are the respective ambient and sample ratios of the 28/29 masses of N_2. Since natural abundance values are small, the following approximation is used by Marriotti et al. (1983):

$$\delta^{15}N = \left(\frac{R_a}{R_s} - 1\right) \times 1000 \tag{6c}$$

A true $\delta^{15}N$ value of 17.00, for example, will give a value of 17.06, which is a negligible error.

With the advent of microchip electronics, which has made signal averaging to a background standard very stable, increased precision in determination of natural abundance ratios may very well raise some basic questions about the "true value" for ambient $^{15}N_2$ (see Marrioti et al., 1983) as well as the acceptable method for sample preparation. Fiedler and Proksch (1975) observed that identical samples prepared by the Dumas method gave consistently higher ^{15}N values than the Kjeldahl method. Whether this is due to differences in chemical isotope fractionations between the two processes, to aerial contamination by ammonia in the Kjeldahl digestion, or to residual ambient N_2 in the Rittenberg reagent is unknown.

When analytical capabilities are sufficiently accurate that differences in single $\delta^{15}N$ units can be measured, the fractionation that occurs during BNF should be taken into account. The fractionation constant is determined empirically by growing legumes axenically and measuring the $\delta^{15}N$, which is always negative (see Kohl and Shearer, 1980; Marriotti et al., 1983; Shearer et al., 1983). Fractionation effects due to differences in *Rhizobium* strains must also be considered (Bergersen et al., 1986).

A major problem with $\delta^{15}N$ methodology is the variability that occurs within depths at specific sites (Rennie et al., 1976; Broadbent et al., 1980; Karamanos and Rennie, 1980; Steele et al., 1981; Ledgard et al., 1984; Wada et al., 1984). This problem is resolved only by comparison of the ^{15}N content to a non-nitrogen-fixing plant. The methods for calculation are the same as those used in ^{15}N isotope dilution studies, which are described in the following section.

B. Isotope Dilution with $^{15}NH_4^+$ or $^{15}NO_3^-$

Inorganic ^{15}N Fertilizer

The addition of $^{15}NH_4^+$ or $^{15}NO_3^-$ to soil is a means of increasing the "$\delta^{15}N$ values" to a degree that analytical detectability presents no problem--even with optical emission mass spectrometry. If the atom % of ^{15}N of the inorganic nitrogen content remained constant, N_2 fixation could be calculated very easily from equation (4a), where atom % ^{15}N is represented by P_1 = soil, P_2 = ambient, and P_3 = plant. Unfortunately, it is not so simple because the mineral atom % ^{15}N is constantly being diluted from mineralization of the large organic nitrogen pool, which contains an ^{15}N content that is much closer to the ambient enrichment than to the fertilizer (Witty, 1983; Chalk et al., 1983).

Since legume uptake of nitrogen is represented by three sources (air, soil, ^{15}N enriched fertilizer), two of which have almost the same ^{15}N enrichment, a nonnodulating isoline is used as a comparison. In cases where nonnodulating isolines are not known, as with many of the cowpea cultivars (e.g., *Vigna, Cajanus, Arachis*), a nonleguminous plant is used instead. The amount of N_2 fixed is calculated by the differences in ^{15}N enrichment between the nitrogen-fixing and non-nitrogen-fixing plant. The underlying assumption is that the ^{15}N

content of the nonfixing plant represents the available ^{15}N content in the soil.

Rennie (1986) points out that the choice of the nonfixing control plant may be important. Comparison of the legume should always be made to a nonnodulating isoline or a nonlegume with as similar a rooting system as possible to ensure that both plants have the same access to the soil; this is particularly true where the rooting system is deep and the ^{15}N fertilizer is banded near the surface.

If we consider equation (4a) and let X = amount of nitrogen taken up from the available nitrogen pool with an atom % ^{15}N content of P_1, Y = amount of nitrogen derived from BNF with an ambient atom % nitrogen of P_2, and Z = X + Y, the total nitrogen in the plant with atom % ^{15}N of P_3, then

$$X = \frac{Z(P_3 - P_2)}{P_1 - P_2} \tag{7a}$$

Since X/Z = the fraction of nitrogen derived from the soil, that derived from BNF (f_f) = 1 - X/Z; thus

$$f_f = 1 - \frac{P_3 - P_2}{P_1 - P_2} \tag{7b}$$

This can also be written as

$$f_f = \frac{P_1 - P_3}{P_1 - P_2} \tag{7c}$$

Equation (7b) simplifies to

$$f_f = 1 - \frac{\Delta P_3}{\Delta P_1} \tag{7d}$$

where ΔP_3 is the excess atom % ^{15}N of the legume and P_1 is the excess atom % ^{15}N of the nonlegume (control), which is assumed to be the same enrichment as the available soil nitrogen. The total amount Y derived from fixation is

$$Y = Z\left(1 - \frac{\Delta P_3}{\Delta P_1}\right) \tag{7e}$$

Equation (7e) is the isotope dilution equation that has been most widely used since its introduction by McAuliffe et al. (1958), who used $^{15}N_2$. Equaiton (7c) is applicable for $\delta^{15}N$ methods, except that P_2 is divided by the isotope correction factor β. Rennie et al. (1978) noted that the failure to correct for isotope discrimination during BNF might lead to overestimates if β is significant. However, Rennie et al. (1976) found no evidence of isotope discrimination in *Vicia faba* grown in the field. As Rennie and Rennie (1983) and Marriotti et al. (1980) point out, isotope discrimination is difficult to prove in the field since it is less with N_2 fixation than with NO_3^- assimilation when the two are compared axenically.

Because ^{15}N is expensive ($40 per gram at 10% ^{15}N) it would be desirable in principle to use ^{15}N-depleted fertilizer (12 cents per gram at 0.010% ^{15}N). In his review of the literature, Knowles (1980) found no cases of this ever being reported in BNF studies. Shortly thereafter, Broadbent et al. (1982) were the first to use ^{15}N-depleted isotopes in greenhouse and field studies to measure BNF. However, they did not add depleted ^{15}N fertilizer to soil, but used instead a soil which was depleted in ^{15}N organic nitrogen as a result of a 5-year study involving ^{15}N-depleted fertilizer additions to cornfields (Broadbent and Carlton, 1978).

The question that is frequently asked is whether or not the relatively small difference between 0.366 and 0.010% ^{15}N (versus 10% ^{15}N used in BNF studies) is sufficient to detect any differences when the isotope is added at the low nitrogen rates (1 to 3 kg/ha) necessary to prevent inhibition of N_2 fixation. If we examine the data from Figure 1 of Ledgard et al. (1985b) for the 18% ^{15}N addition at 1 kg/ha, we can determine a 12.8-fold dilution of nitrogen from mineralization of soil organic matter from equation (4a) [$P_1 = 18$, $X = 1$, P_2 - 0.366, $P_3 = 0.54$ (nonlegume), Y = kg nitrogen from soil]. Using equation (4a) again and using $P_1 = 0.01$, $X = 1$, $P_2 = 0.366$, and $Y = 12.8$, $P_3 = 0.340$, which is sufficiently less than 0.366. Because of experimental errors inherent in all field studies, a difference of 0.026% excess ^{15}N would be marginal for less exacting mass spectrometers and insufficient for optical emission spectrometry. However, ^{15}N-depleted isotopes offer considerable promise for field studies of BNF when high-precision mass spectrometers are available.

A-Value Correction

Considerable attention has been given between comparisons of the difference method, the isotope dilution procedures, and the A-value correction. The difference method does not involve the use of isotopes and is based on the unproven assumption that the difference between total nitrogen in the N_2-fixing plant and the nonfixing control is due to BNF. Comparisons for a control can be made between nonnodulating isolines (Williams and Lynch, 1954; Weber, 1966; Ruschel et al., 1979), inoculated versus uninoculated (Bezdicek et al., 1978), and legume versus nonlegume (Wagner, 1954; Rizk, 1966). Although the first method is clearly superior, a synergistic effect was noted by Ruschel et al. (1979) whereby the nodulated isoline appeared to absorb more total soil nitrogen than did the nonnodulating isoline. The inoculated versus uninoculated method works only if indigenous rhizobia are absent or do not inoculate the legume. Comparison of a legume to a nonlegume is clearly the most inferior method since nitrogen partitioning and total yields cannot be assumed to be equal for the two different cultivars.

In theory and in practice, there is no difference between the A-value correction procedure (Fried and Broeshart, 1975) and the isotope dilution procedure discussed earlier if the same amount and enrichment of ^{15}N is used (Fried and Middleboe, 1977). The reason for using different amounts of nitrogen originated from the rationalization that nitrogen would be a limiting element that could restrict growth of the control plant. Hence the greater mass of legume roots would enable the exploitation of a greater volume of soil nitrogen than the control. This would cause an overestimate of BNF. Consequently, the control should be fertilized with sufficient nitrogen to ensure maximum growth, while the legume should be fertilized with ^{15}N at a low enough rate (1 to 3 kg N/ha) that would not inhibit BNF.

The A value term is essentially a fertilizer-use efficiency factor derived from earlier work of Fried and Dean (1952) and Legg and Stanford (1967). It is based on the same assumption of the isotope dilution technique that if two sources of nitrogen (soil and isotope) are present, the plant will take up each in direct proportion to that which is available, but it is also based on the untenable assumption that the fertilizer-use efficiency for the heavily fertilized control is the same as for the lightly fertilized legume. Broadbent et al. (1982) maintain that the A-value equations can be rearranged mathematically and expressed in the form of differences in unlabeled plant-derived nitrogen such that the A-value correction method reduces to nothing more than the difference method and is thus no improvement on isotope dilution methodology, although Fried and Broeshart (1983) disagree.

Wagner and Zapata (1982) postulated that available S should be in the same relative ratio as available nitrogen and thus conducted experiments using $(^{15}NH_4)_2{}^{35}SO_4$ at 20 and 100 kg N/ha with a S/N ratio of 1.14 in both cases. Uptake ratios of soils to ^{35}S were statistically the same for legume and control at the lower fertilizer rate, but at the higher rate, the ratios between control plants were quite variable and one-third to one-sixth lower than the 20-kg N/ha treatment. This casts considerable doubt over the second assumption inherent in the A-value methodology.

The A-value methodology in terms of theory, experimentation, and mathematical calculations is more complex than the isotope dilution procedure. It is also more prone to errors (Rennie and Rennie, 1983). Inasmuch as many studies over the last 10 years have failed to show that the A-value determination is superior to the isotope dilution procedure, and in fact raise serious doubts about the validity of the fertilizer-use efficiency, it seems pointless to pursue the matter further. Mathematical considerations of the A-value concept can be found in Rennie and Rennie (1983) and Fried and Broeshart (1975). A critical review on the questionable use of this procedure has recently been put forth by Chalk (1985). Thus the concern that legumes might remove more availalbe nitrogen from soil than growth-restricted nonlegumes appears unwarranted. Finally, the question of what constitutes available nitrogen is most eloquently expressed by Rennie and Rennie (1983): "There can be no better estimation of plant-available nitrogen than allowing the plant itself to sample."

Immobilized ^{15}N

Some investigators have incorporated an organic amendment along with ^{15}N into soil either immediately or a few weeks before starting the experiment. The rationale is that large amounts of nitrogen can be added which will not inhibit BNF because of immobilization. As Chalk (1985) points out, this may be true, but the enhancement of immobilization may actually cause a decrease in nitrogen availability and thus an enhancement in N_2 fixation above what would be otherwise observed. Thus the immobilization method offers no advantage over the inorganic ^{15}N isotope dilution method in terms of immediate usage.

C. Humic ^{15}N

Clearly the most desirable method for in situ determination of BNF would be to have the soil organic matter uniformly labeled with ^{15}N. The closest this comes to reality in nature is the natural abundance differences discussed earlier with its inherent analytical difficulties. Innumerable ^{15}N fertility studies directed toward aspects other than BNF have been done over the last 20 years, and there are probably ample sites and opportunities to use these soils for nitrogen-fixation studies. However, one cannot be certain if soils fertilized with ^{15}N 10 or more years ago are uniformly labeled.

Legg and Sloger (1975) added $^{15}NH_4$, sucrose, and cellulose to soil, which they then incubated for 1 year to effect microbial immobilization and subsequent "stabilization" into the organic nitrogen pool. N_2 fixation was determined by the standard isotope dilution methodology using a nonnodulating isoline of soybeans. Because the excess % ^{15}N was only slightly higher in the first two cuttings (1.040%) than in the latter harvests (0.877%) of the control plant, they assumed that most of the organic ^{15}N was uniformly labeled. Since they gave no data regarding the excess % ^{15}N of total Kjeldahl nitrogen it is difficult to verify this conclusion. Nevertheless, Broadbent and Nakashima (1965) have shown that immobilizaiton of residual $^{15}NH_4^+$ became constant after

1 year. Excesses % ^{15}N of -0.078, -0.081, and -0.092 were found in the soil Kjeldahl nitrogen, nonfixing plant shoots, and roots, respectively, using a Hanford soil that had previously been fertilized with ^{15}N depleted $(NH_4)_2SO_4$ for 5 years and then given 1 year of incubation before starting the experiment (Broadbent et al., 1982). This would indicate that uniform labeling of soil had occurred. On the other hand, a soil used by LaFavre and Focht (1983) that had previously been fertilized with $^{15}NH_4SO_4$ more than 6 years ago (Focht and Stolzy, 1978) was not uniformly labeled. The atom % excess ^{15}N of soil Kjeldahl nitrogen was 0.096, while that observed in the 26-day-old plants of *Cajanus cajan* was 0.211 ± 0.006. The excess atom % ^{15}N of available nitrogen content of this soil, as determined from a nonfixing control of California buckwheat (*Eriogonum fasiculatum*) after 11 months growth, was 0.163 ± 0.030 (Poth, unpublished data).

If there can be no assurance that the ^{15}N content of humus is uniformly labeled, it is generally good practice to employ a nonfixing control. Although LaFavre and Focht (1983) did not use a nonfixing control in ^{15}N enriched soil, BNF was so high that the final isotope composition of the plants was very close to ambient. By using ^{15}N content of the soil Kjeldahl nitrogen instead of the plant available nitrogen—as determined from the 26-day-old plants, when BNF was low--they obtained a more conservative estimate of N_2 fixation in the early stages of plant development (Figure 5). Thus the calculations of BNF with regard to ΔP_3 [equation (7d)] being represented by soil Kjeldahl nitrogen as opposed to being represented by a reference plant are important only when (a) N_2 fixation rates are low, (b) plant biomass is low, and (c) ^{15}N has been incorporated recently.

D. N_2-Fixation Inputs into Soil

Contributions of BNF to plants intercropped with legumes can be very significant. Broadbent et al. (1982) conducted a field trial in a soil that contained -0.027, -0.038, and -0.066% excess ^{15}N as a result of a 5-year fertilization of ^{15}N depleted $(NH_4)_2SO_4$ at respective rates of 90, 180, and 360 kg N/ha. When the crops were grown separately, clover (*Trifolium repens*) derived 85% of its nitrogen from BNF as determined from rye grass (*Lolium rigidum*) as a control. When the two were intercropped, 79% of the nitrogen in the grass was derived from root transfer of biologically fixed nitrogen in the clover. Since the use of a "nonfixing" control gives an erroneous measure of BNF if it is intercropped with the legume, it must be grown separately to determine the available ^{15}N content of the soil. Alternatively, a recent method is given by Ledgard et al. (1985a,b) to determine N_2 fixation in intercropping by adding three different isotope enrichments to separate pots or plots. Since the proportion of $\Delta P_3/\Delta P_1$ [equation (7d)] was constant (about 0.46) for all three ^{15}N enrichments (Ledgard et al.,

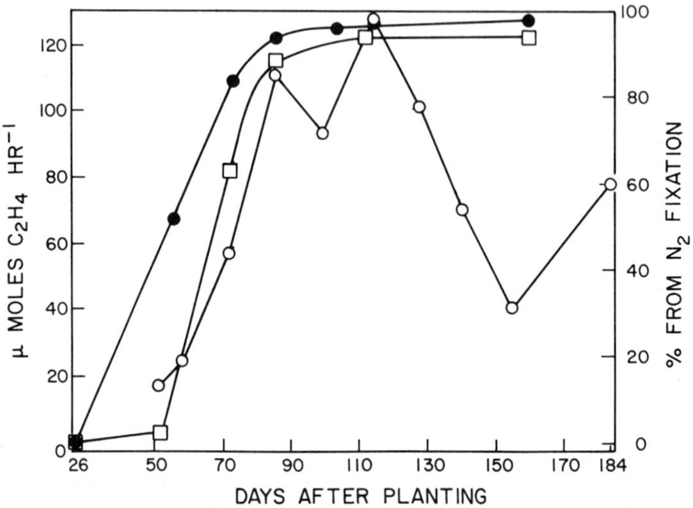

Figure 5 Calculation of BNF from *Cajanus cajan* grown in lysimeters containing ^{15}N-labeled soil from the isotope dilution equation using plant available nitrogen (closed circles) and Kjeldahl nitrogen (squares) for ^{15}N enrichment of soil. Acetylene reduction activity (open circles) was measured in situ. (Redrawn from LaFavre and Focht, 1983.)

1985b) (Figure 1), the slopes of the lines for reference and leguminous plants are different and can be used to differentiate the amount of transfer of fixed nitrogen available nitrogen, and isotope.

Another unique advantage in the use of ^{15}N enriched (or depleted) humus is in the determination of BNF inputs into soil. This can not be obtained from the inorganic isotope dilution method because the atmospheric contributions by N_2 fixation have the same isotopic composition as the soil humus. Although BNF inputs can be estimated from long term (10 years) lysimeter studies, the amount of N_2 that is fixed is frequently too small to be noticeable from the large soil organic nitrogen pool in a single year. This is illustrated in a recent study by Poth et al. (1986), who examined the ^{15}N-enriched soil used by LaFavre and Focht (1983) at the end of their lysimer N_2-fixation experiment: 26 to 32% of the soil Kjeldahl nitrogen was derived from BNF as indicated by a decrease in % ^{15}N from 0.458 ± 0.012 to 0.429 ± 0.003. Nevertheless, the total nitrogen content remained unchanged. The amount of N_2 fixed in the soil was roughly equivalent to that fixed in the crop.

It is understandable that the major focus of BNF has been centered on what goes into the crop since this is a matter of immediate concern and an event that can be quantified by isotope dilution methodology. Yet the contribution of the residual legume root system and its products has a direct bearing on the introduction of a subsequent crop, not only in terms of the surviving inoculum, but also in terms of the available nitrogen content of the soil. Surely it seems reasonable to ask the question: What occurs in the plant-soil ecosystem after the experiment has been completed? When agronomic studies on BNF are viewed on a long-term basis, the use of ^{15}N-labeled humus becomes operationally more realistic.

REFERENCES

Austin, B. M., Galonsky, A., Borting, J., and Wolk, C. P. (1975). *Nucl. Instrum. Methods Phys. Res. 126*: 373.

Bergersen, F. J. (1980). In *Methods of Evaluation Biological Nitrogen Fixation*, F. J. Bergersen, ed. John Wiley & Sons, Inc., New York, p. 65.

Bergersen, F. J., Turner, G. L., Amarger, N., Mariotti, F., and Mariotti, A. (1986). *Soil Biol. Biochem. 18*: 97.

Bezdicek, D. F., Evans, P. W., Abede, B., and Witters, R. E. (1978). *Agron. J. 70*: 865.

Bremner, J. M. (1965a). *Agronomy 9*: 1149.

Bremner, J. M. (1965b). *Agronomy 9*: 1256.

Bremner, J. M., and Mulvaney, C. S. (1982). *Agronomy 9*: 595.

Broadbent, F. E., and Carlton, A. B. (1978). In *Nitrogen in the Environment*, D. R. Nielsen and J. G. MacDonald, eds. Academic Press, Inc., New York, p. 1.

Broadbent, F. E., and Nakashima, T. (1965). *Soil Sci. Soc. Am. Proc. 29*: 55.

Broadbent, F. E., Rauschkolb, R. S., Lewis, K. A., and Chang, G. Y. (1980). *Soil Sci. Soc. Am. J. 44*: 524.

Broadbent, F. E., Nakashima, T., Chang, G. Y. (1982). *Agron. J. 74*: 625.

Burris, R. H. (1942). *J. Biol. Chem. 143*: 509.

Burris, R. H., and Miller, C. E. (1941). *Science 93*: 114.

Chalk, P. M. (1985). *Soil Biol. Biochem. 17*: 389.

Chalk, P. M., Douglas, L. A., and Buchanan, S. A. (1983). *Can. J. Microbiol. 29*: 1046.

Cheng, H. H., Bremner, J. M., and Edwards, A. P. (1964). *Science 146*: 929.

Conway, E. J. (1947). *Microdiffusion Analysis and Volumetric Error*, 2nd ed. Crosby, Lockwood Publishers, London, p. 465.

Danso, S. K. A., Bole, J. B., Zapata, F., eds. (1983). *A Guide to the Use of Nitrogen-15 and Radioisotopes in Studies of Plant Nutrition: Calculations and Interpretation of Data*, Technical

Document. International Atomic Energy Agency, Vienna, p. 64.
Delwiche, C. C., and Steyn, P. L. (1970). *Environ. Sci. Technol.* *4*: 929.
De Polli, H., Matsui, E., Dobereiner, J., and Salati, E. (1977). *Soil Biol. Biochem.* *9*: 119.
Eskew, D. L., Eaglesham, A. R. J., and App, A. A. (1981). *Plant Physiol.* *68*: 48.
Fiedler, R., and Proksch, G. (1972). *Plant Soil* *36*: 371.
Fiedler, R., and Proksch, G. (1975). *Anal. Chim. Acta* *78*: 1.
Fluhler, H. (1973). *Mitt. Schweiz. Anst. Forstl. Versuchswes.* *49*: 125.
Focht, D. D., and Martin, J. P. (1979). In *Agriculture in Semi-arid Environments*, A. E. Hall, G. H. Cannell, and H. W. Lawton, eds. Springer-Verlag, Berlin, p. 119.
Focht, D. D., and Stolzy, L. H. (1978). *Soil Sci. Soc. Am. J.* *42*: 894.
Focht, D. D., Valores, N., and Letey, J. (1980). *J. Environ. Qual.* *9*: 218.
Fried, M., and Broeshart, H. (1975). *Plant Soil* *43*: 707.
Fried, M., and Broeshart, H. (1983). *Agron. J.* *75*: 718.
Fried, M., and Dean, C. A. (1952). *Soil Sci.* *73*: 263.
Fried, M., and Middleboe, J. (1977). *Plant Soil* *47*: 713.
Garteiz, D. A., and Vestal, M. L. (1985). *Liq. Chromatog.* *3*: 334.
Hauck, R. D. (1982). *Agronomy* *9*: 735.
Hauck, R. D., and Bremner, J. M. (1976). *Adv. Agron.* *28*: 219.
Hauck, R. D., and Melsted, S. W. (1956). *Soil Sci. Soc. Am. Proc.* *19*: 871.
Hernandez, B. S., and Focht, D. D. (1985). *Agron. J.* *77*: 867.
Hoering, T. C., and Ford, H. T. (1960). *J. Am. Chem. Soc.* *82*: 376.
Howe, I., Williams, D. H., and Bowen, R. D. (1981). *Mass Spectrometry*. McGraw-Hill Book Company, New York, p. 276.
Jones, K. (1973). In *Comprehensive Inorganic Chemistry*, Vol. 2, J. C. Bailar, H. J. Emeleus, R. Nyholm, and A. F. Trotman Dickenson, eds. Pergamon Press Ltd., Oxford, p. 147.
Junk, G., and Svec, H. J. (1958). *Geochim. Cosmochim. Acta* *14*: 234.
Karamanos, R. E., and Rennie, D. A. (1980). *Can. J. Soil Sci.* *60*: 365.
Kenney, D. R., and Nelson, D. W. (1982). *Agronomy* *9*: 643.
Knowles, R. (1980). In *Methods for Evaluation Biological Nitrogen Fixaiton*, F. J. Bergersen, ed. John Wiley & Sons, Inc., New York, p. 557.
Kohl, D. H., and Shearer, D. (1980). *Plant Physiol.* *66*: 51.
LaFavre, J. S., and Focht, D. D. (1983). *Plant Physiol.* *72*: 971.
LaRue, T. A., and Patterson, T. G. (1981). *Adv. Agron.* *34*: 15.

Ledgard, S. F., Freney, J. R., and Simpson, J. R. (1984). *Aust. J. Soil Res. 22*: 255.

Ledgard, S. F., Morton, R., Freney, J. R., Bergersen, F. J., and Simpson, J. R. (1985a). *Soil Biol. Biochem. 17*: 317.

Ledgard, S. F., Simpson, J. R., Freney, J. R., Bergersen, F. J., and Morton, R. (1985b). *Soil Biol. Biochem. 17*: 323.

Legg, J. O., and Sloger, C. (1975). In *Proceedings of the Second International Conferences on Stable Isotopes*, E. R. Klein and P. D. Klein, eds. Oak Brook, Ill., p. 661.

Legg, J. O., Stanford, G. (1967). *Soil Sci. Soc. Am. J. 31*: 215.

Letey, J., Jury, W. A., Hadas, A., and Valoras, N. (1980). *J. Environ. Qual. 9*: 223.

Mariotti, A. (1983). *Nature 303*: 685.

Mariotti, A. (1984). *Nature 311*: 251.

Mariotti, A., Mariotti, R., Armager, N., Pizzelle, G., Ngambi, J. M., Champigny, L., and Moyse, A. (1980). *Physiol. Veg. 18*: 163.

Mariotti, A., Mariotti, F., and Amarger, N. (1983). In *Nuclear Techniques in Improving Pasture Management*, Proceeding of an Advisory Group Meeting. International Atomic Energy Agency, Vienna, p. 61.

McAuliffe, C., Chamblee, D. S., Uribe-Arago, H., and Woodhouse, W. W., Jr. (1958). *Agron. J. 50*: 334.

McLafferty, F. W., ed. (1983). *Tandem Mass Spectrometry*. John Wiley & Sons, Inc., New York, p. 340.

Muhammad, S., and Kumazawa, K. (1975). *Proceeding of the Second International Conference on Stable Isotopes*, E. R. Klein and P. D. Klein, eds. Argonne National Laboratory, Argonne, Ill., p. 674.

Parkinson, J. A., and Allen, S. E. (1975). *Commun. Soil Sci. Plant Anal. 6*: 1.

Peterson, J. B., Glenister, R. A., and Eskew, D. L. (1982). *Anal. Biochem. 121*: 335.

Porter, L. K., and O'Deen, W. A. (1977). *Anal. Chem. 79*: 514.

Poth, M., LaFavre, J. S., and Focht, D. D. (1986). *Soil Biol. Biochem. 18*: 125.

Rennie, R. J. (1986). *Agron. J. 78*: 158.

Rennie, R. J., and Rennie, D. A. (1983). *Can. J. Microbiol. 29*: 1022.

Rennie, D. A., Paul, E. A., and Johns, L. E. (1976). *Can. J. Soil Sci. 56*: 43.

Rennie, R. J., Rennie, D. A., and Fried, M. (1978). In *Isotopes in Biological Dinitrogen Fixation*, Proceedings of an Advisory Group Meeting. International Atomic Energy Agency, Vienna, p. 107.

Rittenberg, D. (1948). *Preparation and Measurement of Isotopic Tracers*, W. W. Wilson, A. O. C. Nier, and S. P. Reimann, eds.

Science Publishers, Ann Arbor, Mich., p. 31.

Rizk, S. G. (1966). *J. of Microbiol. U.A.R. 1*: 33.

Rose, M. E., and Johnstone, R. A. W. (1982). *Mass Spectrometry for Chemists and Biochemists*. Cambridge University Press, New York, p. 307.

Ruschel, A. P., Vose, P. B., Victoria, R. L., and Salati, E. (1979). *Plant Soil 53*: 513.

Shearer, G. R., Kohl, D. H., and Commoner, B. (1974). *Soil Sci. 118*: 308.

Shearer, G., Kohl, D. H., and Chien, S. H. (1978). *Soil Sci. Soc. Am. J. 42*: 899.

Shearer, G. R., Kohl, D. H., Virginia, R. A., Bryan, B. A., Skeeters, J. L., Nilsen, E. T., Sharifi, M. R., and Rundel, P. W. (1983). *Oecologia 56*: 365.

Silvester, W. B. (1983). In *Biological Nitrogen Fixation in Forest Ecosystems*, J. C. Gordon and C. T. Wheeler, eds. Martinus Nijhoff/Dr. W. Junk Publishers, Dordrecht, The Netherlands, p. 173.

Sprent, J. I. (1979). *The Biology of Nitrogen-Fixing Organisms*. McGraw-Hill Book Company, New York, p. 196.

St. John, R. T., and Hollacher, T. C. (1976). *J. Biol. Chem. 252*: 212.

Steele, K. W., Wilson, A. T., and Saunders, M. H. (1981). *N.Z. J. Agron. 24*: 167.

Tiedje, J. M. (1981). In *Genetic Engineering of Symbiotic Nitrogen Fixation and Conservation of Fixed Nitrogen*, J. M. Lyons, R. C. Valentine, D. A. Phillips, D. W. Rains, and R. C. Huffacker, eds. Plenum Press, New York, p. 481.

Tiedje, J. M., Firestone, R. B., Firestone, M. K., Betlach, M. R., Smith, M. S., and Caskey, W. H. (1979). *Soil Sci. Soc. Am. J. 43*: 709.

Virginia, R. A., and Delwiche, C. C. (1982). *Oecologia 54*: 317.

Wada, E., Imaizumi, R., and Takai, Y. (1984). *Geochem. J. 18*: 109.

Wagner, R. E. (1954). *Agron. J. 46*: 233.

Wagner, G. H., and Zapata, F. (1982). *Agron. J. 74*: 607.

Weber, C. R. (1966). *Agron. J. 58*: 46.

Wellman, R. P., Cook, E. D., and Krouse, H. R. (1968). *Science 161*: 269.

Wesseling, J., and van Wijk, W. R. (1957). In *Drainage of Agricultural Lands*, J. N. Luthin, ed. *Agronomy 7*: 461.

Whalen, M., and Yoshinari, T. (1985). *Nature 313*: 780.

Williams, L. F., and Lynch, P. L. (1954). *Agron. J. 46*: 28.

Witty, J. F. (1983). *Soil Biol. Biochem. 15*: 613.

12

Estimation of Bacterial Nitrogen Fixation Using Acetylene Reduction

ROBERT G. UPCHURCH* *Allied-Signal Corporation, Syracuse, New York*

I. INTRODUCTION

The reduction of dinitrogen to ammonia catalyzed by the enzyme nitrogenase may be determined directly by the measurement of the product ammonia and the products of its further assimilation as described in Chapters 8, 9, and 11, or by indirect means that are based on the ability of the enzyme to reduce chemical analog substrates of dinitrogen. In addition to the reduction of acetylene (C_2H_2), the subject of this chapter, other nonphysiological substrates, such as nitrous oxide,

Present affiliation: North Carolina State University, Raleigh, North Carolina

azide, cyanide, and isocyanide, are all reduced with similar requirements for MgATP and reductant to that for N_2 fixation. With regard to simplicity, the reduction of acetylene has the advantage in that ethylene is the single reduction product. The acetylene reduction assay had its inception in the independent observations of Dilworth (1966) and Schölhorn and Burris (1966, 1967) that C_2H_2-inhibited nitrogen fixation by extracts of *Clostridium pasteurianum*. Dilworth (1966) went on to show by means of mass spectrometry that acetylene was reduced to ethylene. The first reported use of the new assay method was that of Koch and Evans (1966) in their measurement of nitrogen fixation in soybean root nodules. Since then, the assay and the development of its applications have been the subject of rather extensive and regular review (see Burris, 1972, 1974; Stewart, 1975; Nutman, 1976; Hardy and Holsten, 1977), with the most recent being that of Turner and Gibson (1980).

During the last two decades the acetylene reduction assay has been a very significant and stimulating force in the science of nitrogen fixation because of its simplicity and numerous advantages in comparison to other methods. Both acetylene and ethylene are readily measured at low concentrations by gas chromotography using a flame-ionization detector. This method, with its relative simplicity, low equipment cost and resource investment, high sensitivity, and rapidity of use has become a standard procedure in many laboratories. The principal value of the assay is in determining the changes in nitrogenase activity as a result of experimental treatment. With an awareness of certain precautions, the acetylene reduciton assay can be utilized to study the process of nitrogen fixation at diverse levels of biological complexity, from purified enzyme protein components to natural ecosystems. In this discussion we review the basic procedural aspects of the assay and then focus attention on its limitations and problems of procedure and interpretation.

II. THE ASSAY

A. The Gas Phase

Safety precautions should be taken with the use of acetylene because it forms explosive mixtures with air. The explosive mixture limits are between 2.5 and 80% as determined by Perry (1950).

Acetylene is available as a commercial compressed gas. It may also be produced in the laboratory by the addition of water to calcium carbide (Postgate, 1972; Tjepkema and Burris, 1976). It may be necessary to scrub commercial gas to remove impurities, especially if assays are expected to show low levels of C_2H_2 reduction. Impurities commonly found in industrial C_2H_2, such as acetone, H_2, PH_3, NH_3, CO_2, and

C_2H_4, may be removed by buggling the gas first through concentrated H_2SO_4 and then through water. Drechsel bottles are useful for this process.

Acetylene may be introduced into the assay gas phase directly as pure gas or as a gas mix (e.g., 12.5 to 25% argon or helium mix). In either case, the final desired C_2H_2 concentration is usually 10% of the assay gas phase. The argon or helium mixtures can be bought commercially. For safety reasons, gas mixtures containing C_2H_2 with O_2 should not be prepared and transported under pressure. Specific gas mixtures are easily made using a vacuum mannifold, manometer, and pump system as described by Turner and Gibson (1980). If acetylene gas or gas mixtures are not to be added directly to assay vessels in the laboratory, they may be stored and/or transported in reservoirs such as the evacuated mechanical chamber described by Turner and Gibson (1980) or in a rubber football bladder (Stewart et al., 1971). For assaying aerobic systems in a defined atmosphere, industrial oxygen purity is adequate for the gas phase. The gas phase for anaerobic assay utilizes argon or helium or nitrogen. To assay nitrogenase enzyme activity, special precautions as described by Eady (1980) must be followed to remove O_2 from the gas phase. Incubation vessels are either evacuated on the manifold and filled with a gas mixture or with premixed gas (at least 20 volumes are required to reduce the O_2 concentration to 1 volume per minute).

Syringing of C_2H_2 from the reservoir or gas cylinder is accomplished through a rubber tube fitted to the gas outlet or cylinder regulator or beyond the scrubbing bottles if the gas is being scrubbed. If these gas transfers are done in the laboratory or in a confined space, the outflow from the tubing should be voided into a fume hood or through a window as a safety precaution.

After injection of the acetylene into the assay vessel, an equivalent volume of the gas phase is withdrawn to return the pressure to 1 atm. In an aerobic system this equilibration will cause a decline in pO_2 equivalent ot the pC_2H_2 added, and some aerobic systems are sensitive to this decline in pO_2 with significant consequences in the level of activity. To avoid this problem, the vessel pressure should first be reduced to about 0.2 atm and brought back to atmospheric with a prepared gas mixture containing appropriate concentrations of C_2H_2, O_2, and N_2 or Ar and the operation repeated (Turner and Gibson, 1980).

B. Preparation of the Sample

The N_2-fixing activity of many diverse biological systems ranging from purified enzyme components to field and forest soil cores have been studied with the acetylene reduction assay. Details of the assay techniques and adaptations of example systems varying in complexity are described in the references in Table 1.

Table 1 Nitrogen-Fixing Systems Analyzed by the Acetylene Reduction Assay

	System	Reference
I.	Nitrogenase enzyme components	Eady (1980)
II.	Cell-free extracts	Eady (1980), Farnden and Robertson (1980)
III.	Whole cells	Campbell and Evans (1969), Eady (1980)
IV.	*Ex planta* nitrogenase of *Bradyrhizobia*	Agarwal and Keister (1983)
V.	Bacteroids, *Rhizobium*	Evans et al. (1972); Bergersen and Turner (1975), Dipankar and Weaver (1984), Sen and Weaver (1984)
VI.	*Frankia* vesicles	Tjepkema and Winship (1980)
VII.	Detached nodules *Rhizobium*	Bergersen (1970), Dart and Schwinghamer et al. (1970), Sprent (1969), Stephens and Neyra (1983)
	Nodules of Alnus and Hippophae Actinorhizal	Akkermans (1970), Tjepkema et al. (1981)
	Other nonlegume nodules	Stewart et al. (1967), Wheeler (1971); Waughman (1977)
VIII.	Intact roots	
	Intact soybean roots	Fishbeck et al. (1973)
	Grass roots with *Azospirillum*	Dobereiner (1980)
	Rapid soybean plant screen	Wacek and Brill (1976), Main and Bond (1978)
IX.	Nondestructive in situ assay	Balandreau et al. (1978), LaRue and Kurz (1973), Mederaski and Streeter (1977), Ruegg and Alston (1978)
X.	Habitats and ecosystems:	
	Intact soil cores	Lee et al. (1977), Witty and Day (1977)
	Forest soils	Nohrstedt (1982, 1983)
	Field and grasslands	Paul (1975), Rinaudo et al. (1971), Haystead et al. (1980)
	Natural plant communities	Knowles (1980, 1981)

C. Incubation of the Sample

The essential requirement for the incubation vessel is that the container material not be porous or capable of occluding C_2H_2 or C_2H_4. Vessel dimensions and composition are determined by the size and composition of the assay material. It is essential that the volume be adequate to ensure that the gas phase does not change substantially during the assay, especially with regard to the O_2 concentraiton in aerobic systems. Vessels used may be of fixed volume, ranging in size from small vials to large jars, but many other enclosure systems have been used. Plastic bags fitted with inlet ports and sealed with tape or heat sealing have been used under some field conditions (Hardy et al., 1973; Lee and Watanabe, 1977). Balandreau et al. (1978) used metal cylinders driven into the soil and sealed with a transparent cover for in situ estimation of N_2-fixing activity. Soil cores have been incubated in syringes (Hardy et al., 1968) and bottles (Silvester and Bennett, 1973) or within soil borers (Zuberer and Silver, 1978). Other in situ systems employing plastic or glass canopies or plastic buckets have been described by Jordan et al. (1978), Paul et al. (1971), and LaRue and Kurz (1973), respectively. In most instances, a gastight seal that provides access for introducing and removing gases is provided by an implanted rubber serum stopper. For larger samples such as intact roots, Mason jars which have gastight lids that can be ported and sealed with a serum stopper can be used.

The adaptations and variations for incubation chambers are numerous and only a few examples are mentioned below. A continuous type of assay has been reported for chemostat cultures of nitrogen-fixing bacterial (LaRue and Kurz, 1970). Other continuous-type assay systems have been described for nodulated legumes (Fishbeck et al., 1973), and nodulated roots and detached nodules (Minchin et al., 1983). In these systems, a low level of C_2H_2 is introduced into the gas stream passing through the incubation system, or C_2H_2 is pulsed through the system at regular intervals (Mederski and Streeter, 1977) and the effluent is sampled at regular intervals. A significant problem with this type of assay in which the system is continuously exposed to C_2H_2 is in ensuring that the low level of acetylene does not greatly interfere with N_2 fixation. The known growth-regulating effects of C_2H_4 indicate the need for caution in interpreting the data from these experiments. Other types of systems have been developed to assay shoot respiration and nodule N_2-fixing activity simultaneously and continuously (Haystead et al., 1980; Minchin et al., 1977). Nitrogen-fixing systems may be cultured in the assay vessel itself. Modified Pankhurst tubes have been sued for screening large numbers of potential diazotrophs (Campbell and Evans, 1969; Tubb and Postgate, 1973). McCartney bottles have been used for microaerophilic culturing and assay of *Bradyrhizobia* (Gibson et al., 1976). Disposable syringes have been used successfully as incubation chambers for detached nodules, nodulated roots, soil cores, marine samples, and *Rhizobium*-plant tissue culture samples with the gas phase being collected with other syringes or injected into evacuated tubes for storage before C_2H_4 determination (Hardy et al., 1973; Stewart et al., 1971). Special chambers for root enclosure have been used for rhizosphere studies (Pan and Schmidt, 1969; Huang et al., 1975).

D. Sampling the Gas Phase

After the incubation period (usually from 1 to 4 hr) is completed, the gas phase is sampled with any of several syringe types, depending on the assay, and introduced into the gas chromatograph or stored for later C_2H_4 determination. For nodulated soybean roots incubated in 250-ml Erlenmeyer flasks or small Mason jars, we routinely remove 600 μl, allowing 10-sec equilibration and inject 500 μl. If the incubation vessel contains obstacles to the free diffusion of gas, roots, and/or soil, for example, the vessel should be shaken before sampling the gas space. Injection technique should be smooth, fast, and as standardized as possible.

Straight-sided triangular peaks fo equal base and constant retention time are the desired goal of the operator. Disposable 1-ml syringes are frequently used when many samplings are needed for an experiment and short-term storage is needed because the process rate is exceeded. The syringe needles can be stuck into rubber stoppers, and these may be placed under water and kept for several hours before analysis. Fo storage of samples for periods longer than several hours, preevacuated blood sampling tubes (e.g., Vacutainer; Becton-Dickenson and Co., Rutherford, New Jersey) are recommended. The sample is taken from the assay vessel through the replaceable double-ended sampling needle. With more critical timed assays of cell-free extracts or enzyme components where sample storage is not needed, the gas phase is usually sampled with a glass barrel, gastight syringe (e.g., Gastight No. 1705, Hamilton Co., Reno, Nevada). It is important to flush these syringes three to four times with air and then to rinse two to three times with an anaerobic buffer (degased 50 m*M* Tris, 7.5 pH, containing 1 m*M* $Na_2S_2O_4$) between samplings. This procedure will prevent both ethylene carryover that could result in a false-positive determination and O_2 inactivation of the enzyme components.

The gas chromatograph port septa must be monitored for leaking, as repeated use of 21 to 23-gauge needles can cause rapid deterioration. Septa will be far less affected when the fine needles of gastight syringes are used. Precaution must be taken when sampling by syringe or evacuated tube if the ratio of the syringe or tube volume to incubation vessel is greater than 1:20. In this case, consideration should be given to the reduced pressure within the syringe or tube. For example, samples of 250 μl reduce the gas pressure within a 5-ml gas space by 5%, and this should be accounted for in subsequent calculations, especially when a number of consecutive assays are to be made from the same gas space. If the C_2H_2 response by the gas chromatograph remains linear at the high C_2H_2 concentration used in the assay, the pressure loss due to sampling can be determined by use of the C_2H_2 peak as an internal standard as described by Turner and Gibson (1980). With nonrigid containers, such as plastic bags, the ratio volume of gas sample to volume of assay sample is usually small enough that corrections are not necessary.

In most cases, sampling the gas phase is preferable to stopping the reaction by the addition of acid or base, as this may be slow in penetrating the biological material or cause release of C_2H_4 absorbed by rubber caps. On the other hand, this is standard practice and permissible with certain types of nitrogenase enzyme assays. Recently, David et al. (1980) reported the use of ammoniacal silver nitrate to terminate acetylene reduciton. This compound terminated the reaction by quantitatively precipitating acetylene as the carbide salt and did not affect the ethylene formed. The advantage of this process was that it allowed a faster sampling rate of up to 13 samples in 10 min with no wait for C_2H_2 to be purged from the gas chromatographic column between sample injections.

The effects of leghemoglobin on the nitrogenase activity and O_2 consumption by bacteroids and bacteria have been studied in a liquid phase by Bergersen and Turner (1975). In these experiments, solutions preincubated with a gas mixture of C_2H_2, O_2, and argon to achieve the desired dissolved gas concentraiton were injected into the reaction chamber and the headspace displaced. The reaction in solution was stopped by injecting a sample from the reaction vessel into trichloroacetic acid (TCA) or by pumping samples into TCA preloaded syringes. Release of the C_2H_2 and C_2H_4 from the liquid phase was accomplished by injecting the sample into an evacuated tube or by degasing in a Van Slyke apparatus.

E. Analysis and Calculations

Although ethylene has been measured colorimetrically (LaRue and Kurz, 1973) or by mass spectrometry (Dilworth, 1966), the most convenient and sensitive method of detection is by gas chromatography (Hardy et al., 1968; Postgate, 1972). Gas chromatographs equipped with a H_2-flame ionization detector (FID) are used universally for the detection of C_2H_2 and C_2H_4. The highest sensitivity is achieved with nitrogen as the carrier gas, and helium the least sensitive. Argon used as the carrier gas produces lower and slower response than nitrogen, but the peak areas are similar for both carriers. In most cases, measurement of peak heights provides a satisfactory estimate if injection procedures

Table 2 System for the Acetylene Reduction Assay

Detector type	H_2 FID, H_2 25 ml/min, air 200 ml/min, both supplied at 138 kPa (20 psi), temperature 100°C
Column	Stainless steel, 0.45 m × 2 mm ID, Porapak R (100-120 mesh)
Carrier gas	N_2, 276 kPa (40 psi), 30 ml/min
Oven temperature	50°C
Septum type	Supelco Microsep
Retention times	C_2H_4 0.6 min, C_2H_2 1.0 min

have been standardized. The signal is transmitted to a strip-chart recorder of 1-mV input. A reliable system providing quick assays and good peak separations in use in our laboratory and similar to that of Turner and Gibson (1980) is shown in Table 2. The linearity of response to a range of C_2H_4 and C_2H_2 concentrations should be checked before using the gas chromatograph for assays. Subsequently, one or two different gas standards should be run each day provided that the response is linear. Detector fouling should be suspected when changes are observed to a given response. An easily fabricated, portable, gas chromatograph especially useful for field operations has been described by Mallard et al. (1977). This unit has a claimed sensitivity of 10 ppm of C_2H_4 in a 1-ml sample.

Two types of gas standards are used in most laboratories for calibration: one being a premix of ethylene in argon or N_2 (known concentration), and the other a sampling of the gas from an injection of pure ethylene into the reaction vessel of the same volume as those used in the experiment. The former standard requires a knowledge of the assay vessel volumes, as they may vary, in order to calculate the molar quantities of C_2H_4 produced. An assumption is made with the latter approach that all vessels have the same volume and it thus provides a correlation between the molar quantity of C_2H_4 added and peak height. Calculations for both approaches have recently been discussed by Turner and Gibson (1980).

III. PRECAUTIONS AND INTERPRETATION

The acetylene reduction assay is a rapid and sensitive method, but it is also affected by operational and interpretative problems. Therefore, appropriate care must be taken in the mechanics of procedure and in the interpretation of experimental results. As a generality, the further experiments are removed from in vitro studies toward the field situation, the greater are both procedural and interpretative difficulties.

A. Procedural Problems

A number of difficulties caused by gas leakage can be avoided by frequent observation and careful maintenance of conditions. The seals of incubation vessels and syringes used for sampling should be monitored for deterioration and tested periodically for leaks. Syringe needles are occasionally blocked by septa "cores." Depending on needle gauge, injection port leaks can be expected sooner or later from repeated needle injection. Acetylene peaks will be lowered if a blocked needle is used. Both the C_2H_2 and C_2H_4 peaks will be lowered and retention times increased if there is drop in the carrier gas pressure caused by leakage at the injection port, declining carrier cylinder supply, or exhaustion of H_2 or compressed air.

Acetylene is very soluble in water (42.6 m*M* at 1 atm and 20°C). A large volume of water in the assay vessel, including that in the soil or on or in plant material, can cause a significant decline in the C_2H_2 concentraiton in the gas phase. Acetylene and to a lesser extent ethylene are readily absorbed by rubber used in serum stoppers. Where lower C_2H_4 peaks are suspected to be due to gas loss in in vivo systems, the inclusion of an internal gas standard supplied with the C_2H_2 can be used for correction.

While C_2H_2 itself may act as an adequate internal standard, in the presence of a significant aqueous phase other standards, such as methane (Pendersen et al., 1978), ethane (Hansen, 1977), propylene, and propane (Balandreau and Dommergues, 1971) are advised. These gases differ in H_2O solubility and also in the order of elution from commonly used Porapak column packings. The choice of standard should be made with respect to the possibility of metabolism of the standard by the sample and compatibility with the gas chromatographic system (Knowles, 1980). Another aspect of the solubility is the danger that at commonly used pressures of 0.05 to 0.1 atm C_2H_2 may saturate nitrogen-fixing sites to a greater extent than would N_2, leading to an overestimation of activity in assays of whole organisms, symbioses, and ecological systems (Rice and Paul, 1971).

Rubber stoppers, but not serum stoppers, can be used many times without gas leakage. However, as mentioned previously, C_2H_2 and C_2H_4 are absorbed by the rubber (Kavanagh and Postgate, 1970) and their slow release to the atmosphere can cause problems with low-activity assays or where endogenous production of C_2H_4 is being determined. New stoppers may be washed in laboratory detergent and rinsed before use. Old stoppers should be rinsed and allowed to "rest" for 2 to 3 weeks before reuse, ideally in a wire rack under a stream of air. Ethylene is also absorbed by some plastics and this should be considered when constructing incubation chambers of synthetic materials or reusing disposable plastic syringes.

Evacuated tubes or other vessels used for the collection or storage of gas samples should be standardized to a pressure of 0.1 to 0.2 atm. If the volume of the sample taken by syringe, relative to the sampling tube or incubation vessel volume, is greater than 3%, the C_2H_4 determination should be corrected accordingly. Furthermore, the incubation vessel volume should be adequate to minimize changes in conditions during the assay. For instance, a fall in pO_2 due to respiration will cause a decline in the rate of C_2H_2 reduction by aerobic nitrogen-fixing systems (Sprent, 1969).

Acetylene concentration in the gas phase must be sufficient to saturate the nitrogenase of the system under study; otherwise, the rate of reduction will be proportional to the p-C_2H_2 (Hardy et al., 1973). Saturation of N_2-fixing sites is achieved with 0.05 atm p-C_2H_2 in soil of moderate porosity, but 0.2 atm may be required in waterlogged soil or sediment systems. In other in vivo systems (e.g., roots roots or detached nodules), acetylene pressures of 0.10 atm is sufficient. That individual nitrogen-fixing systems differ in response to C_2H_2 was shown by Spiff and Odu (1972) and MacRae (1977), who found increased C_2H_2 by *Beijerinchia* spp. with increasing p-C_2H_2 up to 0.7 atm. These observations indicate the need to determine the optimum p-C_2H_2 for each system to be studied.

B. Interpretative Problems

A major problem in the interpretation of acetylene reduction data where the assay is being used to estimate absolute rates of N_2 fixation is the precise relationship between C_2H_2 reduction and N_2 fixation and the correct conversion ratio to calibrate the assay. In theory, a C_2H_2 to N_2 ratio of 3:1 can be justified (based on electrons needed for reduction), and many researchers have used this ratio to convert results without attempting to validate the system under study. In practice, ratios of 1.5:1 up to 25:1 have been reported (Hardy et al., 1973). As this wide variation in ratio would indicate, cross-checking with ^{15}N is always desirable and can be performed readily with high-activity and/or enzyme systems. On the other hand, difficulty may be encountered in obtaining reliable estimates of N_2 fixation by ^{15}N or total N methods (Paul, 1975) with low-activity-field and natural ecosystems. However, in view of the variability seen in C_2H_2:N_2 ratios within, and between, various diazotrophic systems under laboratory conditions, caution should be exercised in the use of the 3:1 ratio. A number of factors may contribute to these variations. Acetylene competitively inhibits H_2 evolution such that total electron flow through nitrogenase is used to reduce C_2H_2. This fact alone would lead to overestimation of nitrogen-fixation rates based on the assay. It has been shown with soybeans that a rigorous correction of C_2H_2 reduction for H_2 evolved (in air) during N_2 fixation nevertheless leads to a 30 to 40% overestimation of activity compared to direct ^{15}N assay (Saito et al., 1980). This discrepancy may be explained by an effect of C_2H_2 in increasing the total electron flux through nitrogenase (Hageman and Burris, 1980). On the other hand, studies like those of Sprent (1972) in which acetylene reduction data are compared with nitrogen incremental data have shown that C_2H_4:$^{15}NH_3$ ratios decreased with plant age to values in the range of 0.2 to 0.8 at the end of the growing period. In addition, the ratio of H_2 produced to N_2 fixed varies and may be greater than 1:1 in some systems, thus precluding the use of a theoretical 4:1 conversion ratio. Many diazotrophs have an active-uptake hydrogenase that reutilizes the H_2 evolved by nitrogenase, so that attempts to correct C_2H_2 reduction values by measuring nitrogenase-catalyzed H_2 production may be futile. The implications of this finding are not clear. Since C_2H_2 inhibits H_2 evolution by nitrogenase, the additional inhibition of H_2 uptake could be irrelevant; however, inhibition of a conventional bidirectional hydrogenase could deprive a N_2 fixer of an electron sink or of an additional source of reducing power.

A fundamental assumption in the use of the assay is that the rate of nitrogenase activity is not affected by a change in substrate (N_2 versus C_2H_2) or a change in end product (NH_3 versus C_2H_4). A recent study by Minchin et al. (1983) in which a flow-through gas circulating system was used to measure simultaneously acetylene reduction and respiration of clover and other legumes and their detached nodules indicated a marked decrease in CO_2 production rate in the presence of C_2H_2 with a concurrent decline in acetylene reduction activity. A similar decline in H_2 evolution and respiration was observed when N_2 was replaced with argon or helium. Not all symbiotic associations exhibited this effect and its presence or absence did not appear to be correlated with taxonomic grouping, geographic origin, or the presence of an uptake hydrogenase. The decline in CO_2 production would appear to be due to a decrease in respiration coupled to nitrogenase activity, although this has not been established conclusively, and suggests that the cessation of ammonia production is the causal factor in the decline. It is suggested that calculations of nitrogenase rates based on cumulative ethylene production will underestimate actual rates and produce low apparent values for both N_2 fixation and relative efficiencies of H_2 production. These authors conclude that closed incubation vessels might be adequate if the assay is used for comparative purposes, but this is true only if the decline in nitrogenase activity is the same for all treatments. Open, flow-through systems must be employed to determine whether an acetylene-induced decline occurs. If a decline is observed, the short-duration maximum plateau can be determined accurately only by the use of a flow-through system.

There are a number of other side effects that acetylene has on natural habitats that may complicate data interpretation. Exposure of nitrogen-limited systems to C_2H_2 imposes further nitrogen limitations on nitrogen fixers, which therefore do not proliferate (Brouzes and Knowles, 1973). Activity estimates are then low relative to ^{15}N determinations but possibly close to initial rates of fixation. The proliferation of N_2 fixers on combined nitrogen during preincubation of excised grass roots and even during exposure to C_2H_2 (Van Berkum and Sloger, 1979; Van Berkum, 1980) has been one of the controversial aspects of excised grass root assays (Dobereiner, 1978). As previously mentioned, long-term exposure of C_2H_2 has been reported to increase nitrogenase activity either by hyperinduction/hyperderepression or by some conformational change in the enzyme structure (David et al., 1978).

Acetylene inhibits ammonia-supported growth of *Clostridium pasturainum* (Brouzes and Knowles, 1971). This effect may be caused by inhibition of an electron sink hydrogenase in conditions of excess carbon. The implications of this observation should be considered when measuring C_2H_2 reduction in anaerobic habitats. The oxidation of C_2H_4 is inhibited by C_2H_2 (DeBont and Leijten, 1976), but the production of C_2H_4 by aerobic and anaerobic organisms (Goodlass and Smith, 1978) other than N_2 fixers is not inhibited (Witty, 1979). A control without C_2H_2 may not show C_2H_4 accumulation, but the presence of C_2H_2 blocks the further metabolism of C_2H_4 and causes accumulation of C_2H_4 from endogenous sources as well as from C_2H_2 reduction. In these cases there is danger of overestimating activity. Acetylene inhibits both methanogenesis (Raimbault, 1975) and the consumption of methane (DeBont and Mulder, 1976). The latter effect precludes the use of the C_2H_2 reduction assay for measurement of CH_4-supported N_2 fixation by methanotrophs. With the acetylene reduction assay the accumulation of H_2 with its eventual inhibition of nitrogenase is absent. The use of acetylene reduction could thus lead to an overestimation of activity in long-term soil assays (Knowles, 1976).

The measurement of nitrogen fixation in natural habitats by the C_2H_2 reduction assay is complicated by the limitation inherent in the use of a short-term assay to assess overall activity. Variation in activity due to temperature, moisture levels, and light intensity, whether seasonal, day to day, or diurnal, makes the integration of a series of short-term assays difficult. For example, Goh et al. (1978) found different patterns of diurnal variation in studies with field-grown white clover. Several researchers (Sloger et al., 1975; Halliday and Pate, 1976; Ruegg and Alston, 1978) have derived a correction factor from a series of diurnal assays. Halliday and Pate (1976) derived a factor of 0.87, which was used to multiply assay results obtained at 10 a.m. in their monthly assays with field-grown *Macrozamia riedlei*, to obtain an estimated daily rate of nitrogen fixation. Sloger et al. (1975) has suggested that sampling of soybeans at 6 a.m., 8 a.m., 10 a.m., and

12 noon gave good estimates of 24-hr activity. A conclusion from these studies by Sinclair et al. (1976), Goh et al. (1978), and others is that the acetylene reduction assay is probably unsuitable for long-term quantitative N_2-fixation measurement.

Other problems associated with assessing nitrogen fixation in natural ecosystems are organism or ecosystem related. For example, grass roots removed from soil reduce C_2H_2 only after an 8 to 12-hr lag phase (Rinaudo et al., 1971). The reason for this lag has not been clarified and more than one factor is probably involved. As long as this problem remains, the acetylene reduction method should not be used for extrapolation to quantitative estimates with this system. Other difficulties arise in the assay of nonlegume field organisms for acetylene reduction activity. The effects of temperature, interference of other rhizosphere organisms that reduce C_2H_2, and the natural formation of C_2H_4 in forest soil have been discussed by Nohrstedt (1983). It has been suggested that CO together with a container system be used to suppress nitrogenase-catalyzed C_2H_2 reduction in order to estimate background C_2H_4.

These problems, whether associated with the diversity of environments, the complexity of biological systems, or the variation of H_2 produciton, which may or may not be directly related to nitrogen-fixing activity, require detailed analysis of the particular system under study if C_2H_2 reduction is to be used to provide accurate estimates.

REFERENCES

Agarwal, A. K., and Keister, D. L. (1983). *Appl. Environ. Microbiol.* *45*: 1592.

Akkermans, A. D. L. (1971). *Nitrogen fixation and nodulation of Alnus and Hippophae under natural conditions.* Ph.D. thesis, University of Leiden, Leiden, The Netherlands.

Balandreau, J., and Dommergues, Y. (1971). *C. R. Acad. Sci.*, Paris 273: 2020.

Balandreau, J., Ducerf, P., Hamad-Fares, I., Weinhard, P., Rinaudo, G., Millier, C., and Dommergues, Y. (1978). In *Limitations and Potentials for Biological Nitrogen Fixation in the Tropics*, J. Doberiner, R. H. Burris, and A. Hollaender, eds. Plenum Press, New York, p. 275.

Bergersen, F. J. (1970). *Aust. J. Biol. Sci. 23*: 1015.

Bergersen, F. J., and Turner, G. L. (1975). *J. Gen. Microbiol. 89*: *31*.

Brouzes, R., and Knowles, R. (1971). *Can. J. Microbiol. 17*: 1483.

Brouzes, R., and Knowles, R. (1973). *Soil Biol. Biochem.* 5: 223.

Burris, R. H. (1972). *Methods Enzymol. 24*: 415.

Burris, R. H. (1974). In *The Biology of Nitrogen Fixation*, A. Quispel, ed. North-Holland Publishing Company, Amsterdam, p. 9.

Campbell, N. E. R., and Evans, H. J. (1969). *Can. J. Microbiol. 15*: 1342.

Dart, P. J., and Day, J. (1971). In *Biological Nitrogen Fixation in Natural and Agricultural Habitats.*, T. A. Lie and E. G. Mulder, eds. *Plant Soil, Spec. Vol.*: 167.

David, K. A. V., and Fay, P. (1977). *Appl. Environ. Microbiol. 34*: 640.

David, K. A. V., Apte, S. K., and Thomas, J. (1978). *Biochem. Biophys. Res. Commun. 82*: 39.

David, K. A. V., Apte, S. K., Banerji, and Thomas, J. (1980). *Appl. Environ. Microbiol. 39*: 1078.

DeBont, J. A. M., and Leijten, M. W. M. (1976). *Arch. Microbiol. 107*: 235.

DeBont, J. A. M., and Mulder, E. G. (1976). *Appl. Environ. Microbiol. 31*: 640.

Dilworth, M. J. (1966). *Biochim. Biophys. Acta. 127*: 285.

Dipankar, S., and Weaver, R. W. (1984). *J. Exp. Bot. 35*: 785.

Dobereiner, J. (1978). *Ecol. Bull. Stockholm 26*: 343.

Dobereiner, J. (1980). In *Methods for Evaluating Biological Nitrogen Fixation*, F. J. Bergersen, ed. John Wiley & Sons Ltd., Chichester, West Sussex, England, p. 535.

Eady, R. R. (1980). In *Methods for Evaluating Biological Nitrogen Fixation*, F. J. Bergersen, ed. John Wiley & Sons Ltd., Chichester, West Sussex, England, p. 244.

Evans, H. J., Koch, B., and Klucas, R. (1972). *Methods Enzymol. 24*: 470.

Farnden, K. J. F., and Robertson, J. G. (1980). In *Methods for Evaluating Biological Nitrogen Fixation*, F. J. Bergersen, ed. John Wiley & Sons Ltd., Chichester, West Sussex, England, p. 265.

Fishbeck, K., Evans, H. J., and Boersma, L. L. (1973). *Agron. J. 65*: 429.

Gibson, A. H., Scowcroft, W. R., Child, J. J., and Pagan, J. D. (1976). *Arch. Microbiol. 108*: 45.

Goh, K. M., Edmeades, D. C., and Robinson, B. W. (1978). *Soil Biol. Biochem. 10*: 13.

Goodlass, G., and Smith, K. A. (1976). *Soil Biol. Biochem. 10*: 193.

Hageman, R. V., and Burris, R. H. (1980). *Biochim. Biophys. Acta. 591*: 63.

Halliday, J., and Pate, J. S. (1976). *J. Brit. Grassl. Soc. 31*: 29.

Hansen, R. B. (1977). *Appl. Environ. Microbiol. 33*: 596.

Hardy, R. W. F., and Holstein, R. D. (1977). In *A Treastise on Dinitrogen Fixation*, Sec. 4, *Agronomy and Ecology*, R. F. W. Hardy and A. H. Gibson, eds. John Wiley & Sons, Inc., New York, p. 468.

Hardy, R. W. F., Holstein, R. D., Jackson, E. K., and Burns, R. E. (1968). *Plant Physiol. 43*: 1185.
Hardy, R. W. F., Burns, R. C., and Holstein, R. D. (1973). *Soil Biol. Biochem.* 5: 47.
Haystead, A., King, J., and Lamb, W. I. C. (1980). *Grass Forage Sci. 34*: 124.
Huang, C.-Y., Boyer, J. S., and Vanderhoef, L. N. (1975). *Plant Physiol. 56*: 222.
Jensen, B. B., and Cox, R. P. (1983). *Appl. Environ. Microbiol. 45*: 1331.
Jordan, D. C, McNicol, P. J., and Marshall, M. R. (1978). *Can. J. Microbiol. 24*: 643.
Kavanagh, E. P., and Postgate, J. R. (1970). *Lab. Practice 19*: 159.
Knowles, R. (1976). In *Proc. 1st International Symposium Nitrogen Fixation*, W. E. Newton and C. J. Nyman, eds. Washington State University Press, Pullman, Wash., p. 539.
Knowles, R. (1980). In *Methods for Evaluating Biological Nitrogen Fixation*, F. J. Bergersen, ed. John Wiley & Sons Ltd., Chicester, West Sussex, England, p. 557.
Knowles, R. (1981). In *Current Perspectives in Nitrogen Fixation*, A. H. Gibson and W. E. Newton, eds. Elsevier/North Holland, New York, p. 327.
Koch, B., and Evans, H. J. (1966). *Plant Physiol. 41*: 1748.
LaRue, T. A., and Kurz, W. G. W. (1970). Abstract. *X Int. Congr. Microbiol.*, Mexico City, p. 142.
LaRue, T. A., and Kurz, W. G. W. (1973). *Can. J. Microbiol. 19*: 304.
Lee, K.-K., and Watanabe, I. (1977). *Appl. Environ. Microbiol. 34*: 654.
Lee, K.-K., Alimago, B., and Yoshida, T. (1977). *Plant Physiol. 47*: 519.
MacRae, I. C. (1977). *Aust. J. Biol. Sci. 30*: 593.
Main, S., and Bond, G. (1978). *New Phytol. 80*: 187.
Mallard, T. M., Mallard, C. S., Holfeld, H. S., and LaRue, T. A. (1977). *Anal. Chem. 49*: 1275.
Mederski, H. J., and Streeter, J. G. (1977). *Plant Physiol. 59*: 1076.
Minchin, F. R., Neves, M. C. P., Summerfield, R. J., and Richardson, A. C. (1977). *J. Exp. Bot. 28*: 507.
Minchin, F. R., Witty, J. F., Sheehy, J. E., and Muller, M. (1983). *J. Exp. Bot. 34*: 641.
Nohrstedt, H. O. (1982). *Soil Biol. Biochem. 15*: 275.
Nohrstedt, H. O. (1983). *Soil Biol. Biochem. 15*: 281.
Nutman, P. S., ed. (1976). *Symbiotic Nitrogen Fixation in Plants.* Cambridge University Press, Cambridge, p. 307.
Pan, S., and Schmidt, E. L. (1969). *Bacteriol. Proc. 14*: 21.
Paul, E. A. (1975). In *Nitrogen Fixation by Free-Living Microorganisms*, W. D. P. Stewart, ed. Cambridge University Press, Cambridge, p. 259.

Paul, E. A., Myers, R. J., and Rice, W. A. (1971). In *Biological Nitrogen Fixation in Natural and Agricultural Habitats.*, T. A. Lie and E. G. Mulder, eds. *Plant Soil, Spec. Vol.*: 495.

Pendersen, W. L., Chakrabarty, K., Klucas, R. V., and Vidaver, A. K. (1978). *Appl. Environ. Microbiol. 35*: 129.

Perry, J. H. (1950). *Chemical Engineers' Handbook.* McGraw-Hill Book Company, New York, p. 1868.

Postgate, J. R. (1972). *Methods in Microbiology*, Vol. 6B. Academic Press, Inc., New York, p. 343.

Raimbault, M. (1975). *Ann. Inst. Pasteur Microbiol. 126A*: 247.

Rice, W. A., and Paul, E. A. (1971). *Can. J. Microbiol. 17*: 1049.

Rinaudo, G., Balandreau, J., and Dommergues, Y. (1971). In *Biological Nitrogen Fixation in Natural and Agricultural Habitats.* T. A. Lie and E. G. Mulder,eds. *Plant Soil, Spec. Vol.*: 1471.

Ruegg, J. J., and Alston, A. M. (1978). *Aust. J. Agric. Res. 29*: 951.

Saito, S. M. T., Matsui, E., and Salati, E. (1980). *Physiol. Plant 49*: 37.

Schöllhorn, R., and Burris, R. H. (1966). *Fed. Proc. Fed. Am. Soc. Exp. Biol. 24*: 710.

Schöllhorn, R., and Burris, R. H. (1967). *Proc. Natl. Acad. Sci. USA 58*: 213.

Schwinghamer, E. A., Evans, H. J., and Dawson, M. D. (1970). *Plant Soil. 33*: 192

Sen, D., and Weaver, R. W. (1984). *J. Exp. Bot. 35*: 785.

Silvester, W. B., and Bennet, K. J. (1973). *Soil Biol. Biochem. 5*: 171.

Sinclair, A. G. (1973). *N.Z. J. Agric. Res. 16*: 263.

Sinclair, A. G., Hannagan, R. B., and Risk, W. H. (1976). *N.Z. J. Agric. Res. 19*: 415.

Sloger, C., Bezdicek, D., Milberg, R., and Boonkerd, N. (1975). In *Nitrogen Fixation by Free-Living Micro-organisms*, W. D. P. Stewart, ed. Cambridge Unviersity Press, Cambridge, p. 271.

Spiff, E. D., and Odu, C. T. I. (1972). *Soil Biol. Biochem. 4*: 71.

Sprent, J. I. (1969). *Planta Berlin 88*: 372.

Sprent, J. I. (1972). *New Phytol. 71*: 603.

Stephens, B. D., and Neyra, C. A. (1983). *Plant Physiol. 71*: 731.

Stewart, W. D. P., ed. (1975). *Nitrogen Fixation by Free-Living Micro-organisms.* Cambridge University Press, Cambridge, p. 471.

Stewart, W. D. P., Fitzgerald, G. P., and Burris, R. J. (1967). *Proc. Natl. Acad. Sci. USA 58*: 2071.

Stewart, W. D. P., Mague, T. M., Fitzgerald, G. P., and Burns, R. H. (1971). *New Phytol. 70*: 497.

Tjepkema, J. D., and Burris, R. H. (1976). *Plant Soil 45*: 81.

Tjepkema, J. D., and Winship, L. J. (1980). *Science 209*: 279.

Tjepkema, J. D., Ormerod, W., and Torrey, J. G. (1981). *Can. J. Microbiol 27*: 815.

Tubb, R. S., and Postgate, J. R. (1973). *J. Gen. Microbiol. 79*: 103.

Turner, G. H., and Gibson, A. H. (1980). In *Methods for Evaluating Biological Nitrogen Fixation*, F. J. Bergersen, ed. John Wiley & Sons, Inc., New York, p. 111.

Van Berkum, P. (1980). *Soil Biol. Biochem. 12*: 141.

Van Berkum, P., and Sloger, C. (1979). *Plant Physiol. 64*: 739.

Wacek, T. J., and Brill, W. J. (1976). *Crop Sci. 16*: 519.

Waughman, G. H. (1977). *J. Exp. Bot. 28*: 949.

Wheeler, C. T. (1971). *New Phytol. 70*: 487.

Witty, J. F. (1979). *Plant Soil 52*: 165.

Witty, J. F., and Day, J. M. (1977). In *Isotopes in Biological Dinitrogen Fixation*, International Atomic Energy Agency, Vienna, p. 135.

Witty, J. F., Minchin, F. R., and Sheehy, J. E. (1983). *J. Exp. Bot. 34*: 951.

Zuberer, D. A., and Silver, W. S. (1978). *Appl. Environ. Microbiol. 35*: 567.

13

Measuring Ureides

RICHARD A. GLENISTER and THOMAS A. LaRUE *Boyce Thompson Institute for Plant Research, Ithaca, New York*

I. INTRODUCTION

Ureides are compounds that yield urea upon hydrolysis. The term here refers to allantoin and allantoic acid. These compounds occur in many legume and nonlegume plants, and in some species they are a major form in which nitrogen is translocated or stored.

In some legumes, ureides are the predominant form in which newly fixed nitrogen is exported from the nodules to the shoot. The ureides are transported to the shoot via xylem sap, in which the ureide concentration may be very high. The ureide concentration in the xylem is higher if fixation is high. When these legumes are grown in the presence of ample nitrate, the concentration of ureides in xylem sap may be lower, and the dominant nitrogen compounds are asparagine, glutamine, and nitrate. The apparent correlation between nitrogen fixation and xylery ureides suggests that ureide concentrations in xylem or shoot may serve as an indirect indicator of symbiotic nitrogen fixation. In this chapter we describe methods of measuring ureides in legumes.

II. PREPARATION OF ALLANTOIN AND SODIUM ALLANTOATE

A. Allantoin

Allantoin is readily available from many sources. It should always be recrystallized from hot water to remove impurities. To recrystalize allantoin:

1. Add enough hot water (ca. 230 ml) to just dissolve 10 g of allantoin.
2. Filter (keep warm) and then let the filtrate cool slowly to room temperature.
3. Filter to collect the first crop of crystals, and rinse with a small amount of cold (0°C) water.
4. Cool the filtrate (0 to 4°C) to obtain a second crop of crystals. Rinse with a small amount of cold (0°C) water.
5. Dry the crystals in a vacuum desiccator until the weight is constant. The monoclinic crystals melt at 238 to 240°C [molecular weight (MW), 158.2 Molecular formula (MF), $C_4N_4O_3H_6$].

B. Sodium Allantoate

Sodium allantoate-H_2O, a stable convenient form of allantoic acid, is prepared by the method of Hermanowicz (1948).

1. Combine 10 g of allantoin and 20 ml of 30% potassium hydroxide solution, and let stand 17 to 20 hr at 25°C.
2. Dilute with approximately 60 ml of water, cool (0 to 4°C), and

then add cold 2 *N* sulfuric acid until the pH is 2.6 to 2.8.

3. Filter, and wash the precipitate with cold water.
4. Dissolve the precipitate in a small quantity of water and titrate with 1 *N* NaOH to pH 8.0.
5. Back titrate with 1 *N* acetic acid to pH 7.0 and filter.
6. Add 8 to 10 volumes 95% ethanol to the filtrate and allow it to stand until crystals form.
7. Collect the crystals by filtration, and dry in a vacuum desicator or vacuum oven at 50°C to constant weight. The yield is approximately 7 g; the amorphous crystals melt at 168°C.

Sodium allantoate prepared in this way contains 1 mol of water; MW, 216.2; MF, $C_4N_4O_4H_7Na \cdot H_2O$.

III. SAMPLE PREPARATION

A. Plant Tissue

Drying and Grinding

Plant stem tissue can be harvested, dried, and stored with only small losses (about 5%) of ureides. Dry samples at 65 to 75°C in a forced-air oven for 2 days. Grind the dry plant tissue to pass a 60-mesh (1.0-mm) screen. Store in a dry place before extraction.

Leaf samples will lose 15% or more of the ureides when they are dried at 65°C. Therefore, quick freezing in liquid nitrogen or an acetone-dry ice bath is preferred for leaf samples. Immediately upon harvest place leaves in a plastic bag and immerse them in the liquid nitrogen or an acetone-dry ice bath. Store at -20°C or freeze-dry in a lyophilizer.

Alkaline Extraction Procedures for Total Ureides

To determine the total ureide content of dried ground stem tissue, simultaneous extraction and hydrolysis with dilute sodium hydroxide is used (see Section IV.A, "*Alternative Procedure for Total Ureides*").

If only an estimate of total ureides (allantoin plus allantoic acid) is required, fresh stem tissue can be extracted in dilute sodium hydroxide solution as follows:

1. Slice or chop the stems into pieces 5 mm or smaller.
2. Weigh a sample of chopped stem tissue (approximately 1 g) and place in test tube.
3. Add 3 ml of 0.5 *N* sodium hydroxide, heat 5 to 10 min in a hot water bath (90 to 95°C), then cool to room temperature.
4. Transfer 0.5 ml to test tubes, dilute with water to 3.5 ml, and continue the assay for total ureides in Section IV.A, "Alternative Procedure for Total Ureides," step 2).

Alcoholic Extraction Procedure for Ureides

If allantoin and allantoic acid contents are to be determined separately (Section IV.A, "*Differential Analysis of Allantoin and Allantoic Acid*") or if total ureides are measured by the modified Berthelot assay (Section IV.B), extract plant material using buffered alcohol-water solution as follows:

1. Extraction solution. Combine equal volumes of ethanol with 0.1 *M* potassium phosphate buffer, pH 7.0. Prepare the buffer by dissolving 17.4 g of dibasic potasium phosphate in approximately 950 ml of water, titrate to pH 7.0 with concentrated HCl, and dilute to 1.0 liters.
2. Combine 0.20 to 0.50 g of ground plant tissue and 10 ml of extraction solution in a large test tube, heat in a 80°C water bath for 5 min or more, then cool.
3. Filter to remove tissue residue, or centrifuge at low speed and decant.
4. Store samples in tightly closed vials at -20°C.

B. Xylem Sap

Bleeding Sap

Cut the shoot below the first node with a very sharp blade. Lubricate the stem below the cut with stopcock grease. Place plastic tubing 2 to 4 cm long with a diameter slightly smaller than the stem over the cut end to act as a reservoir. Because bleeding depends on root pressure, keep the soil surrounding the root saturated with water. Discard the first few drops of exudate to minimize contamination from the cut cells of the stem. Remove the accumulated exudate every few minutes with a small pipette and place it in a vial kept on ice. A mature soybean plant will yield about 500 μl of xylem sap in 1 hr, and the ureide concentration is stable during this time (McClure et al., 1980). Store samples frozen until analysis.

Xylem Extraction

Xylem sap can be obtained by vacuum extraction of the stem (Bollard, 1953). Cut the stem above ground level and remove the leaves. A small trap (2 to 3 ml) is connected in a flexible line between a vacuum pump and the basal end of the shoot. A small hand-held vacuum pump is sufficient and is convenient. With the vacuum applied each time, cut short segments (1 to 2 cm) from the upper stem. A small amount of sap will be pulled from the stem after each cut. The total volume will rarely exceed 500 μl. Ureide concentrations in xylem collected by vacuum extraction are lower than in xylem obtained by root bleeding (Herridge, 1984).

C. Phloem Sap

With a sharp blade, cut the petiole while it is held under EDTA. (Dissolve 7.60 g of EDTA and 0.37 g of HCl in 900 ml of water, adjust the pH to 7.0 with 1 *N* KOH, and make up to 1.0 liter.) To collect petiole phloem exudate, place the cut end of the petiole with attached leaflets into 2.0 ml of 20 m*M* EDTA, 5 m*M* KCl, pH 7.0. Keep the detached leaves in a plastic bag to maintain high humidity.

To obtain fruit tip phloem from attached pods, cut the distal end under the EDTA solution, and place the cut tip in 2.0 ml of the 20 m*M* EDTA solution. For petiole and pod samples, the collection time is from 5 to 10 hr.

IV. COLORIMETRIC ANALYSIS

A. Ureides as Glyoxyl Phenylhydrazones

Allantoin and allantoic acid are hydrolyzed to glyoxylic acid. The selective hydrolysis of allantoic acid, but not allantoin, by weak acid provides a way of measuring each separately. The assay is based on the oxidative decarboxylation of the phenylhydrazone of glyoxylic acid to form a highly colored dibenzyl formazan (Vogels and Van der Drift, 1970). This reaction permits the determination of glyoxylate with phenylhydrazine even if other phenylhydrazones are formed (see the section "Notes and Precautions," note 7). The high molar extinction coefficient of the formazan (ca. 50,000) provides high sensitivity. The practical limit of detection is as little as 1 nmol. Under careful conditions the coefficient of variation is 5 to 10%.

There is a method by which ureides are measured by converting the glyoxyl phenylhydrazone into a chromophore in alkaline solution, avoiding the oxidation-decarboxylation step (Borchers, 1977; van Berkum and Sloger, 1983). van Berkum and Sloger showed that interference from glyoxylic acid or other keto acids was negligible in extracts from dried soybean tissue but was significant in extracts from fresh tissue. This interference could be eliminated by including phenylhydrazine in the alkaline hydrolysis step for total ureide. This method for total ureides can be automated.

Reagent Solutions

1. *0.5 N NaOH solution.* Dissolve 20 g of sodium hydroxide in approximately 950 ml of water, and dilute to 1.0 liter.
2. *0.65 N HCl solution.* Dilute 108 ml of 37% hydrochloric acid to 1.0 liter.
3. *0.4 M* K_2PO_4 *pH 7.0.* Dissolve 69.7 g of dibasic anhydrous potassium phosphate in 950 ml of water, adjust the pH to 7.0 with concentrated HCl, and dilute to 1.0 liter.

4. *Phenylhydrazine solution.* Dissolve 0.10 g of phenylhydrazine HCl in 30 ml of water a few minutes before use.
5. *Ferricyanide solution.* Dissolve 0.50 g of potassium ferricyanide in 30 ml of water. Prepare on the day of use.
6. *Cold concentrated (37%) hydrochloric acid.* Cool to 0°C before use.

Assay for Total Ureides (Allantoic Acid plus Allantoin)

1. *Sample dilution.* Dilute the sample extract with water so that 3.0 ml contains approximately 10 to 250 nmol of ureide. Prepare duplicate samples for a sample blank if necessary (see "Notes and Precautions," note 1).
2. *Conversion of allantoin to allantoate.* To the 3.0-ml sample add 1.0 ml of 0.5 *N* NaOH, mix, and heat at 90 to 100°C for 5 min in a boiling-water bath.
3. *Conversion of allantoate to glyoxylic acid.* To the sample from step 2, add 1.0 ml of 0.65 *N* HCl, mix, and heat at 90 to 100°C for 10 min.
4. *Neutralization of hydrolyzed sample.* To the sample from step 3, add 1.0 ml of 0.4 *M* potassium phosphate buffer pH 7.0 (see the section "Notes and Precautions," note 2).
5. *Conversion of glyoxylate to glyoxyl phenylhydrazone.* Within 2 hr, add 1.0 ml of phenylhydrazine solution to the neutralized sample and let stand at room temperature for 5 min.
6. *Conversion of glyoxyl phenylhydrazone to dibenzyl formazan (colored product).* Cool the samples in an ice bath, and with samples still in the ice bath, add 5 ml of precooled concentrated HCl, stir, then add 1.0 ml of ferric cyanide solution and stir (see the section "*Notes and Precautions*," notes 3, 4, and 5).
7. *Color development and measurement.* Remove samples from the ice bath and after 15 min at room temperature read the absorbance at 535 nm.

Differential Analysis of Allantoin and Allantoic Acid

Allantoic acid determination. To determine allantoic acid, delete step 2 in the procedure above and add 1.0 ml of 0.5 *N* NaOH solution immediately before step 3.

Allantoin determination. Allantoin concentration is determined from the difference between the total ureide content and allantoic acid content.

Alternative Procedure for Total Ureides

If only the total ureide content is desired, simultaneous ureide extraction and hydrolysis of allantoin is achieved by extracting samples with an alkaline solvent.

1. *Extraction and hydrolysis.* Place 50 to 100 mg ground dry plant tissue in a test tube, add 2.0 ml 0.25 *N* NaOH, stir, and heat at 80 to 100°C for 10 min.
2. *Allantoate to glyoxylic acid.* To the sample from step 1, add 1.0 ml 0.65 *N* HCl, mix, and heat at 95 to 100°C for 10 min.
3. *Neutralization and centrifugation.* Neutralize the sample with 1.0 ml 0.4 *M* potassium phosphate buffer pH 7.0, cool the samples to room temperature in a water bath, and centrifuge for 5 min. at 400 g to sediment plant tissue.
4. *Conversion of glyoxylate to glyoxyl phenylhydrazone.* Remove a 1.5 ml aliquot of the supernatant from step 3 to a test tube and add 0.25 ml of phenylhydrazone solution, and let stand at room temperature for at least 5 min.
5. *Conversion of glyoxyl phenylhydrazone to dibenzyl formazan (colored product).* Cool the samples in an ice bath, and with the tubes still in the ice bath add 1.25 ml precooled concentrated HCl, stir, then add 0.25 ml ferric cyanide solution. (Notes 3, 4, and 5.)
6. *Color development and measurement.* Remove samples from the ice bath, and after 15 min. at room temperature read the absorbance at 535 nm.

Notes and Precautions

1. Prepare sample blanks by adding the required volume of phosphate buffer, NaOH solution, and HCl solution in that order, but omitting the heating steps. This blank measures the free glyoxylic acid in tissue. Blank values for soybean stem tissue are usually low (10 to 20 μmol/g) and constant, but the glyoxylic acid content of leaves may be high. If only an approximation of total ureide content is required, the sample blanks may be omitted.
2. With the exception of steps 4 and 5, the time interval between steps is not critical but should be as short as possible.
3. Ensure that the concentrated HCl is cold before use.
4. Mixing the sample vigorously while adding the ferric cyanide solution in step 6 yields more reproducible results.
5. Provide eye protection and adequate ventilation, particularly when using concentrated HCl.
6. Use deionized or distilled water for extraction and reagent solutions.
7. Steps 5 through 7 of the section "Assay for Total Ureides" can be automated using a Technicon AutoAnalyzer, which provides more consistent results. An AutoAnalyzer flow diagram is outlined in Figure 1.

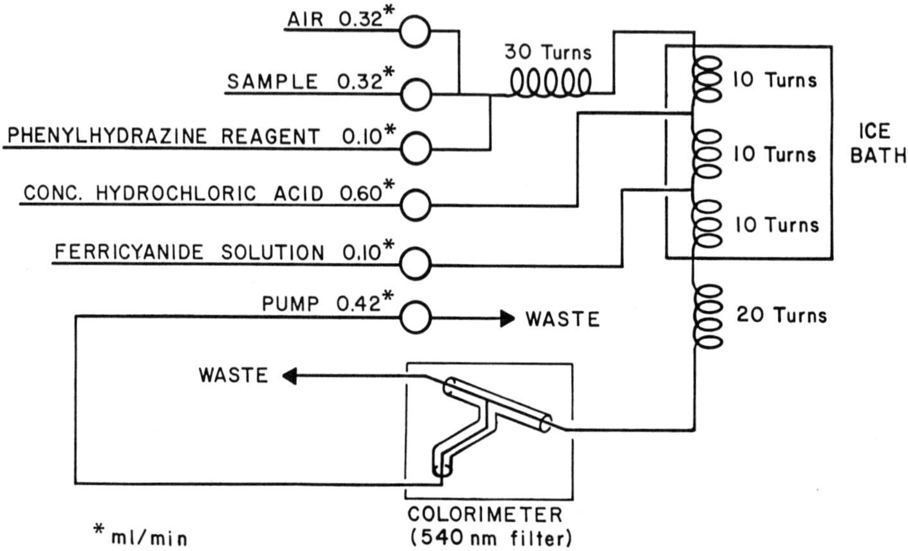

Figure 1 AutoAnalyzer flow diagram for a semiautomated analysis of ureide.

B. Modified Berthelot Assay

This is based on the Berthelot assay for ammonia, in which ammonia reacts with hypochlorite and alkaline phenol to form indolphenol. Domnas (1961) observed that compounds containing amide nitrogen would react with hypochlorite in acidic solutions to form intermediates, which would then react with alkaline phenol to form indolphenol.

In addition to ureides, other nitrogen compounds, including amino acids, will react in this assay. Because the modified Berthelot assay is not specific for ureides, the plant extract is treated with a cation-exchange resin to remove basic amino acids (Patterson et al., 1982).

Reagent Solutions and Materials

1. Dowex HCR-S, a strongly acidic cation-exchange resin (obtained from Sigma Chemical Co., St. Louis, Missouri). The resin (approximately 1 kg) is prepared for use as follows:
   a. Wash with 4 volumes of water.
   b. Equilibrate with 2 volumes of 0.5 *N* NaOH, and rinse with water until the pH of the wash is neutral.
   c. Equilibrate with 2 volumes of 1.0 *N* HCl, and rinse until the pH of the wash is neutral.
   d. Remove excess water with vacuum filtration.

   e. Equilibrate the resin in 50% aqueous ethanol.
   f. Remove excess solution by vacuum filtration. Do not allow the resin to dry completely.
2. 0.2 *M* potassium phthalate solution. Dissolve 40.8 g of potassium hydrogen phthalate in approximately 950 ml of water. Adjust the pH to 4.0 with dilute HCl or NaOH, and dilute the solution to 1.0 liter.
3. Bleach solution. Dilute 3 volumes of commercial household bleach (5.25% active chlorine) with 7 volumes of water just before use.
4. Phenol-methanol solution. Combine methanol with an equal volume of chromatography-grade liquid phenol (phenol saturated with water) or phenol crystals dissolved in excess water and allowed to equilibrate (remove excess water). Store in a brown bottle.
5. 20% NaOH solution. Dissolve 20 g of NaOH in 80 ml of water.

Procedure

1. Place approximately 2 g of exchange resin and 10 ml of sample extract in a 20-ml screw-top vial, and agitate the vials vigorously for 30 min on a shaker. Sample the solutions immediately or decant and store for later use (see the note below).
2. Dilute sample aliquots containing 50 to 1000 nmol of ureide to 1.0 ml with water in 18 mm × 150 mm test tubes.
3. Add 1.0 ml of 0.2 *M* phthalate solution and 0.5 ml of bleach solution, mix vigorously, then let stand at room temperature for 5 min.
4. Add 2.0 ml of phenol reagent, mix vigorously, and let stand 10 min.
5. Add 5.5 ml of water, mix, and measure the absorbance at 625 nm.

Note: Because the resin will slowly hydrolyze the ureides, the sample should be removed from the resin immediately after step 1.

V. SPOT TEST AND CHROMATOGRAPHY

The spot test is based on the acid-catalyzed Schiff reaction of aldehydes with amides. The test estimates the ureide content of the sap expressed onto filter paper from segments of legume petioles. By comparison with sodium allantoate standards applied to the paper before color development, the approximate ureide content of the petioles can be visually estimated.

The scoring system is based on our observation that twofold differences in allantoate concentration can be differentiated reliably.

A. Reagents and Apparatus

1. 0.05 *M* dimethylaminocinnamaldehyde stock reagent. Dissolve 0.875 g of dimethylaminocinnamaldehyde in 95 ml of absolute ethanol. Dimethylaminocinnamaldehyde can be obtained from Aldrich Chemical Company, Milwaukee, Wisconsin.
2. Concentrated (37%) hydrochloric acid.
3. Acidified ethanol. Combine 1 ml of concentrated (37%) hydrochloric acid and 125 ml of dry 100% ethanol. Store in a tightly closed container.
4. Prepare the spray reagent just before use. Combine 1 part concentrated HCl with 19 parts of 0.05 *M* stock reagent.
5. Sodium allantoate standards, 1, 2, 4, 8, 16, 32, and 64 m*M*. Dissolve 0.124 g of sodium allantoate in 9.5 ml of water and dilute to 10.0 ml for 64 m*M* stock. Make serial twofold dilutions for 32, 16, 18, 4, 2, and 1 m*M*.
6. Plexiglass sample holder, punch, and clipboard (figure 2).
7. Chromatography-grade filter paper: Whatman No. 3, SS-295, or equivalent.

B. Sample Collection

1. Place a sheet of filter paper on the clipboard under the plexiglass sample holder. Cut the petioles with a sharp razor and place the distal portion of the petiole, with the leaflets still attached, into a hole in the plexiglass sample holder. Cut the petiole again at the surface of the plexiglass holder, leaving a short segment (6 mm) of the petiole in the sample holder. After sampling 10 to 12 petioles, and before they have dried too much, push the punch into the hole, crushing the petiole segment and expressing sap onto the filter paper. After all samples are collected, remove the filter paper from the clipboard and carefully remove any pieces of petiole remaining attached to the paper. Label each sample with a pencil (see Section V.D, note 1).
2. Allow the filter paper to dry completely before storage or processing (see Section V.D, note 2).

C. Color Development and Estimations

1. Apply 5 μl of each of the sodium allantoate standards to the same paper and dry for approximately 5 min at 60°C.
2. Gently agitate the sample for approximately 10 sec in the acidified alcohol to remove interfering pigments and to bleach the chlorophyl (see Section V.D, note 3).

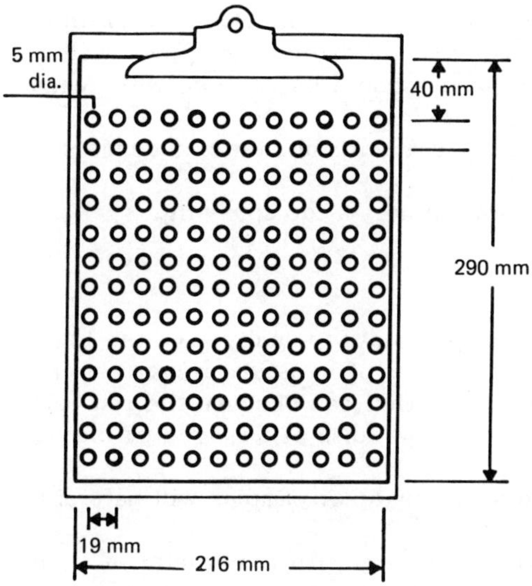

(a)

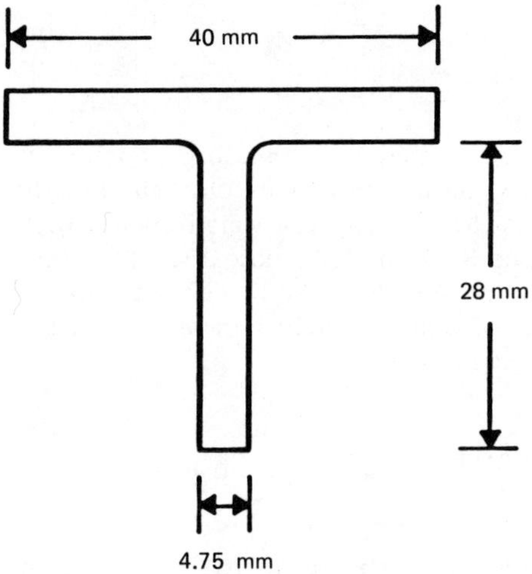

(b)

Figure 2 (a) Plexiglass sample holder on clipboard. A matrix of 5-mm-diameter holes are drilled into a sheet of 6.75-mm-thickness plexiglass, with dimensions and spacings shown above. (b) Punch. Two pieces of 4.75-mm-diameter steel rod are brazed or welded in the form of a tee.

3. Suspend the sheet and allow excess wash solvent to drain off.
4. Remove remaining wash solvent by drying for approximately 5 min at 60 to 65°C. The paper should be sprayed with reagent within a few minutes after drying.
5. Spray the paper with spray reagent until the standards show adequate color intensity.
6. Score samples within a few minutes after spraying.

D. Notes and Precautions

1. The petioles of fully developed leaves usually give the highest response. Extracts from petioles of immature leaves sometimes produce a purple interference that can obscure the red color from ureides. Do not harvest petioles of damaged or diseased leaves.
2. When stored dry at room temperature, samples will be stable for several weeks.
3. The rinsing time required will vary with the amount of chlorophyll and with the thickness of the paper. Preliminary trials may be necessary to determine the optimum rinsing time.
4. The limit of detection is approximately 5 nmol of sodium allantoate per spot.

E. Chromatography

Allantoic acid and allantoin can be separated by ascending paper chromatography with ethanol:water (7:3) or by thin-layer chromatography with butanol:acetic acid:water (12:3:5). Spray the dried sheets with freshly prepared spray reagent (see Section V.A, note 4). This technique should be applied to novel plant samples, for it will not only indicate the relative concentrations of allantoic acid and allantoin, but will detect potential interferring compounds.

VI. RELATIONSHIP OF UREIDE CONTENT TO NITROGEN FIXATION BY LEGUMES

In greenhouse-grown soybeans, there is a strong positive correlation between xylary ureides concentration and nitrogen fixation as estimated by acetylene reduction, ^{15}N dilution, or nitrogen accumulation (McClure et al., 1980; Herridge, 1982a; McNeil and LaRue, 1984; van Berkum et al., 1985).

In field-grown soybeans, the relationship is not so clear. A portion of the ureide nitrogen in xylem arises from reduction of nitrate in the roots. The ureide concentration in the shoot is a function of its source (cotyledons, nodules, and root) and its sinks (developing shoot, leaves, and pods). It is unlikely that environmental factors affect sources and sinks in the same way, and the relationship of ureide content to symbiotic fixation is not always statistically significant (Patterson and LaRue, 1983a).

Nitrate in soil decreases symbiotic nitrogen fixation but contributes in part to the ureide content of xylem. When the soybean is obtaining nitrogen from nitrate, it is necessary to determine the relative abundance of ureides, determined as the ratio of ureide N to total N (McClure et al., 1980; Herridge, 1982b; Herridge et al., 1984) or the ratio of ureide N to ureide N + nitrate N (Patterson and LaRue, 1983b). There is a stronger correlation between measures of fixation and relative abundance of ureides than with ureide concentration alone.

The relative abundance of ureides in vacuum-extracted xylem sap of soybean was estimated nine times during the growing season, and estimated nitrogen fixed in each interval was summed to provide an estimate of seasonal fixation (Herridge et al., 1984). The estimate of 90 kg of fixed nitrogen per hectare was close to the estimate of 112 kg/ha determined from the difference of nitrogen content between inoculated and uninoculated plants.

The highest correlations in soybean are obtained during early pod fill, the R2 and R3 stages. The analysis of ureides in young shoot tissue harvested at R2 was used to rank soybean cultivars, but ureide concentrations could not distinguish among cultivars in the same maturity group (Patterson and LaRue, 1983a,b). In another study, van Berkum et al. (1985) did not find a cultivar difference, but detected differences associated with strains of *Bradyrhizobium japonicum*. When the spot test was applied to field-grown soybeans, it correlated well with the colorimetric assay for ureides and could be used to identify poorly fixing plants (Glenister and LaRue, 1986).

There are many potential advantages to using ureides to estimate fixation. There is no diurnal variation, only a portion of the plant need be harvested, samples are easily sorted, and assays are sensitive and do not require expensive chemicals or apparatus. If nitrate or total nitrogen in the plant must also be measured to determine the relative ureide abundance, however, the procedure is then much more time consuming than are acetylene reduction assays.

Detailed studies have been conducted only with one species, and it is not known how widely applicable the ureide analysis will be. The potential use of ureide analysis for comparing cultivars or rhizobial strains deserves further research.

ACKNOWLEDGMENT

Research at the Boyce Thompson Institute described here was supported by a grant from the United Nations Development Plan to the International Institute for Tropical Agriculture at Ibadan, Nigeria.

REFERENCES

Bollard, E. G. (1953). *J. Exp. Bot. 4*: 363.
Borchers, R. (1977). *Anal. Biochem. 79*: 612.
Domnas, A. (1961). *J. Biochem. Tokyo 50*: 46.
Glenister, R. A., and LaRue, T. A. (1986). *Exp. Agric. 22*: 405.
Hermanowicz, W. (1948). *Rocz. Chem. 22*: 159 [from *Chem. Abstr. 44*: 1418h (1950)].
Herridge, D. F. (1982a). *Plant Physiol. 70*: 1.
Herridge, D. F. (1982b). *Plant Physiol. 70*: 7.
Herridge, D. F. (1984). *Crop Sci. 24*: 173.
Herridge, D. F., Roughley, R. J., and Brockwell, J. (1984). *Aust. J. Agric. Res. 35*: 149.
McClure, P. R., Israel, D. W., and Volk, R. J. (1980). *Plant Physiol. 66*: 720.
McNeil, D. L., and LaRue, T. A. (1984). *Plant Physiol. 74*: 227.
Patterson, T. G., and LaRue, T. A. (1983a). *Crop Sci. 23*: 819.
Patterson, T. G., and LaRue, T. A. (1983b). *Crop Sci. 23*: 825.
Patterson, T. G., Glenister, R., and LaRue, T. A. (1982). *Anal. Biochem. 119*: 90.
van Berkum, P., and Sloger, C. (1983). *Plant Physiol. 73*: 511.
van Berkum, P., Sloger, C., Weber, D. F., Cregan, P. B., and Keyser, H. H. (1985). *Plant Physiol. 77*: 53.
Vogels, G. D., and Van der Drift, C. (1970). *Anal. Biochem. 33*: 143.

14

Evaluation of Nitrogen Fixation by Legumes in the Greenhouse and Growth Chamber

ALAN H. GIBSON *Division of Plant Industry, Commonwealth Scientific and Industrial Organisation, Canberra, A.C.T., Australia*

I. INTRODUCTION

To propose that laboratory-based evaluation of biological nitrogen fixation by legumes is an artform rather than a scientific exercise would be incorrect, but the plethora of techniques described in recent years lends credence to the proposal. Not very long ago, most experiments were conducted on agar in cotton-plugged tubes (Thornton, 1930) or in sand or vermiculite in the cumbersome but relatively safe Leonard jars (Leonard, 1944). Over the past 15 years, both systems have been modified by numerous researchers and a wide range of new systems proposed. These extend from petri dishes and plastic growth pouches, through virtual anaerobic plastic tubes, to large-scale tanks for evaluating host rather than endosymbiont performance. Most of these developments have undoubtedly facilitated the research programs, either through increased speed in setting up, through more efficient use of space, through the use of cheaper or more readily available materials, or by adapting systems with which the researcher has familiarity in other fields of study. The extent to which these developments have affected the standard of the research is a subject rarely considered in polite circles.

The first criterion in selecting or developing a system is the purpose of the program. For example, where it is the intention to use a "plant dilution most probable number" approach to count rhizobia in soils, inoculants, or culture suspensions, there need be *less* concern over physical growth conditions than over bacteriological control. Similar considerations apply to the determination of a strain's ability to form nodules on a particular host. However, in both cases, neglect of the physical conditions could have disastrous consequences, as outlined later. For evaluation of the symbiotic capability of strains known to nodulate particular hosts, it is more realistic to provide optimal growth conditions to achieve full expression of symbiotic potential, even at the risk of low-level bacterial contamination. In all cases, it is a matter of priority, backed by a sound understanding of the material under consideration, the objectives of the program, and of course, the prudent and judicious use of adequate controls. Under no circumstances is expediency a satisfactory alternative to the adoption of sound scientific methods.

In this chapter the principal features of environmental influences on symbiotic development and plant growth are reviewed, particularly as they are relevant to the evaluation of nodulation and nitrogen fixation, and various systems for plant culture are presented.

II. ENVIRONMENTAL CONSIDERATIONS

Temperature, light, and moisture are obvious environmental parameters to consider, but O_2 and CO_2 levels, mineral nutrition, and pH must be taken into account when establishing cultural conditions. There are specific effects of most environmental variables on one or more stages of symbiotic development and conditions adequate for the growth of plants supplied with mineral nitrogen may not be suitable for nodulated plants.

A. Temperature

The optimal root temperature for infection and for nodule development and function varies depending on the characteristic being studied, and on the host species. For species derived from temperate regions, root temperatures in the range 20 to 24°C provide ideal conditions for orderly nodule development. Although it is possible to hasten infection by increasing the temperature to 30°C (Gibson, 1967a), subsequent nodule development may be disrupted severely (Pankhurst and Gibson, 1973). Infection of temperate species is progressively retarded with decreasing temperature to approximately 7°C, where it ceases (Roughley and Dart, 1970), although nodulated plants transferred to this root temperature will fix nitrogen, albeit at a greatly reduced rate (Gibson, 1967b). For species of tropical and subtropical origin, root temperatures in the range 25 to 30°C appear optimal for nodulation and nitrogen fixation (Jones and Tisdale, 1921). The upper limit for nodulation varies between 35 and 40°C (see Gibson, 1977; Munevar and Wollum, 1981, 1982), an effect that could be associated with poor growth and survival of the rhizobia as much as inherent inability of the plants to nodulate at these temperatures. Below 15°C, nodulation may not occur, but if it does, the delay in providing nitrogen may still cause the death of plants growing in a nitrogen-free medium.

Shoot temperature has a strong effect on assimilate flow to the roots (Schweitzer and Harper, 1980; Eckart and Raguse, 1980). Care should be exercised in doing acetylene reduction assays at or near the growing temperature of the plants (Gibson, 1976).

For most forms of experimentation, the selected temperature conditions should be optimal for the species, and if this information is not available in the literature, preliminary studies should be made to ascertain the appropriate conditions. Black or dark pots will elevate soil temperatures and these may be detrimental to nodulation (Meyer and Anderson, 1959), while the temperature in closed containers (e.g., cotton-plugged tubes) will be 2 to 4°C higher than ambient temperatures during the light period. Shading of greenhouses by whitewash, of which there are a number of suitable formulations, by shade cloth, slatted blinds, or the use of adjustable louvres (Hely, 1959) is essential in many areas, while another approach is to place sealed pots in cooled-water baths (Rogers, 1969). Insulated metal boxes fitted with copper-covered heating elements, fans, and thermostat can provide supplementary heating without the expense of unnecessarily raising greenhouse temperatures (Gibson, 1980a) (see Section III.A).

B. Light

The quantity (irradiance), quality (spectral composition), and photoperiodic duration can each influence nodulation, either directly, or indirectly through a general effect on plant growth. Irradiance levels strongly affect nodule tissue development (Gibson, 1976) but reports vary as to their effect on the specific nitrogenase activity of the nodules (e.g., Wahua and Miller, 1978; Sprent, 1973). Legumes are particularly sensitive to spectral composition and nodulation is affected by far-red wavelengths, presumably mediated by phytochrome (Lie, 1969a). The inclusion of tungsten filament lamps, either as conventional globes or strip lamps, in the light bank is frequently beneficial for legumes, although many modern fluorescent tubes have improved spectral composition for plant growth. To achieve plant development similar to that found under field conditions, scientists at the Plant Growth Laboratory, University of Reading, England, grow species to be studied under a range of light sources before beginning experimentation (Summerfield et al., 1977b).

The flowering of many legumes is influenced by daylength; for example, the Fabeae (*Vicia, Pisum, Cicer*) and Genisteae (*Lupinus*) are long-day plants, whereas the Phaseoleae (*Glycine, Vigna, Phaseolus*) flower as the days shorten (Evans and King, 1975). As many species show a change in N_2-fixing activity during pod fill (e.g., Mague and Burris, 1972; LaRue and Kurz, 1973; Lawn and Brun, 1974a), due presumably to competition for assimilates between the developing pods and the nodules, consideration must be given to daylength and changes in daylength during the course of an experiment.

Under most greenhouse conditions, the quality and quantity of light is adequate for normal plant development. During the winter months in regions of high latitude, supplementary light may be essential.

High-intensity lamps such as mercury vapor lamps, well spaced to minimize shading, should be used. Within artificially lit controlled environment cabinets, irradiance levels rarely exceed 30 to 40% of brigh sunshine, although the uniform irradiance throughout the light period has a compensatory effect. Irradiance levels are reduced by approximately 8% on passage through clear glass and continuing efforts should be made to keep glass surfaces free of dust.

Where possible, roots should be shielded from light to minimize harmful effects on nodulation. In addition to Lie's studies (1969a) and those of Grobbelaar et al. (1971), various personal communications from others attest to reduced nodulation of illuminated roots. However, the effects on specific stages of infection and nodule development are poorly recorded and more poorly understood. For these reasons, caution should be exercised in the use of systems that permit root illumination (e.g., petri dishes, growth pouches, plugged tubes in greenhouses) until it can be demonstrated that the characteristic under study is not affected adversely, either qualitatively or quantitatively. In some species (e.g., *Glycine max, Crotalaria juncea, Sesbania rostrata*) light can induce chlorophyll development in the nodule cortex, but little is known of the effect that this has on nitrogenase activity or nodule longevity. In one system, light was found to promote nodulation of *Trifolium subterraneum* by strains that would not nodulate darkened roots, as in soil (Gibson, 1968). Wooden racks with close-fitting holes, or paper cylinders wrapped around the lower part of test tubes, can reduce light effects on roots of plants grown entirely within tubes.

Plant spacing is an important consideration in some types of experimentation. For symbiotic effectiveness studies, mutual shading should be minimal if symbiotic potential is to be realized. Growth rates are reduced after the leaf area index (ratio of leaf area to experimental area) exceeds 4. Alternatively, in his studies with pasture species, Silsbury (1981) used close plant spacing and allowed full canopy cover to be achieved before imposition of treatments or commencing manipulation of the material; the canopies were confined within aluminum borders to prevent lateral spread.

C. Water

Water stress may be due to deficiency or excess, and for nodulated legumes, the optimum level of supply covers a more limited range than plants supplied with adequate mineral nitrogen. Deficiency decreases infection of clovers (Worrall and Roughley, 1976) and severely retards nodule development (Zablotowicz et al., 1981). Water-deficient nodulated plants exhibit a marked decline in nitrogenase activity, with recovery dependent on the degree and length of stress and the species; meristemmatic-type nodules tend to show better recovery than spherical nodules (Sprent, 1976; Aparicio-Tejo et al., 1980; Ismaili et al., 1983; Albrecht et al., 1984). Opinions vary as to whether the effect is due to nodule dehydration, reduced photosynthesis, or lowered rates of assimilate supply to nodules.

The effect of excess water is primarily one of lowered oxygen supply, and has the effect of retarding nodulation (Bond, 1950; Minchin et al., 1978) and of decreasing nitrogenase activity. In peas, the decline in nitrogenase activity appears to be associated with a decrease in nitrogenase component II (the Fe-Mo protein, or dinitrogenase reductase) (Bisseling et al., 1980). Under conditions of a high maintained water table, soybeans show very extensive nodule development above the free water level (Hunter et al., 1980; Nathanson et al., 1984).

Essential to experimentation with water stress is the measurement of soil and plant water potentials. For leaves and larger nodules, a pressure chamber is used (Scholander et al., 1965), the pressure being increased slowly until water appears at the cut surface of the petiole or nodule extruding from the chamber. Ismaili et al. (1983) used a Wescor HR-33T microvoltmeter and C-52 sample chamber to determine the water potential of both leaves and nodules, while dew point hygrometers and thermocouple psychrometers have also been used (Nelsen et al., 1978). Soil water potential is measured by psychrometers, by tensiometers, or by oven-drying soil and relating water content to previously determined soil moisture release curves. With regard to experimentation on moisture stress, Begg and Turner (1976) point out that stress under laboratory conditions is induced more rapidly than in the field, where moisture reservoirs are larger and there is opportunity for a degree of adaptation by the plants.

Various systems have been developed to maintain predetermined soil moisture levels. Hack (1971) placed auto-irrigator cells around each plant, with each cell supplied from barostat reservoirs to provide equal hydrostatic pressure throughout the pot. Polyethylene glycol can be used to regulate water supply, but it is toxic to roots and must be separated from the growth medium by a semipermeable membrane (Johnson and Asay, 1978). Although these systems have not been used with legumes, they should be adaptable for studies on nodulation and nitrogen fixation.

For most experimental purposes, the principal concern is the maintenance of optimal moisture conditions. The pots should be free draining, and a free-draining substrate such as vermiculite, perlite, coarse sand, or special commercial preparations selected. A 1:2 mixture of vermiculite and perlite, or 1:1 sand and vermiculte, has been used successfully in my laboratory. Fine sand is likely to cause waterlogging. Depending on pot size, plant size, and the temperature conditions, one or two waterings each day will suffice, although under hot conditions (above 27°C), three waterings will be necessary for well-grown plants. The placement of pots in large trays is satisfactory under some circumstances, but is not recommended for strain studies due to contamination risks, or for nutritional studies. Although the pots should be free draining to prevent waterlogging, great care should be exercised to prevent run-through where different bacterial strains are used. When run-through cannot be avoided, the use of drainage pots fitting around the base of the plant pot is recommended. Alternatively, the use of capped, 10 to 15-mm watering tubes, extending 60 to 100 mm into the growth medium, can minimize the risk of contamination. The Leonard jar assembly (see Section IV.F) obviates many of these problems. Finally, care should be exercised to water the pots with water or nutrient solution at the same temperature as the greenhouse or growth chamber. Cold water, as from many water supply systems, may seriously retard growth even though the temperature effect is transient (Brockwell and Gault, 1976).

D. Carbon Dioxide

The close relationship between photosynthesis, nitrogen fixation, and plant growth is widely accepted and changes in CO_2 concentration can markedly influence the symbiotic association involving legumes. Increasing CO_2 levels in the canopy will promote nitrogen fixation (Wilson et al., 1933; Havelka and Hardy, 1976), although the basis for the increase (increased specific nodule activity, increased nodule growth) varies with species (e.g., Masterson and Sherwood, 1978). Apart from effects on photosynthesis, increased CO_2 in the root zone can affect symbiotic performance. Nodulation of root cultures of *Phaseolus vulgaris* is inhibited by high CO_2 (Grobbelaar et al., 1971) but elsewhere, the effect on nodulation and nitrogen fixation is generally stimulatory (Mulder and van Veen, 1960). A possible explanation lies in the presence of phosphoenolpyruvate carboxylase, responsible for dark CO_2 fixation in many legume nodules (Christeller et al., 1977; Coker and Schubert, 1981; Vance et al., 1983).

In controlled environment cabinets, the fresh-air flow into cabinets should be four to six times greater than the maximum CO_2 demands of the plants (Morse and Evans, 1962) if ambient CO_2 levels are to be maintained. This point is frequently neglected in assessing cabinets, especially those constructed in local workshops. Plant growth and nitrogen fixation in cotton-plugged test tubes may be limited by CO_2 diffusion into the tubes, with the result that differences in symbiotic performance by different strains of rhizobia may be masked (Gibson, 1967c). The use of *some* foam push-in caps, or loose-fitting "bacteriological" caps, restricts diffusion six to eight times more than cotton plugs, and they should be avoided unless the purpose is for bacterial enumeration in plant dilution assays. The slow growth of plants in tape-sealed petri dishes (see Section IV.C) is probably a consequence of poor CO_2 diffusion and low light intensity.

E. Mineral Nutrition

Many elements have been implicated in the successful formation and function of legume nodules, with calcium, phosphorus, molybdenum, boron, cobalt, copper, and iron each having a particular role (see Munns, 1977). More recently, nickel has been implicated in the hydrogen uptake enzyme (see Harker et al., 1984).

Table 1 provides the formulations for a number of commonly used plant nutrient solutions. The list is by no means exhaustive, but analysis of the formulations indicates a basic similarity in the concentration of the component ions. Most of the elements are provided in excess and nutrient disorders are likely to arise from toxicity rather than deficiency. Phosphate toxicity is the most likely source of trouble, and downward adjustment, at the risk of loss of buffering capacity, may be necessary. Common symptoms include leaf margin scorch, leaf wrinkle, deformed leaves, and chlorosis, but in many instances, nutrient disorders will be reflected only in reduced yield. The diagnosis of nutrient disorders for many legumes is reviewed by Tisdale and Nelson (1966). Caution is advised in discarding experiments at an early stage due to chlorosis. This may be due to a delay in the nodules meeting the plant's requirements for fixed nitrogen when the plants are grown under otherwise ideal conditions in nitrogen-free media. Such plants should green rapidly within several days although I have observed plant death (e.g., in some *Vigna* species) where the onset of nitrogen fixation has been delayed by 3 to 4 days.

For pot-grown plants, it is advisable to water once each day with nutrient solution and on other occasions with water. This minimizes "salt" injury. Nutrient solutions are normally prepared from stock solutions 100+ times more concentrated than the final nutrient. To avoid precipitation of essential salts, especially trace elements, these stocks should be added to the final volume of water rather than mixed then diluted.

F. Combined Nitrogen

The literature on the effects of combined nitrogen on nodulation and nitrogen fixation is as extensive as that on any other topic involving the symbiosis. Most of the reports concern inhibition of infection, nodule development, or nitrogenase activity, or increased rates of nodule senescence. Over longer periods, stimulation of nodulation and nitrogen fixation is common, an effect due to enhanced plant growth prior to nodulation and the depletion of supplied mineral nitrogen (see

Dart, 1977; Munns, 1977; Harper and Gibson, 1984). The extent of the inhibitory effects is influenced by the host plant, the bacterial strain, and the form and concentration of the combined nitrogen source. There is poor understanding of the physiological effects of combined nitrogen on nodulation, despite numerous hypotheses, although there is reason to believe that the influence of nitrate is external to the plant (Gibson and Harper, 1985). Photosynthate deprivation (Oghoghorie and Pate, 1971) and the inhibition of nitrogenase by nitrite (Trinchant and Rigaud, 1984) are two popular explanations for the inhibition of nitrogenase activity following nitrate application. Recent unpublished results from this laboratory suggest that both factors could be involved.

Most laboratory and greenhouse media, apart from soil, contain little or no combined nitrogen. However, if tap water is used to prepare the medium or to water the growing plants, this water should be analyzed for nitrate prior to experimentation. Depending on the nature of the experiments, as little as 4 ppm NO_3-N can influence results (Gibson and Nutman, 1960).

For some types of experiments, the use of nitrogen-free media can be questioned. Since the significance of legume nodules was realized 100 years ago, it has been customary to use nitrogen-free media. Assessment of nitrogen fixation is facilitated and the interpretation of the results is not complicated by possible differential inhibition of nodulation by mineral nitrogen. However, most soils have low-to-moderate levels of available nitrogen at the time of sowing and the inclusion of 10 to 20 ppm NO_3-N in the medium at sowing would be realistic. This approach can be justified for strain trials and would benefit the cultivation of nodulated legumes, where the principal purpose is the preparation of large quantities of nodule material. Recent studies in this laboratory using the stem-nodulating species *Sesbania rostrata* have been facilitated by supplying the plants with mineral nitrogen for 3 weeks, then eluting the nitrogen from the medium at the time of inoculation. Initial growth of this small-seeded species is very slow, but a word of caution is required. Between 3 and 8 weeks from germination, the plants increase in height from 30 to 220 cm under favorable growing conditions.

G. pH

As a general rule, pH should be maintained in the range 6.0 to 7.0. Temperate species tend to have a higher minimum pH for nodulation than the tropical species. For example, *Medicago* spp. show poor nodulation below pH 6.0 and develop few nodules below pH 5.2; this reduction in nodule number may be compensated by an increase in the size of individual nodules. For *Trifolium* species, pH 4.7 tends to be the lower limit, although Small (1968) has recorded nodules on *T. africanum*

Table 1 Composition of Nitrogen-free Mineral Nutrient Solutions (mg/liter) Used for Tube, Pot, and Hydroponic Culture of Nodulating Legumes[a]

	Tube culture			Pot culture		Hydroponics	
Chemical	Thornton (1930)	Jensen (1942)	Fåhraeus (1957)	McKnight[c] (1949)	Gibson (1980a)	Munns (1968)	Harper and Gibson (1984)
K_2HPO_4	500	200	—	—	130	—	30
KH_2PO_4	—	—	100	200	—	140	11
NaH_2PO_4	—	—	150	—	—	—	—
$CaHPO_4$	—	1000	—	—	—	—	—
$Ca_3(PO_4)_2$	2000	—	—	—	—	—	—
$CaCl_2$	—	—	100	—	6	—	56
$CaSO_4 \cdot 2H_2O$	—	—	—	1500	—	1088	350
$MgSO_4 \cdot 7H_2O$	200	200	120	200	110	230	250

K_2SO_4	–	–	–	–	–	–	210
NaCl	100	200	–	–	–	–	–
KCl	–	–	–	300	110	0.7	–
$FePO_4$	1000	–	–	–	–	–	–
$FeCl_3$	10	140	–	140	8[d]	3[d]	e
Fe citrate	–	–	5	–	–	–	–
pH	6.6	6.6	6.5	7.0	6.8	6.5	6.5[f]

[a]The trace element solution can be used with all formulations. H_3BO_3, 2.86 g; $MnSO_4 \cdot H_2O$, 2.08 g; $ZnSO_4 \cdot 7H_2O$, 0.22 g; $CuSO_4 \cdot 5H_2O$, 0.08 g; Na_2MoO_4, 0.11 g. Prepared in 1000 ml, used at 1.0 ml/liter final nutrient.

[b]Agar 1.2 to 1.6% (w/v) depending on batch, brand, and firmness required. 1.6% preferred for aluminum-foil-capped tubes.

[c]Also used in tube culture with vermiculite as base.

[d]$FeCl_3$ (10 g) or $FeSO_4 \cdot 7H_2O$ (17.2 g) is sequestred with ethylenediametetraacetic acid (22.8 g) in 1000 ml of H_2O containing KOH (10 g). Aerate mixture vigorously overnight.

[e]Sequestrene Fe-300 chelate (Ciba-Geigy, Greensboro, North Carolina) at 22 mg/liter.

[f]MES at 21.0 g/100 liter, with KOH (4.4 g/100 liter).

as low as pH 4.0. While tropical legumes may nodulate down to pH 4.0, many show a strong increase in nodule number associated with liming (Munns et al., 1977). Some tropical species (e.g., *Stylosanthes* spp.) show a decline in nodulation above pH 5.5, but this is unusual. Alkaline pH has received less attention than acid conditions, but the lowered availability of Fe, B, Zn, and phosphate above pH 6.0 to 6.5 should be taken into consideration in formulating any medium.

All plants cause a drop in the pH of the medium due to H^+ excretion during cation uptake and to the excretion of organic acids. Soils have, to a greater or lesser extent, a natural buffering capacity, but sand culture and other artificial media depend on the nutrient medium supplied. With agar medium, appropriate mixtures of K_2HPO_4 and KH_2PO_4 can be used to establish and maintain the required pH for 20 to 30 days depending on volume of medium, plant species, and growth conditions. However, phosphorus toxicity can result in excessive use of these salts. In hydroponic systems, the use of a cation-exchange system such as Amberlite IRC50, in association with a nutrient-recirculating system, minimizes pH changes (Lahav et al., 1976). With *Glycine max*, the inclusion of MES [2-(*N*-morpholino)ethanesulfonic acid] at the rate of 21.0 g per 100 liters of final nutrient solution, following adjustment with 8.8 g KOH to raise the pH to 6.5, does not interfere with nodulation and reduces the need for adjustment (Imsande and Ralston, 1981). However, the effect of MES on the nodulation of other species is variable and should be examined when establishing a hydroponic system (Harper and Gibson, personal experience). The micaceous substrate, vermiculite, may have a pH as high as 9.0, depending on source. Thorough washing overnight in acidified water (pH 5.0) is recommended. Vermiculite has an inherent buffering capacity that renders it unsuitable for pH studies, as it can revert to pH 6.0 to 7.0 (from 5.0, or from 8.7) in 2 days (D. C. Jordan, personal communication). Nitrogen control systems behave differently from no-nitrogen nutrient media. With NO_3^- uptake, the pH rises and with NH_4^+ uptake, it falls, and in hydroponic culture in particular, regular amendment of pH is required. The use of NH_4NO_3 can minimize these changes.

Low pH affects root hair infection rather than later stages of nodule development. Advantage of this situation may be taken to increase nodule size, at the expense of nodule number, where nodule material is required for biochemical or physiological studies on plants normally producing many small nodules (Munns, 1968; Lie, 1969b). This is achieved by lowering the pH of the medium 4 to 7 days after nodules first appear. However, experimentation is required to achieve a pH that minimizes any adverse effects on nutrient availability. Lie and Brontonegoro (1969) have used Fe(III)EDTA with peas to reduce nodule number and to increase nodule size.

III. GREENHOUSES AND GROWTH CHAMBERS

The necessity for adequate temperature and lighting was discussed in Section II, and in this section, consideration will be given to some means of achieving suitable conditions for growth of inoculated plants. A feature not considered previously was hygiene, which is vital for experiments in both greenhouses and growth chambers.

Legumes are particularly susceptible to many fungal pathogens, to predatory and/or virus spreading insects, and to some arachnids. To minimize insect attack, fine mesh should be placed over evaporative cooler entry ports and ventilation louvres in the side or top of the greenhouse, and over air-exchange ports in growth chambers. They should be cleaned and checked regularly. A double-door system, providing an atrium (insect-proof vestibule) to the greenhouse, reduces the potential danger of insect and fungal entry. An atrium also enables workers to clean or change their footwear before entry and to cover their clothes with a clean laboratory coat or overall; this is particularly important for workers moving among different greenhouses (e.g., those responsible for watering). Combing the hair, cranial and facial, helps to reduce both spore loads and insect transmission. Ideally, the greenhouse should have a cement floor that can be hosed down each day. All dead leaves should be removed from benches regularly, as they provide a haven for insects and fungi. Should aphids or red spider (*Tetranychus urticae*) appear, fumigation by experts should be done. In fact, before embarking on a program with particular species, preliminary trials with recommended dosages of fungicides, aracnicides, and insecticides should be made to ensure that they are not toxic to the nodulated plants. Recent experience in my laboratory has indicated that some of the more exotic legumes (e.g., *Sesbania rostrata*) may defoliate with normal applications of insecticides or aracnicides; it is advantageous to have a safe, but reliable battery of pesticides at your disposal in the event of trouble. Under some circumstances, biological control can be achieved, such as the use of the predaceous mites *Phytoseiulus persimilis* or *Typhodromus occidentalis* to control the ubiquitous red spider (*Amblyserus fallacis* is recommended for higher temperatures), or the parasitic chalcid wasp *Encarsia formosa* (Perifleur Ltd., Woodlands Avenue, Littlehampton, Sussex, England) to control white fly (*Trialeurodes vaporariorum*).

The second requirement for a high level of sanitation concerns experiments involving different bacterial strains. The aerial load of rhizobia must be kept to a minimum. Air circulation in greenhouses and growth chambers is vital; hence it is essential that control be exercised through adequate watering procedures that minimize splash and minimize run-through from pots. Drainage saucers beneath pots provide a continuing source of contamination, but plastic bags around the pots serve to catch run-through and prevent dispersal of the rhizobia. Essential for these experiments are the inclusion of uninoculated control plants, sufficient to permit frequent sacrificial harvests to determine the level of contamination.

A. Greenhouses

A severe problem in greenhouses is overheating during the summer months, especially where plant culture is done in test tubes, although I have recently seen unacceptably high temperatures (30°C+) in an uncooled, poorly ventilated greenhouse when external ambient temperatures were approximately 10 to 14°C. Louvres (Hely, 1959), blinds, or whitewash help keep the heat load to an acceptable level, while evaporative coolers may be used under low-humidity conditions. Top ventilation, preferably controlled by a thermostat, is essential. Flowing water over the roof and/or regular hosing of the floor may be necessary under very hot conditions. Recent Australian research into greenhouse construction has resulted in 30 to 40% savings in operation through the use of new materials and innovative design. An inner lining of transparent polythene film beneath the glass, or cladding with double-sheet acrylic with at least 5 mm between the sheets, can reduce fuel consumption by 30%. Rock piles, stored adjacently and/or beneath benches, are used to store heat (or "coolth"). Other features are applicable to hot environments with high solar radiation year round, but are also relevant in temperate regimes. Details can be obtained from the CSIRO Centre for Irrigation Research, Griffith, N.S.W., 2680, Australia.

Under some circumstances, such as the culture of legumes of tropical or semitropical origin, it may be essential to provide heating, especially at night. These species may perform poorly when temperatures fall below 20 to 25°C, particularly during the nodule formation stage. A procedure adopted by F. J. Bergersen to ensure adequate nodulation of *G. max* in the autumn (fall) and early spring is to place the pots with germinating seedlings into holes in the top of a heated box (Bergersen and Turner, 1970; Gibson, 1980a) (Figure 1). The air in the box is heated with copper-covered soil heating cables, distributed by fans in both ends of the box, and thermostatically controlled to maintain 26°C. A prime advantage of this system is that heating costs are reduced greatly as the remainder of the greenhouse is held on a 24°C (day)/18°C (night) regime. This procedure has extended by 4 months the period under which suitable bacteroid preparations can be made. The system has also proved invaluable in obtaining nodulation of *Casuarina* spp. by *Frankia* sp., these plants seemingly having a minimum temperature for nodulation above 20°C.

Figure 1 Insulated root temperature boxes used to promote early nodulation of tropical-type legumes and casuarinas. Copper-covered insulated soil heating wires, thermostatically controlled, are located 60 mm above the base of the box. Pots rest neatly in holes in the plywood cover (not shown). Note also the fine mesh over the window, to exclude insects. This greenhouse is fitted with adjustable louvres.

B. Controlled Environment Chambers

Supplied by numerous manufacturers, controlled environment cabinets come in a range of sizes and with various degrees of sophistication to achieve required conditions. Alternatively, suitable chambers can be constructed in situ, but the services of experienced refrigeration and electrical engineers should be sought at the design and construction stages. Underestimation of cooling requirements is a common problem; some alleviation of a potential heating problem can be achieved by locating the fluorescent-tube transformers (ballasts) in a well-ventilated position external to the chamber but having consideration for the need to keep the envelope temperature of the tubes in the range 35 to 40°C for maximum efficiency. In any system, humidity control may not be a luxury if it reduces watering with its attendant risk of contamination.

Root temperature control is more important than shoot temperature control, and various systems for maintaining root temperature under both warm (e.g., Rogers, 1969) and cool (e.g., Gibson, 1980a) conditions have been described. For the culture of small-seeded legumes, with the roots growing on agar slopes in aluminum-foil-covered test tubes and the shoots freely exposed above the aluminum foil (Section IV.D), I have used a specially constructed cabinet with four temperature-controlled tanks, an independently controlled shoot temperature system and a light bank providing 300 microeinsteins per square meter per second (Gibson, 1980a). For larger plants, insulated tanks 25 cm width, 30 cm deep, and 200 cm long have been constructed. Drain plugs fitted to the base of 8-cm plant culture pots screw into drainage lines. An automatic watering system delivers sterile nutrient solution every 2 hr to the pots in which the growth medium is perlite. The nutrient solution is circulated within stainless steel pipes within the tanks to achieve temperature equilibration before delivery. A refrigerated water bath provides coolant for the tanks, each of which has a thermostatically controlled heater pump circulating the water in the tank. These tanks can be used in growth chambers or in the open greenhouse.

IV. MICROBIOLOGICALLY CONTROLLED SYSTEMS

The purpose and nature of the planned experimental program will determine the degree of microbiological control necessary. Where the symbiotic capability of a strain, be it a field isolate, a known laboratory strain, or a genetic construct, with a particular host is not known, strict control is essential to ensure that the inoculated strain is responsible for the observed response. Similar considerations apply where it is known that the inoculated strain is slow in nodulating the host, or if that strain has low symbiotic effectiveness with the chosen host. Some relaxation of these controls *may* be tolerated when the inoculated strain is known to nodulate promptly and form a highly effective symbiosis with the host, as in environmental studies or where the intention is to obtain a large amount of nodule material for biochemical or physiological studies. However, such relaxation can be acceptable only where the experimenter has a sound knowledge of the symbiotic attributes of host and rhizobia and is aware of the dangers of contamination when interpreting the results. The inclusion of adequate uninoculated controls cannot be overemphasized.

The safest systems are those where surface-sterilized seeds are planted into enclosed systems (e.g., cotton-plugged test tubes). Such systems have the additional merit of requiring minimal watering and hence reduced risk of contamination. However, they are suitable only for small-seeded (up to 10 mg) legumes; other systems are required for larger-seeded species. Furthermore, distinction between degrees of symbiotic effectiveness is difficult in plugged-tube systems (Gibson, 1967c).

Various small-seeded species can be used to substitute for larger-seeded species under certain circumstances, for example, counting populations of rhizobia, where both have the same specificities with respect to nodule formation. Brockwell (1980) lists various alternate hosts, such as *Glycine ussuriensis* Regal and Maack for *G. max*, (*Bradyrhizobium japonicum*), *Macroptilium atropurpureum* (DC) Urb. for various tropical legumes (*Bradyrhizobium* sp.), and *Vicia hirsuta* (L.) S.F. Gray, for a range of legumes nodulated by *Rhizobium leguminosarum*; this latter observation was confirmed and extended to include other *Vicia* spp. by van Brussel et al. (1982). (Further alternate hosts are listed in Section V.C.) However, in the absence of published information, it is essential to demonstrate that the small-seeded surrogate host has the same affinities for strains of rhizobia as the species under examination. For example, although *M. atropurpureum* nodulates with most strains of *Bradyrhizobium* sp., many tropical legumes exhibit a degree of specificity (e.g., Gibson, 1980b; Date and Halliday, 1980). Symbiotic effectiveness is frequently more specific than nodulation, and extreme caution should be exercised in extrapolating from results with surrogate hosts to those expected with other species.

A. Seed Sterilization and Pregermination

The seed sample should be cleaned and all broken seed removed. For symbiotic effectiveness studies, the seed sample should be sieved and graded visually or weighed to provide a sample of relatively uniform size in order to reduce plant-to-plant variation.

The chosen method of sterilization depends on seed hardness. *Arachis hypogea*, with a soft seed coat, can be treated with 2% solution of sodium hypochlorite, the principal ingredient in bleach, for 2 to

5 min, followed by several rinses in sterile water. Mercuric chloride [4 ml of saturated solution (28 g of $HgCl_2$ per 100 ml of H_2O at 80°C) in 100 ml of H_2O plus 0.5 ml of concentrated HCl], preceded by a 10-sec rinse in ethanol, is a common treatment. Exposure for 2 to 5 min, depending on subsequent germination and incidence of contaminants, followed by 10 rinses in sterile water, is normally adequate. The seeds should stand 15 to 60 min after the last washing, and followed by two further rinses at 15-min intervals. For hard-coated seeds, concentrated H_2SO_4 is a suitable sterilant and softener. The seeds, no more than 600 per 50-ml flask, should be desiccated overnight to remove surface moisture. Exposure to acid should be adequate to soften the seed coat (5 to 30 min, depending on hardness and subsequent germination). The acid is drained *thoroughly* into a waste beaker before adding a large volume of water, with vigorous shaking, to minimize local heating. This is followed by a further nine rinses in sterile water, drained into another waste beaker, after which the seed is allowed to imbibe before a final wash. An alternative procedure for hard seed coats is to scarify the seeds with fine emery cloth or sand paper, or to scratch the seeds with a pin. The sodium hypochlorite treatment is then adequate to secure surface sterilization. With all systems, it is vital to ensure that all internal surfaces of the vessel, and the sides of the rubber stopper in contact with the vessel, be moistened by sterilant from the time of addition.

Despite the foregoing procedures, contaminants below the seed coat can be very difficult to kill. Although the contaminants are unlikely to be rhizobia, *Klebsiella* spp. (Evans et al., 1972) and, in the case of *Medicago sativa, Erwinia herbicola* (Handelsman and Brill, 1985) may survive and cause problems in rhizosphere or competition studies. Plating the seeds on a nutrient medium will detect many organisms, but contaminating microaerophiles or anaerobes may not be detected.

Imbibed seeds may be sown directly to the culture medium (twice the number required, then thinned after germination) or plated on 20-ml sterile 0.9% agar in petri dishes. With the latter procedure, the number of seeds per plate will vary between 20 and 100 depending on seed size, as the capacity of the agar to supply water for germination is limited. Inversion of the plates encourages the development of straight radicles, which is advantageous when sowing on to agar. Some species have a cold requirement (e.g., many clovers, medics), and 48 hr at 4°C prior to incubation at 20 to 25°C produces more even germination.

Under conditions where it is known that the inoculant rhizobia will nodulate the host promptly and effectively, and other conditions will provide minimal cross-contamination, surface sterilization with 70 to 95% ethanol, followed by five rinses in sterile water, may be adequate. This procedure has been used for larger-seeded species sown directly into vermiculite/perlite medium in pots; it must always be accompanied by the use of adequate uninoculated controls as a safeguard.

B. Plugged Tubes

The culture of plants entirely within plugged tubes is the safest way of ensuring the integrity of the symbiotic associations under examination (Figure 2a). Thornton (1930) developed the system for small-seeded legumes, using a slope of seedling nutrient agar to support the seedling. Tube size will vary with seed size, but 20 mm × 150 mm tubes, with 8 ml of agar, are adequate for species with seeds up to 8 mg; to 15 mg seed weight, 25 mm × 150 mm or 25 mm × 200 mm tubes, with 15 ml of agar, are preferred. The agar (1.2 to 1.6%, depending on inherent strength) is melted in the required volume of solution, the nutrients as stock solutions added, and the pH adjusted (normally 6.6 to 6.8) before dispensing. Automatic-filling, hand-operated syringes (e.g., Cornwall syringes, Becton-Dickinson, Rutherford, New Jersey) or automatic pipetting machines, greatly facilitate the task.

Traditionally, the tubes have been stoppered with rolled cotton wool plugs, although we have used 25-mm lengths of cylindrical, lenticulated Styrofoam successfully in recent years. The cotton wool plugs are rolled from 30-mm strips cut from a large roll with a bandsaw.

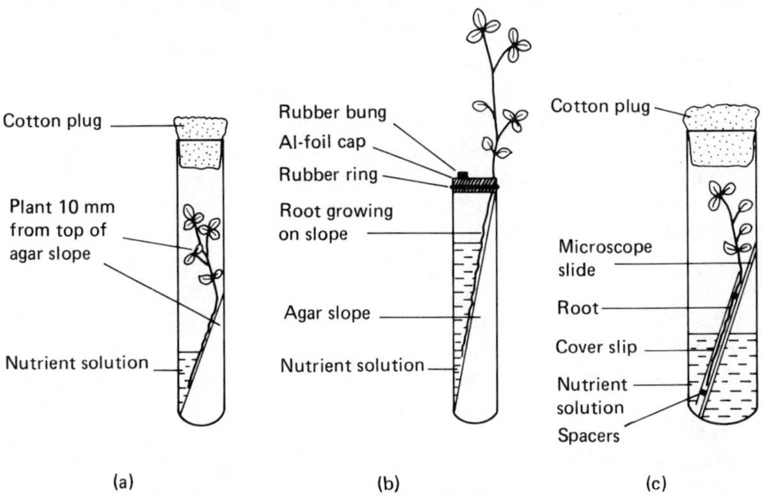

Figure 2 Three systems of tube culture: (a) cotton-plugged tubes (Thornton, 1930); (b) semienclosed tubes, with the shoots extending above an aluminum-foil cover and the roots growing down an agar slope; (c) Fåhraeus slide technique with the roots growing between a microscope slide and a coverslip.

This cotton wool should be nonabsorbent, and the plugs firmly, but not tightly rolled to preserve their shape and form during removal and replacement in subsequent operations. Metal or plastic bacteriological caps should be avoided (Gibson, 1967c). After autoclaving (15 lb, 20 min), the tubes are sloped such that the agar reaches halfway up the tube. Agar deeps should be avoided, as they minimize aeration of the roots.

Various nutrient solutions have been used for tube culture (Table 1), but it is advisable to examine one or more experimentally before adopting one for a species not used previously. A reduction in phosphate levels, at the risk of losing some pH buffering capacity, can reduce leaf scorch with some species. Jensen's medium is satisfactory for the common clovers and some medics, although some species do better in vermiculite medium than in an agar-based medium (Brockwell and Hely, 1966). Standard procedure is to add vermiculite to a depth of 50 mm in the tubes, then 6 to 7 ml (20-mm-diameter tubes) or 12 ml (25-mm-diameter tubes) of McKnight's solution (Table 1) before autoclaving.

For agar culture, seeds may be germinated in situ, near the top of the slopes, or germinated seedlings transferred to the slopes when the radicles are 7 to 15 mm. An inoculation needle bent into the shape of a shepherd's crook (S shaped) is ideal for transferring seedings, and a pair of long-armed, lightly sprung forceps facilitates seed placement. For vermiculite culture, seedlings are preferred to seeds. The seedling is buried 2 mm beneath the surface, with the radicle against the wall of the tube, and the top of the vermiculite is compressed with a flat, sterile rod.

After sowing seedings onto agar slopes or following germination of seeds planted in situ, 2 or 3 ml of nutrient solution, one-fourth the strength of that used in the agar, is added to the tubes. Pipettes or hand syringes can be used, although a burette fitted with a three-way stopcock and a bell-shaped hood around the delivery tip of the burette minimizes contamination. The burette is filled from an overhead reservoir (1-liter conical flask, fitted with a two-hole rubber bung, one hole being occupied by a breathing tube extending to the bottom of the flask and the open end plugged with cotton or surrounded by a cotton-plugged tube; the other hole contains a glass tube fitted with rubber or plastic tubing that leads to the burette). The plugged burette and reservoir fitting, and the stopcock (in a tube), are autoclaved separately and then assembled in a clean, still atmosphere. The flask of sterile nutrient can be replaced as needed.

Plugged tubes are ideal where the aim of the study is to determine nodule-forming ability, to conduct microbiological counts on soils or inoculants by the plant dilution test (Brockwell, 1980), or to ascertain whether a particular symbiotic combination is capable of nitrogen fixation. However, great caution is advised in using this system to ascertain differences in the level of symbiotic effectiveness. Restricted CO_2 diffusion into the tubes limits the expression of potential differences (Gibson, 1967c). Using 20-mm-diameter tubes, exponential growth of *Trifolium repens* (seed weight, 0.7 mg) is retarded at 28 days, while for *Medicago sativa* (2.7 mg) and *Trifolium subterraneum* (10 mg), the retardation in growth commences at 14 and 8 days, respectively. After these periods, the increase in dry weight, and total nitrogen, is linear, and this minimizes potential strain differences. The use of larger tubes delays the onset of linear growth by 5 to 6 days.

Fåhraeus Slide Technique

This is a modified version of the plugged-tube technique, in which the agar slope is replaced by a glass microscope slide; it is used to examine root hair development and infection (Fåhraeus, 1957) (Figure 2c). A large rectangular coverslip is held away from the surface of the slide by two to four chips of cover glass cemented together, with two or three such "spacers" on each edge of the slide. Nutrient solution (Table 1) is added (20 ml) to 25 mm × 150 mm stoppered (cotton or foam) tubes and the assembly autoclaved. Before sowing germinated seeds, sterile soft agar (0.3% w/v) is run into the space and allowed to set (in a sterile petri dish). The germinated radicle of the seed is inserted into the space between the coverslip and slide and a drop of dilute inoculum touched to the root. At various intervals the slide can be removed from the tube for microscopic examination of the roots (e.g., Nutman, 1959, 1962).

C. Petri Dish Method

A procedure favored by many geneticists to screen a large number of isolates for loss of ability to form nodules is the petri dish method (Rolfe et al., 1980). This involves sowing germinated seedlings of small-seeded legumes on nutrient agar (30 ml) in 99-mm petri dishes. The lower half of the plate is streaked with the appropriate culture prior to sowing. The plates are left horizontal for 60 min after sowing, to enable the seedlings to adhere to the agar, and a 400 mm × 25 mm strip of Nesco film (Nippon Shoji Kaisha Ltd., Osaka, Japan) is used to seal the two halves of the dish before placing the dish in a vertical position. Two slits are made in the film, using sterile forceps run across the top of a group of plates, to facilitate gas exchange. Many plates can be located beneath a light bank in a controlled-environment cabinet. Assessment of nodulation can be made regularly. A useful modification to the procedure is to use square (100 mm × 100 mm) dishes.

The technique has value in determining whether or not a strain is capable of nodulating a given host and whether or not nitrogen fixation occurs. However, other techniques are advised where the intention is to determine the degree of symbiotic effectiveness as the conditions, with limited gas exchange and low irradiance within the plates, do not permit full expression of symbiotic potential.

D. Semienclosed Tubes

This system was developed to provide good bacteriological control for plants growing without restriction due to inadequate CO_2 or irradiance, to allow frequent observation of nodulation of the growing plant, and incidentally, to permit studies on the effects of root and shoot temperatures (Gibson, 1965; Vincent, 1970) (Figure 2b). In this system, the roots grow down an agar slope extending to the top of the tube, which is covered by aluminum foil, and the shoots grow freely above the foil cap. Plants from seed weighing 1.0 to 15.0 mg grow well in this system—species grown successfully include *Trifolium subterraneum*, *T. pratense*, *Medicago sativa*, *M. truncatula*, *Macroptilium atropurpureum*, *Stylosanthes humilis*, *Terramnus uncinatum*, *Desmodium* spp., and *Glycine ussuriensis*.

Agar medium (12.5 ml in 20 mm × 150 mm tubes, 25 ml in 25 mm × 150 mm tubes) is prepared as described in Section IV.B. Circles of aluminum foil (45 to 50 mm in diameter) are stamped from sheets of foil folded to 50 mm width and interleaved with newspaper (this aids separation of the disks). Thin foil may have to be doubled before folding. The disk is held in place by a rubber band or preferably, a rubber Elastrator ring normally used for marking and tailing lambs. This ring is applied with Elastrator pliers (Elastrator Pty Ltd., 414 Collins Street, Melbourne, Australia, or stock agents in North America and Europe). Two operators can cap and ring 1000 tubes in 2 hr. After autoclaving, the tubes are sloped with the agar just touching the foil cap. This operation can be facilitated by the use of metal racks, in which the tubes are autoclaved, fitted with two screws protruding 7 mm from the side of the rack 130 mm above the base.

Surface sterilized seeds, germinated in inverted petri dishes, are sown through a hole made in the foil cap with a sterile needle of the same diameter as the radicles. This hole should be 3 mm from the edge of the tube and immediately above the agar slope. Sterile forceps are used to grasp the radicle just below the cotyledons and to insert the radicle into the hole *away from the slope*. The cotyledons are then moved toward the center of the tube so that the radicle lies on the agar surface. To prevent the seed coat from drying out and stopping the cotyledons from separating, a 25 mm × 50 mm tube containing moistened cotton wool is placed over the tube (resting on the rubber ring). To keep the covers in place, wire mesh covers are placed over a closely packed rack or basket of tubes. An easier-to-use but slightly less effective system involves covering the sown seedlings with moistened paper tissues (e.g., Kleenex, Scotties), and the rack or basket of tubes placed in a plastic bag containing free water. Provided that the seeds are well germinated (radicles 10 to 15 mm), two experienced operators can sow 1000 tubes in 2 to 3 hr, quite comparable with rates achieved with other microbiologically controlled systems. The tubes are then placed beneath lights in a controlled environment chamber or in a greenhouse.

The covers are removed after 3 days and any adhering seed coats gently taken off by forceps or a needle. A sterile sharpened probe of 5 mm diameter is used to make a hole in the foil cap diagonally opposite the seedling. Sterile nutrient solution, one-fourth strength of that used in the agar, is added to within 15 mm (*Trifolium* spp.) or 40 mm (other species) of the top of the tube. The hole is fitted with a sterile (rinsed in ethanol, dried 80°C) rubber stopper 4 to 7 mm in diameter, 14 mm long. The inoculum may be added in the watering solution, or as a few drops of suspension (10^7 to 10^8 organisms/ml) by pipette. In both cases, the nutrient is gently agitated over the roots to provide uniform dispersion of the bacteria. Watering is facilitated by an automatic dispensing machine, sterilized by filling the delivery system with ethanol for 5 min, then flushing through with 2 liters of hot sterile water. Ideally, the roots should be darkened during growth, either by a thick paper sleeve or by suspending the tubes into a darkened box (or water bath).

With the consequent improved plant growth, watering is more frequent than in a closed system. The first watering can be delayed until 14 to 16 days, when the water level has dropped to near the lower parts of the roots, but will become more frequent until harvest, which is normally 28 to 35 days from sowing. For longer-term experiments, the plants should be well spaced to ensure that the leaf area index (ratio of leaf area to subtending surface area) does not exceed 4.0, at which time the growth will fall below exponential.

This technique is not recommended where absolute bacteriological control is essential, such as in ascertaining whether a particular strain or isolate is capable of nodulation, or in performing a plant dilution count. Contamination of uninoculated control plants is rare (less than 1 in 100 tubes) and little trouble is encountered with other contaminants. For many effectiveness tests or even environmental studies, this technique is preferred to those involving plugged tubes.

A satisfactory modification of the technique involves replacement of the agar by a 1:1 mixture of vermiculite and perlite. The tube is filled with lightly packed substrate, 15 ml of nutrient solution is added with agitation and packing, and the tube is filled with vermiculite and perlite substrate. The tube is then capped and ringed as above. J. Brockwell (personal communication) has modified the procedure by sowing surface-sterilized seeds through a hole in the cap (15 mm deep) and inserting a rubber bung. After 2 to 3 days, the shoot is gently pulled through this hole and the bung replaced after the tube is inoculated. If these tubes containing vermiculite/perlite are to be placed in a water bath, the inclusion of one or two glass marbles above a shallow layer of vermiculite/perlite during tube preparation will reduce the tendency for the tubes to float. The disadvantages of this system are that watering is more frequent and that nodulation cannot be observed without harvest. However, for those species growing poorly on agar (e.g., *Trifolium resupinatum*, some *Medicago* spp.) it is a most valuable technique.

E. Plastic Pouches

Disposable pouches were introduced by Porter et al. (1966) and modified by Weaver and Frederick (1972) as a plant culture method for dilution counts on legume inoculants and soils, especially where the host plant is large-seeded. The system has been used, with minor modifications, for basic studies of the infection zone on the roots of *Glycine max* (Bhuvaneswari et al., 1980). When stacked carefully, as in a record rack, many pouches can be located in a controlled-environment chamber.

The system comprises a sheet of sterile absorbent paper, with a V-shaped trough, in a 120 to 150-mm square plastic bag. Surface-sterilized seeds are placed in the trough and 40 ml of nutrient solution containing the appropriate dilutions of rhizobia are added through a plastic drinking straw located at one side of the bag. For dilution counts, the assumption is that one viable cell added to the pouch will lead to nodule formation. Again it is essential to include uninoculated controls to maintain a check on the integrity of the system. The system should not be used to assess the level of symbiotic effectiveness of particular host × strain combinations, especially where large-seeded legumes are involved. Nor is the technique recommended for strains whose nodulating behavior is not known or who are known to nodulate slowly (unless other checks of strain identity such as antibiotic resistance or serological characterization are made).

For basic studies of infection, the pouches are marked at the growing point of the root at the time of inoculation. Under some circumstances, a small flap in the side of the bag is created to inoculate the root at a specific point; this flap is then closed and sealed with tape. Anion-exchange beads, 106 to 108 μm in diameter, are used to indicate the point of spot inoculation with 2 to 8 nl of bacterial suspension (Turgeon and Bauer, 1982, 1983).

The pouches may be purchased (American Hospital Supply Corp., Moline, Illinois) or produced in the laboratory. Blotting paper folded to have the V trough is autoclaved in sealed foil (30 min, 121°C) to prevent wetting, and inserted into polythene sandwich bags, under aseptic conditions, immediately before use.

F. Leonard Jar Assemblies

For 40 years, the Leonard jar technique (Leonard, 1944), with some modifications, has been the standard method for growing large-seeded legumes under bacteriological control. The assembly contains two parts, an upper vessel with sand or supporting medium, and this sits on a second vessel containing sterile nutrient solution (Figure 3). The two are connected by a wick. Round, brown beer or spirits bottles of 750-ml capacity make very convenient media containers. The bottom is removed in one of several ways, the easiest being to mark a line around

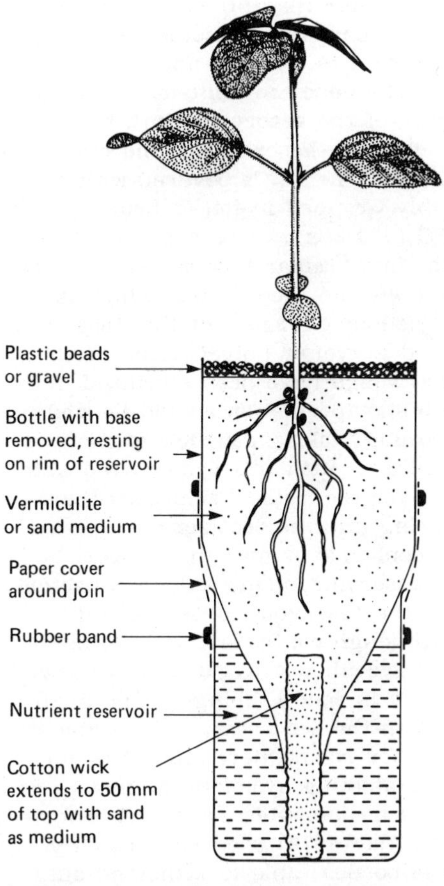

Figure 3 Leonard jar assembly.

the bottle with a diamond-point glass knife, 10 mm above the base. The bottle is rotated slowly in a cradle with the very flat flame from a porcelain-tipped fishtail burner directed at the mark. After four turns, the bottle is plunged into cold water while covering the neck opening with a finger. Sharp edges can be ground down on an emery wheel. A glassblower's abrasive wheel can also be used to remove the bottoms.

A cotton wool strip or lamp wick (thoroughly washed in detergent and rinsed) is used to plug the neck before filling the bottle with vermiculite, well-washed medium-coarse river sand (containing 1 g K_2CO_3/kg), or any other reasonably aerated growth medium. If using sand, the wick should extend to within 50 mm of the top of the medium, to ensure adequate nutrient movement, but with vermiculite, a much shorter wick (50 mm into bottle) is adequate. If a check of the pH of the vermiculite shows it to be high (greater than pH 8), it should be washed in very dilute acid overnight before use. Moisten the medium with nutrient. The reservoir should be 1 to 2 liters, and a square preserving jar is ideal, as it has less tendency to fall over. The medium bottle rests securely on the mouth of the reservoir, with the bottom of the bottle, or at least the wick, close to the bottom of the reservoir, which is filled with nutrient (Table 1). The top is covered with a glass petri dish half and the assembly wrapped in foil or brown paper before autoclaving for 2 hr at 121°C (150 kPa).

Under still, clean conditions, or in a laminar flow bench, surface-sterilized seeds, or seedlings, are sown into the medium, which is recovered with the lid until the cotyledons emerge. At this time, the medium is covered with dry-sterilized gravel or polyethylene beads (e.g., Alkathene, I.C.I., Australia) which have been sterilized in 95% ethanol, dried, and allowed to cool. Inoculation should be made into the holes made at the time of sowing, taking care not to allow inoculum near the surface of the medium. With this system, top watering is eliminated, and the covering over the medium minimizes the chance of casual contamination entering into the root zone.

The units can be placed in a greenhouse or growth chamber. Fresh nutrient can be added to the reservoir, using standard bacteriological procedures, if necessary, or the reservoir bottle exchanged for a fresh, sterilized reservoir. Although contamination by other rhizobia is rare, especially if normal greenhouse hygiene is preserved, other bacteria will be found frequently in root washings. The likely source of contamination is from beneath the seed coat, an important point to consider if using the method for rhizosphere studies or for purposes not solely associated with evaluating symbiotic response.

Various minor modifications to the basic system have been effected in different laboratories. Wacek and Alm (1978) used high-density polyethylene bottles (2 liters); these bottles, able to withstand autoclaving, were cut with a knife just below the shoulder. The two halves served as medium container and nutrient reservoir, but again "joined" by a wick.

Devine and Reisinger (1978) modified the Leonard jar technique to study the symbiotic response of a large number of plants, often derived from a breeding programme, to inoculation with a single strain of rhizobia. Plexiglas trays were filled with sterilized growth medium and lowered onto nutrient reservoirs; the nutrient solution moved into the growth medium through tubes in the bottom of the trays. They used 36 soybean plants per tray, with two trays fitting onto a 80-liter reservoir in which the nutrient was renewed every 2 weeks. Before reuse, the trays were disinfected with 70% ethanol or other disinfectants.

G. Open Pots

Depending on the purpose of the experiments, the cleanliness of the greenhouses, and the degree of contamination that can be tolerated, it is frequently feasible to grow the plants in open pots, with watering of the top surface. The pots must be thoroughly washed between use, rinsed with disinfectant, then rinsed again, preferably immediately before use. A perlite/vermiculite mixture (2:1) provides good aeration as well as satisfactory water-holding capacity of the medium. Sterilization of the medium in autoclavable plastic bags prior to filling the pots is optional. The vermiculite may require neutralization. The medium should be firmly, but not tightly packed and the pots should have a drainage hole to prevent waterlogging. To minimize the potential danger of contamination arising from run-through, the drainage water can be caught in a plastic bag held in place by rubber bands. This does not relieve the responsibility for not overwatering, as excess water will cause anaerobic conditions in the pot. When sowing, either seeds or seedlings, form a hole(s) in the medium and inoculate into the bottom of the hole with 2 ml of bacterial suspension (10^7 bacteria/ml); this minimizes the possibility of emerging seedlings carrying inoculant rhizobia above the surface. Care should be taken not to cross-contaminate the pots when closing the holes and firming the medium around the hypocotyl of seedlings or over the hole. This procedure usually ensures heavy crown nodulation, which is greatly appreciated by "nodule pickers" for physiological studies.

Control procedures may be less stringent when only one strain is in use, as in many physiological-biochemical studies, if only for the reason that the selected host × strain combination is known to form many effective nodules promptly. For example, in studies with soybean nodules, Bergersen and Turner (e.g., 1970) used a 1:1 mixture of washed river sand and vermiculite in 250-mm-diameter pots, each with 10 (summer) or 6 (winter) plants per pot. The nitrogen-free nutrient is McKnight's (Table 1), diluted 1:1 with tap or distilled water. For 2 weeks after germination, the pots are placed in heated boxes at 26 to 28°C to encourage prompt nodulation (Figure 1). To provide the necessary aeration in the medium for their lupin studies, Robertson et al. (1975) preferred sterile pumice, placed in stainless steel trays in the glasshouse or growth cabinet (23/20°C, 12-hr light period). For both soybeans and lupins, the seeds are sown at 40-mm depth, but smaller seeds should be sown nearer the surface.

An interesting adaptation from the horticultural industry has proven valuable for special purposes. This is the Core-tainer Single Cell System (R. Leach, Canby, Oregon), in which plastic tubes, 30 to 100 mm in diameter, fitted into specially constructed racks, are used for growing individual plants (S. G. Pueppke, personal communication). The tubes are filled with vermiculite, coarse sand, or even steamed soil, after treatment with a bacterial agent (e.g., Chlorox) and thorough washing. There is a drain hole in each tube, and watering is done carefully to minimize splashing and to prevent waterlogging. The system is ideal for studying the ability of individual plants to nodulate or to form effective nodules, and as a rough screen of the abilities of individual isolates of rhizobia, in genetic studies. The tubes can be reused many times. Due to their arrangement, very effective utilization is achieved, although care must be exercised to prevent overcrowding (i.e., the leaf area index for the area should not exceed 4).

Provided that the local water supply is free of, or very low in, mineral salts, and especially nitrates and carbonates, tap water can be used to water the plants. If the quality of this water is unacceptable, a demineralizer can be installed, feeding into a header tank as supply rates during watering would exceed the supply of demineralized water. Where water supplies are drawn from undergrounds wells, with the potential for contamination, there is merit in installing an in-line ultraviolet water sterilizer (e.g., Aquafine Corporation, Burbank, California).

Unexplained inhibitory effects on plant growth have been observed where pots are watered with low temperature (10° or less) tap water, even though the fall in temperature in the medium is small and the effect transient (less than 1 hr) (Brockwell and Gault, 1976). The problem can be minimized by installing a storage tank in the greenhouse, or by construction of a mobile, heated tank, with pump and thermostatic control, that is filled each day after use.

For the open pot system to be used, there must be clear recognition of the experimental purposes and the problems that could arise in the interpretation of results if even a low proportion of the nodules are formed by contaminating rhizobia. The efficacy of the system depends on the establishment of a high rhizosphere population of inoculant bacteria and their influence on minimizing nodulation by any contaminating rhizobia. The inclusion of uninoculated controls (5 to 8% of total pots) spread throughout the experimental area is essential; the experiment should be terminated as soon as these plants show evidence of nodulation. Provided that adequate time has elapsed since the establishment of the pots, normal harvest could be appropriate. Although generalizations can be dangerous, the development of nodules on uninoculated controls will often precede the nodulation of inoculated plants by contaminating rhizobia, due to the influence of the rhizosphere population of the latter group.

H. Hydroponics

Large-vessel hydroponic culture has had limited use in nodulation studies, due mainly to problems associated with contamination, the need for good aeration, and concern over pH control. Munns (1968) used 3-liter beakers and 22-liter steel drums, each lined with polyethylene bags, to study pH and combined nitrogen effects on the nodulation of *Medicago sativa*. Fishtank aerators were used to disperse filtered air through each container. The pH was adjusted daily. With these small-seeded plants, it was necessary to germinate the seeds for 7 to 8 days before transferring seedlings with 50 to 60-mm roots to trays with holes for the roots. The nutrient solution was kept to within 25 mm of the top of the container by daily adjustment.

A similar system for studying the nodulation of soybeans has been employed in this laboratory (Harper and Gibson, 1984; Gibson and Harper, 1985). Surface-sterilized seeds were germianted in trays of vermiculite and perlite, and at 6 days, transferred to holes in the lids of 2.5-liter plastic pots. Corks were center drilled with a 3-mm bit and lateral access made through a longitudinal V cut. The hypocotyl was placed in the channel at the base of the V and held in position by a small piece of foam. There were four seedling holes per lid, plus a 1.5-mm hole for the plastic piping supplying air to the nutrient solution. The nutrient solution (Table 1) is amended with MES [2-(*N*-morpholino)ethanesulfonic acid, 21.0 g per 100 liters plus 8.8 g of KOH; pH 6.5]. This had no detectable effect on the nodulation of *G. max*, but trials with other species (e.g., *Lupinus, Vigna*) were inconclusive (see Section II.G). With four plants per pot, daily pH amendment was necessary to keep the pH in the range 6.0 to 6.5; without MES, twice-daily adjustment was necessary. The alternative was to use higher levels of phosphate, with a concomitant increase in the danger of phosphate toxicity. Aeration was provided from a compressed air line, at the rate of 2 to 3 liters per minute. For studies of the effect of combined nitrogen on nodulation, inoculation was delayed until the seedlings were transferred from the germinating trays, but for other studies, the seedlings were inculated in the trays 3 days from setting up and before they had emerged. For experiments longer than 20 days, the plants were reduced to two per pot and the nutrient solution changed completely at 20 days.

A hydroponic culture system for growing single plants to be used in nondestructive acetylene reduction assays has been described by Imsande and Ralston (1981). For plants very susceptible to waterlogging, Zobel et al. (1976) have described a simple, inexpensive aeroponic system for studying nodulation, although to the present time, this technique has had limited use in various experimental programs.

I. Other Systems

The foregoing systems provide a range of techniques that as described or with slight modification, should meet most of the requirements in any studies of the legume–*Rhizobium* association to be made under bacteriologically controlled conditions. However, some specialized techniques have been developed for particular purposes. These are outlined briefly, with further details available in the references cited.

Tissue Culture

Tissue culture systems have been used to study the infection process (e.g., Holsten et al., 1971) but more commonly, to elucidate the factors responsible for the induction of nitrogenase in rhizobia (Child and LaRue, 1974; Phillips, 1974; Gibson et al., 1976). An interesting approach to the problem of induction has involved the culture of plant cells and bacteria in separate containers connected by a tube with one to three bacterial filters to permit the passage of chemical compounds only (e.g., Reporter and Hermina, 1975; Bednarski and Reporter, 1978). Although the results indicate a ping-pong effect (i.e., a bacterial factor is essential to elicit the production of a promotory factor by the plant cells), progress toward defining the compounds involved has been very slow.

Excised Root Culture

Raggio et al. (1957) first described a system in which young roots are excised from the seedling and the cut end inserted into a vial containing organic nutrients and KNO_3. The root then grows on an agar surface, on sand, or into liquid, each of which is supplied with inorganic nutrients (except combined nitrogen). Bunting and Horrocks (1964) modified the technique by including a portion of the hypocotyl when excising the root; this improved greatly the level of nodulation. Despite the apparent promise of this procedure, whereby the contribution of the plant can, in theory, be mimicked by additions to the "organic nutrient vial," the use of the technique has had little impact on our understanding of the symbiosis between legumes and rhizobia.

Propagules

With heterozygous species, plant-to-plant variability often creates severe analytical difficulties in determining the host's genetic influence on nodulation. With *Trifolium repens*, tips of young, actively growing stolons, each with a single apical growing point and an incipient root, have been used (Joshi et al., 1967; Mytton, 1976). Mytton observed that the most effective propagation of the "ramets" was achieved by including one folded and one expanded leaf on each stolon tip.

Stem cuttings have been used to produce populations of genetically identical *Medicago sativa* (Lindsay and Jordan, 1976; L. R. Mytton and A. H. Gibson, unpublished data). Selected donor plants were potted and cut back severely every 4 to 5 weeks to increase the number of new stems. Just before flowering, the stems were cut to produce pieces with 20 mm of stem below a node and 10 mm above the node. The lower end of the root was dipped in talc containing 0.1% w/v 3-indolebutyric acid and the cutting placed in a 50-mm-diameter pot containing sterile sand or perlite-vermiculite mixture. The pots were placed in a humidified atmosphere after watering with a complete nutrient solution, and after 14 days, they were inoculated. A similar procedure has been used with this species by Seetin and Barnes (1977).

Grafting

Various investigators have been interested in distinguishing the influence of the shoots and the roots on nodulation (Hely et al., 1953; Tanner and Anderson, 1934; Lawn and Brun, 1974b; Gibson et al., 1981; Bedmar and Phillips, 1984). The most common procedure is to use a V graft in the hypocotyl region. The excised shoot is pared to reveal 3 to 4 mm of cambial tissue, and when inserting this scion into the V cut into the hypocotyl, care should be taken to ensure that cambial tissue of both stock and scion are in contact. The join is covered (e.g., a plastic drinking straw, 10 to 20 mm long, split, and rolled around the join) and sprayed with water before placing the plant in a plastic bag to retain high humidity. Carbon dioxide can be injected into the bag each day until the graft takes. The plants are hardened slowly, at first by making a cut in the corner of the bag, then opening the bag the next day, before removing the bag the following day. Better than 90% success can be achieved with normally compatible stocks and scions.

V. NONSTERILE SYSTEMS

As a distinguishing title from the previously described systems, this is a misnomer, if only because those systems are not sterile. In addition to the added rhizobia, contaminating bacteria from beneath the seed coat (Evans et al., 1972; Handelsman and Brill, 1985) provide a population of bacteria that may or may not interfere with nodulation by the inoculant bacteria. However, in this section, attention is directed to the use of soil, collected from the field for various reasons, for experiments in the laboratory or greenhouse. The various reasons include a comparison of inoculation responses on different soil types or following fertilizer amendment, comparisons of different populations of indigenous rhizobia, or the sampling of populations of rhizobia in soil at a time when it is impractical to grow plants in the field (for seasonal reasons or by virtue of distance between sites or from the laboratory). Despite the fact that the soils may have a population of rhizobia, this does not remove the necessity to exercise due care and attention in preventing cross-contamination, or contamination from sources within the laboratory or greenhouse. Thought and common sense are required to maintain the populations as sampled.

A. Pot Culture

Surface soil, collected to a depth of 100 to 150 mm, is allowed to air dry before being passed through a 5-mm screen to remove stones, large clods, and root pieces. By using a finer screen, aeration problems during plant growth can occur due to loss of soil structure. The soil may be transferred directly to pots, although the use of polyethylene bags as liners reduces the risk of contamination from the pots and facilitiates handling. The amount of soil is constant for each pot, and to reduce watering difficulties, the pots too are tared to the same weight by the addition of sand or gravel to the base.

As appropriate, basal dressings of fertilizer (e.g., superphosphate, lime, K_2SO_4, trace elements) are applied to the surface of the soil and incorporated, although a more common approach is to mix thoroughly the fertilizer with the soil before adding to the bags or pots. Mineralization of soil nitrogen is promoted when the profile is disturbed. As this could delay nodulation, heavy seeding of the pots with oats or barley for a 4- to 6-week growth period, followed by top removal at soil level (Simpson and Gibson, 1970), can reduce greatly the available mineral nitrogen.

Unless essential for experimental purposes, soil sterilization by steam, chemical treatment, or irradiation is to be avoided. In addition to increasing the availability of various nutrients directly, it will reduce (rarely eliminate) the other elements of the soil microflora that could be vital in affecting the response expected in the field, for example, V-A mycorrhizae can be beneficial (Smith et al., 1979), whereas some actinomycetes (Damirigi and Johnson, 1966) and soil fungi (Angle et al., 1981) may be inhibitory. The same considerations apply to watering as with microbiologically controlled pots, with the further proviso that the water level in the bags should be kept at, or near, field capacity. This is determined by adding water to soil in bags, or in columns where the level of compaction is the same as in the bags, such that soil at the bottom is still dry after a period. The surface soil is removed and the water content determined; the water content of the soil after free drainage is known as the field capacity.

B. Soil Cores

A modification of the above is that cores are removed directly to empty cans 80 mm or more diameter (Vincent, 1970). Food cans, purchased sealed, are ideal. Just prior to use, they are opened at both ends with a can opener that turns the cut edges inward, reinforcing the rim. A metal cap is placed over the can to protect its shape while it is driven into the ground. The can is then dug from the soil and placed in a plastic bag for transport to the laboratory; this minimizes desiccation and soil loss during transport. In the glasshouse, the can and bag are fitted into the top of a jar that holds the can firmly near the base, after first piercing the bag to allow drainage. The top of the bag is left closed over the can after seeding, watering, and so on, and only opened once the seedlings have established. This procedure minimizes lateral windborne contamination and desiccation. Watering is done as required, it being advisable to slowly add water to slight excess, as seen by drainage to the jar.

C. Enumeration of Rhizobia

The most common technique for enumerating rhizobia in a mixed soil population is the plant infection test, a most probable number (MPN) method (Brockwell, 1980). This technique has various modifications in practice, but the essential principal is that in a dilution series, the presence of one organism in the diluent applied to a single plant will lead to nodulation of that plant. The remainder of the exercise is statistical, aided by consultation of tables, showing the probable number of bacteria present in the initial suspension. The procedure most commonly adopted in my laboratory is to have a series of 25 plants and six bottles containing 4.0 ml of water (or dilute inorganic nutrient to provide an osmoticum). From the initial suspension (10 g of soil in 90 ml of water), or a further 10- to 10,000-fold dilution of this suspension, depending on the anticipated number of rhizobia present, a 1-ml sample is taken to the first bottle in the series. Using a fresh pipette, the suspension is drawn into the pipette, and expelled, five times before transferring 1 ml to the next bottle in the series and 4 ml $\times$ 1 ml to four plants. This procedure is repeated until the sixth bottle, when all of the 5 ml is used to inoculate five plants. This procedure covers a range of 1.5×10^4. Should it not be possible to estimate, beforehand, the probable number as falling within this range, further groups of four plants can be added, increasing the range fivefold for each group. Alternatively, a series of 10-fold dilutions can be used with some loss of accuracy.

A number of aspects of the MPN technique should be considered before making the determinations. These include:

1. The plant culture system should be secure against accidental contamination. Cotton-plugged tubes, with the plants entirely within the tube, is the safest approach, although pouches have been used (Weaver and Frederick, 1972), especially for larger-seeded legumes.
2. With agar or liquid medium, the rhizosphere is effectively the complete volume of the medium. However, some legumes grow poorly on agar and a vermiculite or vermiculite-perlite medium is preferred. With low concentrations of bacteria, the physical constraints imposed by the medium could lead to the failure of these bacteria to multiply and extend into the root zone, resulting in an underestimate of the number of rhizobia in the original suspension (Brockwell, 1963; Grassia and Brockwell, 1978).
3. Under some circumstances, fungi in the soil may attack the seedlings and prevent nodulation, leading to "skips" in the series. The incorporation of mycostatin (100 ppm) into the plant culture medium effectively eliminates this problem, which is most common in soils with a relatively low count of rhizobia (Robinson, 1968).
4. In some soils, the rhizobia remain firmly attached to the soil particles, with the consequence that the size of the population will be underestimated. Kingsley and Bohlool (1981) observed that the use of ammonium phosphate as a diluent, rather than sodium phosphate, various chlorides, or water, increased the counts of rhizobia in some soils by 10-fold.
5. Where possible, the host species used should be that which is relevant to the purposes for making an estimate of the number of rhizobia present. With large-seeded legumes, this presents the problem of a culture system secure against accidental contamination. Brockwell (1980) has listed a number of small-seeded species that are suitable alternate hosts for some of the common large-seeded legumes. These include *Vicia hirsuta* (L.), *V. sativa* (L.) or *V. atropurpurea* Desf. for *Pisum sativum* (L.) and *V. faba* (*Rhizobium leguminosarum*); *Ornithopus sativus* Brot., or *O. compressus* (L.) for *Lupinus* spp. (*Bradyrhizobium* sp./*R. lupini*); *Macroptilium atropurpureum* (DC) Urb, *M. lathyroides* (L.) Urb. or *Teramnus uncinatus* (L.) Sw. for many large-seeded tropical-type legumes such as *Vigna* spp., *Cajanus cajan*, and *Lab-lab pur-*

pureum (*Bradyrhizobium* sp., formerly known as the cowpea-type *Rhizobium* sp.; Jordan, 1984), while for *Leucaena leucocephala* (Lam.) de Wit, *Desmanthus virgatus* (L.) Willd. can be used; *Glycine ussuriensis* Regal and Maack for *G. max* (L.) Merr. (*Bradyrhizobium japonicum*).

6. Leonard jar assemblies can also be used for estimating numbers of rhizobia appropriate for large-seeded legumes, although the greater time to prepare the assemblies and increased glasshouse space requirements mitigate against its use for counting large numbers of samples.

Although the plant infection test is used primarily to determine the number of rhizobia in soil, it is also used for enumerating rhizobia in inoculants, especially where such inoculants are prepared in nonsterile carrier materials, and from any situation in which rhizobia are only a component of a mixed microflora. Further details on the application of this method, including appropriate tables for estimating the MPN, can be obtained from various publications, including Brockwell (1963, 1980), Grassia and Brockwell (1978), and Weaver and Frederick (1972).

Other methods for enumerating rhizobia generally involve plate counting. Although media have been described as being selective for rhizobia (e.g., Graham, 1969), or with the potential for being developed into a selective medium (Pattison and Skinner, 1974), there is no popular medium with this capability at the present time.

The foregoing methods provide an estimate of all strains in a soil or inoculant capable of nodulating a particular host. Obviously, these methods have little value in counting particular strains. To do this, it is necessary that these strains have an antibiotic resistance marker, or markers, or that they have been characterized by serological means for ready identification by the ELISA or immunofluorescent techniques (see Chapters 5 to 7). Antibiotic resistance can be used in three ways. One is to determine the intrinsic pattern of resistance of a particular strain against a range (up to 12 to 15) of antibiotics at particular concentrations (e.g., Beynon and Josey, 1980; Sinclair and Eaglesham, 1984; Stein et al., 1982). Although useful, this approach requires that the rhizobia be isolated by conventional procedures from nodules on plants in plant infection tests. Unless they are the predominant organism at the end of the dilution series, there is always the doubt that other strains in the soil may be more competitive and that the absence of the characterized strain from the nodule population *in the test* is not necessarily an indication of its absence or proportionately low representation in the soil.

The second procedure is to select isolates of strains for tolerance to higher levels of antibiotics (e.g., Brockwell et al., 1977), and subsequent testing to ensure there has been no loss of symbiotic effectiveness (Pankhurst, 1977; Pain, 1979). This too may require isolation from nodules before testing against the relevant antibiotics, although direct plating may be feasible where it has been shown that the soil microflora does not possess the same levels of resistance. The third procedure is essentially an extension of the second in that the high level of antibiotic resistance is achieved through transposon insertion. Such transposons possessing genes for resistance to high levels of antibiotics (e.g., kanamycin at 250 μg/ml). The danger of this approach is that the transposon may insert such that it mutates a gene important in nodulation or bacterial survival; there is also the possibility that it will transfer within the bacteria from a harmless position to a gene relevant to nodulation and survival. For further consideration of the use of antibiotic resistance, see the review by Schwinghamer and Dudman (1980).

VI. EXPERIMENTAL OBSERVATIONS AND THEIR EVALUATION

The first step in evaluating results occurs in the planning stages of the experiment. At that stage, there should be a clear understanding of the aims of the experiment and the limitations imposed on the interpretation of the results by the experimental constraints (e.g., physical growth conditions, the adequacy of the included controls, and other constraints referred to in the preceding pages). Then there will be the impact of problems that may have arisen during the course of the experiment, ranging from equipment malfunction, through loss of pH control in the medium or insect or pathogen infestation, to "inexplicable" nodulation of those uninoculated controls. Only the individual scientist can assess their significance and make the decision whether to proceed or start again.

For many of the techniques used here, the purpose of the study is straightforward. For example, what are the nodulation capabilities of a number of strains with "x" hosts? What is the level of symbiotic effectiveness of selected strains? Are selected strains competitive in nodule formation against other strains from the laboratory collection or in a soil population? Is mineral nutrition, including pH, an important consideration in the nodulation of plants grown in particular soils? Basically, the questions are simple, but should it be necessary to explain responses, obviously more detailed investigation is required. In the following sections, some particular techniques in general laboratory/geeenhouse studies are described and the assessment of data discussed.

A. Root Hair Infection

There may be many reasons why a plant fails to nodulate and often it is adequate to note that nodules did not appear. With an increasing interest in the genetics of nodulation, close observation of the roots to ascertain the point of breakdown in nodule formation is essential. The Fåhraeus slide technique (Section IV.B) is ideal for continuing studies of the infection of small-seeded legumes, but the short depth of focus of most microscopes and the extensive nature of the roots limits this technique's usefulness for species with a seed weight in excess of 10 mg. Bright-field illumination, with 30× objectives and 8× eyepieces, give optimal magnification for routine searching (Nutman, 1959, 1962). Doubtful infections can be examined at higher magnification and/or with the use of phase-contrast illumination.

Various staining procedures have been described for studying root hair infection. Phillips (1971) fixed the roots in Carnoy's solution ($C_2H_5OH:CHCl_3:CH_3COOH$, 6:3:1) for 24 hr, and after thorough rinsing, stored them in 70% ethanol. Following further rinsing, the roots were dipped in 0.05% toluidine blue at pH 6.8 for 3 sec; the root hairs stained blue and the infection threads red. Recently, Vasse and Truchet (1984) examined a range of stains and staining procedures, and concluded that 0.01% (w/v) methylene blue, in filtered Fahraeus medium, applied for 3 to 15 min in a humid chamber, gave the best results. Furthermore, they indicated that the material could, after observation, be returned to the growth medium for subsequent development and study. Phase-contrast microscopy has been used in infection studies (Callaham and Torrey, 1981; Turgeon and Bauer, 1982); interference-contrast microscopy has also been used (Turgeon and Bauer, 1982; Pueppke, 1983). For their elegant studies of the pattern of infection development in soybeans, Bhuvaneswari et al. (1983) have described a computer-based system, with a graphics tablet and pen mounted under a dissecting microscope. The program for data acquisition, storage, and analysis is available on request from the authors.

B. Nodulation

Nodule development is the summation of a number of physiological processes and the time taken for nodules to appear has been used as a measure of environmental effects (physical, nutritional, pH) on the establishment of the symbiosis (Gibson, 1967a; Munns, 1968) as well as on symbiotic compatibility (Nutman, 1967; Halversen and Stacey, 1984, 1985, albeit using the location of the developing nodules as an indication of the time of infection). Distinction between young nodules and lateral root initials requires experience but as a guide, the former tend to be rounded, whereas the lateral initials are usually pointed. With plant culture in tubes, plastic growth pouches, or hydroponics, continuing observations are possible and doubtful observations can be subsequently verified or discarded. Pseudonodules (i.e., proliferations of cortical tissue dependent on the presence of rhizobia but white and devoid of bacteria internally; Allen et al., 1953) have been observed commonly on *Medicago sativa* inoculated with rhizobia not of the species *R. meliloti*, but they are best described as "fluffy excrescences," often in the axils of lateral roots. More recently, structures more akin to nodules have been observed when *M. sativa* is inoculated with various mutants or engineered strains of *R. meliloti*. These structures are devoid of bacteria internally and lack leghemoglobin, and can be identified with certainty only by microscopic examination (Hirsch et al., 1984).

Many published papers are dependent on nodule number data for the interpretaiton of results. However, caution is advised in the use of this parameter, as successful infections frequently cause the cessation of further infection in their immediate vicinity (Nutman, 1962). For this reason, nodule number is not a good indication of the number of infections. In some host × strain combinations, large nodules develop rather than many smaller ones; this may be genetic (Nutman, 1969) or environmental (Gibson, 1967a; Munns et al., 1977). At the other end of the scale, mutant *G. max* plants may develop many nodules (Carroll et al., 1985), although the nitrogen-fixing capacity of these individual nodules is less than that of nodules of similar size on wild-type parental plants. Elsewhere, delaying inoculation may lead to high numbers of nodules (Dart and Pate, 1959), while the application of combined nitrogen that will delay nodulation may lead eventually to more nodules (Gibson and Nutman, 1960) and an increase in total nodule weight relative to nonfertilized controls. Hence the use of nodule number data must be used with caution and with definite objectives in mind when collecting this information. In many cases, as in inhibition experiments, the relative ranking of treatments based on nodule number may change with time (e.g., Gibson and Nutman, 1960) and single determinations may be very misleading. For most purposes, the determination of nodule weight is more relevant than assessment of nodule number.

C. Plant Growth

The most common method for assessing symbiotic response is to determine plant dry weight and/or total plant nitrogen. The plant material should be dried in a fan-equipped oven at 75 to 80°C, to constant weight; 48 to 72 hr should be sufficient. Volatilization of some nitrogen compounds can occur at higher temperatures. After drying, the material should be stored in a desiccator or a plastic bag containing silica gel or anhydrous calcium chloride.

The Kjeldahl digestion of the plant material for total nitrogen determination depends on the reduction of all nitrogen in the tissue to ammonium, under the action of concentrated sulfuric acid. In this laboratory we prepare the digestion mixture in advance by adding Se

(1 g/liter) as catalyst and K_2SO_4 (100 g/liter) to raise the boiling point; the mixture is heated until it turns green, then allowed to cool. Normally, we restrict the amount of tissue in the digestion, in 100-ml flasks, to 300 mg. For greater quantities, the tissue is ground in a Wiley-type hammer mill, or preferably a coffee-grinder type of mill in which the blades at the base of the chamber rotate at 26,000 rpm. Where possible, nodules should be separated from the roots as the high level of nitrogen in these discrete structures can lead to unevenness in the distribution of nitrogen through the material. During digestion, temperatures inside the flask should be kept below 320°C, even though this prolongs the digestion period. An alternative system gaining popularity is a heated aluminum block, with holes to receive 100-ml digestion tubes. The material is digested until clear, then another hour to ensure the conversion of all nitrogen to ammonium ions. Ammonia is determined by distillation into 2% (w/v) boric acid and titration with 0.02 *N* HCl, or colorimetrically with a Technicon AutoAnalyzer. For further details, see Bergersen (1980).

D. Symbiotic Effectiveness

"Symbiotic effectiveness" is the term used to describe the ability of a nodulated plant to fix nitrogen, and can be expressed qualitatively (high, moderate or intermediate, ineffective) or quantitatively. Quantitative expression is dependent on a comparison of dry weight, or total plant nitrogen, with plants receiving adequate combined nitrogen or with plants inoculated with a standard strain of known performance. There has been an increasing tendency to use the term "efficiency" in place of "effectiveness" (Caldwell and Vest, 1977), but this is incorrect and should be discouraged. Efficiency is properly expressed as a qualified rate (e.g., mg N_2 fixed per gram of nodule weight, per milligram of bacteroid protein or weight, or per milligram of carbohydrate utilized in the nodules). Highly effective associations are not always efficient (Gibson, 1966, 1969; Carroll et al., 1985). The two terms have a different basis of expression and it is important that each be used in its correct context.

As a general principle, tests for symbiotic effectiveness should continue until the most effective strains have fixed at least five times more nitrogen, and preferably more, than was initially present in the seed of the host (Gibson, 1980a). This should occur within 28 to 35 days from inoculation.

Nitrogen controls provide a measure of the potential growth of the species under the conditions of culture. Expression of the growth of the inoculated plants as a percentage of the growth of the nitrogen controls provides an indication of "absolute" symbiotic effectiveness. However, it is important that these controls be realistic, and for this reason the addition of combined nitrogen (as KNO_3, or as NH_3NO_3 to minimize the pH shift) should be delayed until nodules appear. At that time, a low level of nitrogen is supplied while the nodules on the inoculated plants develop nitrogenase activity, with a further, larger addition when nitrogen fixation starts; this is usually 3 to 5 days after nodule appearance. Care should be taken to avoid damage by excessive combined nitrogen, and to this end, weekly additions are advised. These should be based on the expected growth of the plants in that period. For example, plants weighing 40 mg and having a relative growth rate of 14% (see below) would increase in weight to 100 mg in 7 days. With 4% nitrogen in the tissue, this will require 2.4 mg of nitrogen; a 50% increase is advised, requiring the addition of 3.6 mg of nitrogen. The amount added varies with the species, the stage of growth, and the growth conditions, but at all times it should be adequate but not excessive. In pots, some allowance should be made for possible leaching and denitrification. However, the main consideration is that the additions should not commence too early, as this will accelerate the early growth of the controls and provide an unrealistically low estimate of symbiotic effectiveness. One method to overcome the problem of timing is to inoculate the potential nitrogen controls with a known effective strain, and to begin the nitrogen additions as soon as uninoculated controls show the first sign of nitrogen deficiency. As nitrogen control plants may take up, but not necessarily assimilate, more nitrogen than they can utilize for maximum dry weight increase, the percent nitrogen in the tissues of the nitrogen control plants may be excessive. Hence the comparison of the inoculated plants is most appropriately made on a dry weight basis rather than on total nitrogen data.

$$\text{Symbiotic effectiveness (S.E.)} = \frac{\text{dry weight, inoculated}}{\text{dry weight, N controls}} \times 100$$

Some expressions of S.E. involve subtracting the dry weight of the uninoculated controls from both the inoculated and nitrogen control values. However, the uninoculated controls may grow to dilute the plant nitrogen to a level much lower than that in inoculated plants, and consideration of the weight of the uninoculated controls in deriving S.E. is inappropriate in that it will lead to unrealistic depression of the values for moderately effective associations.

Another approach is to compare the strains under test with the performance of a known standard strain or a suite of standard strains. This approach has been used to determine the effectiveness of field isolates (Gibson et al., 1975). For the calculation of S.E., total plant nitrogen can replace dry weight in the calculation. Provided the initial requirement that the best plants have fixed more than five times the seed nitrogen, the use of total nitrogen instead of dry weight is unlikely to make a great change in S.E.

Relative Growth Rate

In the foregoing, only one harvest is made and the results reflect the integration of effects due to speed of nodulation, rate of nodule development, and induction of nitrogenase activity, as well as inherent nitrogen-fixing ability. Many of the early events may not be related to nitrogen-fixing ability, and as laboratory tests are essentially short term in nature, a guide to the latter characteristic can be achieved by measuring relative growth rate (R_W) or relative nitrogen assimilation rate (R_N) over the last week or 10 days of the experimental period (Gibson, 1965, 1966). This involves two harvests, and the information provided includes:

1. Overall symbiotic effectiveness
2. Relative speed of establishment of the symbiosis (first harvest)
3. R_W or R_N, which are predictors of subsequent growth and nitrogen fixation

For the most efficient use of this approach, the plants within a treatment are "paired" on the basis of size, leaf development, and root development at the time of the first harvest, when one of each pair is taken. Relative growth rate, in mg/mg per day, is calculated as

$$R_W = \frac{\log_e W_2 - \log_w W_1}{t_2 - t_1}$$

where W_1 and W_2 are total plant dry weights at harvest times t_1 and t_2. The relative nitrogen assimilation rate R_N is calculated in the same way, except that total plant N data is used. A typical value of 0.11 mg/mg per day indicates that the plant increases in weight by 11% each day. The doubling time is 6.5 days (for any situation where exponential increase in involved, $R_W \times$ doubling time = 72).

Experimental Design

For genetically homogeneous material, fewer replicates are required than for genetically heterogeneous material (*Trifolium subterraneum*, *G. max*, compared to *Trifolium pratense*, *T. repens*, or *Medicago sativa*). For the former, 8 to 10 replicates of singly grown plants is normally adequate, but for the latter, 15 to 20 replicates are essential to get a reasonable estimate. Plant material should be completely randomized before inoculation (as in tube culture experiments) to minimize location or time-of-planting effects. After inoculation, the treatments should again be randomized into blocks, one replicate per block. This is important for statistical analysis, which can help minimize location effects, be they in the greenhouse or growth cabinet.

For grain legumes, where the principal criterion is grain yield, as opposed to dry matter yield in pasture species, there are considerable difficulties in culturing plants to maturity in either the greenhouse or growth cabinet. Under some circumstances, plants with the highest dry weight and nitrogen content do not always give the highest yield of grain (Summerfield et al., 1977a). The question of adequately assessing the symbiotic effectiveness of grain legumes is one that few researchers have attempted to answer. The only practical answer is to take the best strains from the greenhouse and growth cabinet trials and to field test them before inclusion in inoculants. Ideally, the latter tests should be done at a number of representative sites where environmental variables (physical, biological, nutritional), often impossible to simulate in the greenhouse, could influence the performance of some strains. An example of this is the poor nodulation of *G. max* by USDA 123 at high pH and by USDA 135 in acid soils (Damirgi et al., 1967).

Acetylene Reduction Assay

Since the observation that nitrogenase reduced acetylene to ethylene (Dilworth, 1966; Schöllhorn and Burris, 1966), the technique has become the basis for many experimental observations in the laboratory, the greenhouse, and the field. The prime advantages of low relative cost for equipment and materials, comparative simplicity, speed, and high sensitivity have enabled studies that would otherwise be laborious and expensive, and in other cases, impossible. But as with all experimental procedures, the value of the technique is no greater than its limitations, which essentially involve the validity of the assay as a true measure of nitrogenase activity.

Acetylene reduction is a measure of electron flow through nitrogenase, and although the theoretical $C_2H_2:N_2$ ratio is 4:1, values as high as 6:1 are acceptable where the system is producing a high level of H_2, even though this H_2 may not appear due to use by an active-uptake hydrogenase system. In an examination of a number of studies with legumes soon after the technique went into popular use, Hardy et al. (1973) found a mean ratio of 3.9:1, with a range from 1.5 to 8.4. In my laboratory, using ^{15}N and C_2H_2 in short assays or increase in total plant nitrogen over a period during which several C_2H_2 reduction assays were made, a similar range and mean value to those reported above have been obtained. Various researchers have been very critical of the assay (e.g., Minchin et al., 1983), whereas others (e.g., Imsande and Ralston, 1981) have been satisfied that the assay provides a reasonable measure of true nitrogen fixation. The variability in the ratio and in the degree to which the assay provides a measure of true nitrogenase activity requires that appropriate experiments be run before deciding to utilize, or discard, the technique in any program.

A number of systems have been devised for nondestructive assays on plants, some of which do not require disturbance of the plants. These systems have been designed for specific purposes, such as estimating respiration associated with nitrogen fixation, or to examine the effect of various environmental variables on nitrogenase activity (see Huang et al., 1975; Mederski and Streeter, 1977; Minchin et al., 1983).

On various occasions, researchers interested in determining the symbiotic effectiveness of different strains of rhizobia, or different host × strain combinations under nitrogen-free or low-nitrogen conditions have requested information on acetylene reduction procedures, intending to use the assay instead of plant growth characteristics (weight, total nitrogen). The assay, at best, provides an estimate of nitrogenase activity during the period of the assay; in no way should it be considered as an alternative to the more direct measurements. The assay can be used to complement other estimates, but in the final analysis, the integrated values obtained from plant harvest procedures are those on which the assessments of symbiotic effectiveness must be based.

VII. CONCLUDING REMARKS

At the risk of being repetitive, expediency is no alternative to the adoption of sound scientific technique. In many laboratories, the production of nodulated plants is the tiresome task frequently handed over to a technician, trained or otherwise, while the more interesting procedures on this material are done by the researcher. Although this is satisfactory under some circumstances, without an understanding and awareness of possible shortcomings (e.g., contamination, pathogen or insect infestation, restrictions imposed by the growth conditions), interpretation of the results can be at risk. By following the principles outlined in this chapter, potential problems should be minimized.

REFERENCES

Albrecht, S. L., Bennett, J. M., and Boote, K. J. (1984). *Field Crops Res. 8*: 61.

Allen, E. K., Allen, O. N., and Newman, A. S. (1953). *Am. J. Bot. 40*: 429.

Angle, J. S., Pugashetti, B. K., and Wagner, G. H. (1981). *Agron. J. 73*: 301.

Aparicio-Tejo, P. M., Sánchez-Díaz, M. F., and Peña, J. I. (1980). *Physiol. Plant. 48*: 1.

Bedmar, E. J., and Phillips, D. A. (1984). *Plant Physiol.* *75*: 629.

Bednarski, M. A., and Reporter, M. (1978). *Appl. Environ. Microbiol.* *36*: 115.

Begg, J. E., and Turner, N. C. (1976). *Adv. Agron.* *28*: 161.

Bergersen, F. J. (1980). In *Methods for Evaluating Biological Nitrogen Fixation*, F. J. Bergersen, ed. John Wiley & Sons Ltd., West Sussex, England, pp. 65-110.

Bergersen, F. J., and Turner, G. L. (1970). *Biochim. Biophys. Acta* *214*: 28.

Beynon, J. L., and Josey, D. P. (1980). *J. Gen Microbiol.* *118*: 437.

Bhuvaneswari, T. V., Mills, K. K., Crist, D. K., Evans, W. R., and Bauer, W. D. (1983). *J. Bacteriol.* *153*:443.

Bhuvaneswari, T. V., Turgeon, B. G., and Bauer, W. D. (1980). *Plant Physiol.* *66*: 1027.

Bisseling, T., van Staveren, W., and van Kammen, A. (1980). *Biochem. Biophys. Res. Commun.* *93*: 687.

Bond, G. (1950). *Ann. Bot.* *15*: 95.

Brockwell, J. (1963). *Appl. Microbiol.* *11*: 377.

Brockwell, J. (1980). In *Methods for Evaluating Biological Nitrogen Fixation*, F. J. Bergersen, ed. John Wiley & Sons Ltd., Chichester, West Sussex, England, pp. 417-488.

Brockwell, J., and Gault, R. R. (1976). *Aust. J. Exp. Agric. Anim. Husb.* *16*: 500.

Brockwell, J., and Hely, F. W. (1966). *Aust. J. Agric. Res.* *17*:885.

Brockwell, J., Schwinghamer, E. A., and Gault, R. R. (1977). *Soil Biol. Biochem.* *9*: 19.

Bunting, A. H., and Horrocks, J. (1964). *Ann. Bot.* *28*: 229.

Caldwell, B. E., and Vest, H. G. (1977). In *A Treatise on Dinitrogen Fixation*, Sec. 3, *Biology*, R. W. F. Hardy and W. S. Silver, eds. Wiley-Interscience, New York, pp. 557-576.

Callaham, D. A., and Torrey, J. G. (1981). *Can. J. Bot.* *50*: 1647.

Carroll, B. J., McNeil, D. L., and Gresshoff, P. M. (1985). *Proc. Natl. Acad. Sci. USA* *82*: 4162.

Child, J. J., and LaRue, T. A. (1974). *Plant Physiol.* *53*: 88.

Christeller, J. T., Laing, W. A., and Sutton, W. D. (1977). *Plant Physiol.* *60*: 47.

Coker, G. T., III, and Schubert, K. R. (1981). *Plant Physiol.* *67*: 691.

Damirgi, S. M., and Johnson, H. W. (1966). *Agron. J.* *58*: 223.

Damirgi, S. M., Frederick, L. R., and Anderson, I. C. (1967). *Agron. J.* *59*: 10.

Dart, P. J. (1977). In *A Treatise on Dinitrogen Fixation*, Section 3, *Biology*, R. W. F. Hardy and W. S. Silver, eds. Wiley-Interscience, New York, pp. 367-472.

Dart, P. J., and Pate, J. S. (1959). *Aust. J. Agric. Res.* *12*: 427.

Date, R. A., and Halliday, J. (1980). In *Advances in Legume Science*, R. J. Summerfield and A. H. Bunting, eds. Royal Botanic Gardens, Kew, pp. 597-601.

Devine, T. E., and Reisinger, W. W. (1978). *Agron. J. 70*: 510.
Dilworth, M. J. (1966). *Biochim. Biophys. Acta 127*: 285.
Eckart, J. F., and Raguse, C. A. (1980). *Agron. J. 72*: 519.
Evans, L. T., and King, R. W. (1975). In *Report of the Technical Advisory Committee on the Biology of Yield of Grain Legumes.* TAC Secretariat, Food and Agriculture Organization of the United Nations, Rome, 17pp.
Evans, H. J., Campbell, N. E. R., and Hill, S. (1972). *Can. J. Microbiol. 18*: 13.
Fåhraeus, G. (1957). *J. Gen. Microbiol. 16*: 374.
Gibson, A. H. (1965). *Aust. J. Biol. Sci. 18*: 295.
Gibson, A. H. (1966). *Aust. J. Biol. Sci. 19*: 499.
Gibson, A. H. (1967a). *Aust. J. Biol. Sci. 20*: 1087
Gibson, A. H. (1967b). *Aust. J. Biol. Sci. 20*: 1105.
Gibson, A. H. (1967c). *Aust. J. Biol. Sci. 20*: 837.
Gibson, A. H. (1968). *Aust. J. Agric. Res. 19*: 907.
Gibson, A. H. (1969). *Aust. J. Biol. Sci. 22*: 829.
Gibson, A. H. (1976). In *Symbiotic Nitrogen Fixation in Plants*, P. S. Nutman, ed. Cambridge University Press, Cambridge, pp. 385-403.
Gibson, A. H. (1977). In *A Treatise on Dinitrogen Fixation*, Sec. 4, *Agronomy and Ecology*, R. W. F. Hardy and A. H. Gibson, eds. Wiley-Interscience, New York, pp. 393-450.
Gibson, A. H. (1980a). In *Methods for Evaluating Biological Nitrogen Fixation*, F. J. Bergersen, ed. John Wiley & Sons Ltd., Chichester, West Sussex, England, pp. 139-184.
Gibson, A. H. (1980b). In *Advances in Legume Science*, R. J. Summerfield and A. H. Bunting, eds. Royal Botanic Gardens, Kew, pp. 69-75.
Gibson, A. H., and Harper, J. E. (1985). *Crop Sci. 25*: 497.
Gibson, A. H., and Nutman, P. S. (1960). *Ann. Bot. 24*: 420.
Gibson, A. H., Curnow, B. C., Bergersen, F. J., Brockwell, J., and Robinson, A. C. (1975). *Soil Biol. Biochem.* 7: 95.
Gibson, A. H., Child, J. J., Pagan, J. D., and Scowcroft, W. R. (1976). *Planta, Berlin 128*: 233.
Gibson, A. H., Dreyfus, B. L., Lawn, R. J., Sprent, J. I., and Turner, G. L. (1981). In *Current Perspectives in Nitrogen Fixation*, A. H. Gibson and W. E. Newton, eds. Australian Academy of Science, Canberra, p. 373.
Graham, P. H. (1969). *Appl. Microbiol. 17*: 769.
Grassia, A., and Brockwell, J. (1978). *Soil Biol. Biochem. 10*: 101.
Grobbelaar, N., Clarke, B., and Hough, M. C. (1971). In *Biological Nitrogen Fixation in Natural and Agricultural Habitats*, T. A. Lie and E. G. Mulder, eds. (*Plant Soil, Spec. Vol.*). Martinus Nijhoff, The Hague, pp. 203-214, 215-223.
Hack, H. R. B. (1971). *J. Exp. Bot. 22*: 323.
Halvorsen, L. D., and Stacey, G. (1984). *Plant Physiol. 74*: 84.

Halvorsen, L. D., and Stacey, G. (1985). *Plant Physiol. 77*: 621.
Handelsman, J., and Brill, W. J. (1985). *Appl. Environ. Microbiol. 49*: 818.
Hardy, R. W. F., Burns, R. C., and Holsten, R. D. (1973). *Soil Biol. Biochem. 3*: 47.
Harker, A. R., Xu, L.-S., Hanus, F. J., and Evans, H. J. (1984). *J. Bacteriol. 159*: 850.
Harper, J. E., and Gibson, A. H. (1984). *Crop Sci. 24*: 797.
Havelka, U. D., and Hardy, R. W. F. (1976). In *Proceedings of 1st International Symposium on Nitrogen Fixation*, W. E. Newton and C. J. Nyman, eds. Washington State University Press, Pullman, Wash., pp. 456-475.
Hely, F. W. (1959). *J. Agric. Eng. Res. 4*: 133.
Hely, F. W., Manil, C., and Bonnier, P. (1953). *Nature 171*: 884.
Hirsch, A. M., Wilson, K. J., Jones, J. D. G., Bang, M., Walker, V. V., and Ausubel, F. M. (1984). *J. Bacteriol. 158*: 1133.
Holsten, R. D., Burns, R. C., Hardy, R. W. F., and Hebert, R. R. (1971). *Nature 232*: 173.
Huang, C.-Y., Boyer, J. S., and Vanderhoef, L. N. (1975). *Plant Physiol. 56*: 222.
Hunter, M. N., de Jabrun, P. L. M., and Byth, D. E. (1980). *Aust. J. Exp. Agric. Anim. Husb. 20*: 339.
Imsande, J., and Ralston, E. J. (1981). *Plant Physiol. 68*: 1380.
Ismaili, M., Briske, M. M., and Weaver, R. W. (1983). *Agron. J. 75*: 649.
Jensen, H. L. (1942). *Proc. Linn. Soc. N.S.W. 67*: 205.
Johnson, D. A., and Asay, K. H. (1978). *Crop Sci. 18*: 520.
Jones, F. R., and Tisdale, W. B. (1921). *J. Agric. Res. 22*: 17.
Jordan, D. C. (1984). In *Bergey's Manual of Systematic Bacteriology*, Vol. 1, N. R. Krieg and J. G. Holt, eds. Williams & Wilkins, Baltimore, pp. 234-244.
Joshi, H. U., Carr, A. J. H., and Mytton, L. R. (1967). *J. Gen. Microbiol. 47*: 139.
Kingsley, M. T., and Bohlool, B. B. (1981). *Appl. Environ. Microbiol. 42*: 241.
Lahav, E., Harper, J. E., and Hageman, R. H. (1976). *Crop Sci. 16*: 325.
LaRue, T. A., and Kurz, W. G. W. (1973). *Can. J. Microbiol. 19*: 304.
Lawn, R. J., and Brun, W. A. (1974a). *Crop Sci. 14*: 11.
Lawn, R. J., and Brun, W. A. (1974b). *Crop Sci. 14*: 22.
Leonard, L. T. (1944). Method of testing bacterial cultures and results of tests of commercial inoculants. *U.S. Department of Agriculture Circular 703.*
Lie, T. A. (1969a). *Plant Soil 30*: 391.
Lie, T. A. (1969b). *Plant Soil 31*: 391.

Lie, T. A., and Brotonegoro, S. (1969). *Plant Soil 30*: 339.
Lindsay, C. R., and Jordan, D. C. (1976). *Can. J. Soil Sci. 56*: 495.
Mague, T. H., and Burris, R. H. (1972). *New Phytol. 71*: 275.
Masterson, C. L., and Sherwood, M. T. (1978). *Plant Soil 49*: 421.
McKnight, T. (1949). *Queensl. J. Agric. Sci. 6*: 61.
Mederski, J. H., and Streeter, J. G. (1977). *Plant Physiol. 59*: 1076.
Meyer, D. R., and Anderson, A. J. (1959). *Nature, London 183*: 61.
Minchin, F. R., Summerfield, R. J., Eaglesham, A. R. J., and Stewart, K. A. (1978). *J. Agric. Sci. Cambridge 90*: 355.
Minchin, F. R., Witty, J. F., Sheehy, J. E., and Muller, M. (1983). *J. Exp. Bot. 34*: 641.
Morse, R. N., and Evans, L. T. (1962). *J. Agric. Eng. Res. 7*: 128.
Mulder, E. G., and van Veen, W. L. (1960). *Plant Soil 13*: 265.
Munevar, F., and Wollum, A. G., II. (1981). *Soil Sci. Soc. Am. J. 45*: 1113.
Munevar, F., and Wollum, A. G., III. (1982). *Agron. J. 74*: 138.
Munns, D. N. (1968). *Plant Soil 28*: 129.
Munns, D. N. (1977). In *A Treatise on Dinitrogen Fixation,* Sec. 4, *Agronomy and Ecology*, R. W. F. Hardy and A. H. Gibson, eds. Wiley-Interscience, New York, pp. 353-391.
Munns, D. N., Fox, R. L., and Koch, B. L. (1977). *Plant Soil 46*: 591.
Mytton, L. R. (1976). *Ann. Appl. Biol. 82*: 577.
Nathanson, K., Lawn, R. J., de Jabrun, P. L. M., and Byth, D. E. (1984). *Field Crops Res. 8*: 73.
Nelsen, C. E., Safir, G. R., and Hanson, A. D. (1978). *Plant Physiol. 61*: 131.
Nutman, P. S. (1959). *J. Exp. Bot. 10*: 250.
Nutman, P. S. (1962). *Proc. Roy. Soc. London Ser. B 156*: 122.
Nutman, P. S. (1967). *Aust. J. Agric. Res. 18*: 381.
Nutman, P. S. (1969). *Proc. Roy. Soc. London Ser. B 172*: 417.
Oghoghorie, C. G. O., and Pate, J. S. (1971). In *Biological Nitrogen Fixation in Natural and Agricultural Habitats*, T. A. Lie and E. G. Mulder, eds. (*Plant Soil, Spec. Vol.*). Martinus Nijhoff, The Hague, pp. 185-202.
Pain, A. N. (1979). *J. Appl. Bacteriol. 47*: 53.
Pankhurst, C. E. (1977). *Can. J. Microbiol. 23*: 1026.
Pankhurst, C. E., and Gibson, A. H. (1973). *J. Gen. Microbiol. 74*: 219.
Pattison, A. C., and Skinner, F. A. (1974). *J. Appl. Bacteriol. 37*: 239.
Phillips, D. A. (1971). *Physiol. Plant. 25*: 482.
Phillips, D. A. (1974). *Plant Physiol. 54*: 654.

Porter, F. E., Nelson, I. S., and Wold, E. K. (1966). *Crops Soils 18*(8): 10.

Pueppke, S. G. (1983). *Can. J. Microbiol. 29*: 69.

Raggio, M., Raggio, N., and Torrey, J. G. (1957). *Am. J. Bot. 44*: 325.

Reporter, M., and Hermina, N. (1975). *Biochem. Biophys. Res. Comm. 64*: 1126.

Robertson, J. G., Farnden, K. J. F., Warburton, M. P., and Banks, J. M. (1975). *Aust. J. Plant Physiol. 2*: 265.

Robinson, A. C. (1968). *Aust. J. Exp. Agric. Anim. Husb. 8*: 327.

Rogers, V. E. (1969). *Aust. CSIRO Div. Plant Ind. Field Stn. Rec. 8*: 37.

Rolfe, B. G., Gresshoff, P. M., and Shine, J. (1980). *Plant Sci. Letts. 19*: 277.

Roughley, R. J., and Dart, P. J. (1970). *Plant Soil 32*: 518.

Scholander, P. F., Hammel, H. T., Bradstreet, E. D., and Hemmingsen, E. A. (1965). *Science 148*: 339.

Schöllhorn, R., and Burris, R. H. (1966). *Proc. Fed. Am. Soc. Exp. Biol. 24*: 710.

Schweitzer, L. E., and Harper, J. E. (1980). *Plant Physiol. 65*: 51.

Schwinghamer, E. A., and Dudman, W. F. (1890). In *Methods for Evaluating Biological Nitrogen Fixation*, F. J. Bergersen, ed. John Wiley & Sons Ltd., Chichester, West Sussex, England, pp. 337-365.

Seetin, M. W., and Barnes, D. K. (1977). *Crop Sci. 17*: 783.

Silsbury, J. H. (1981). *Plant Physiol. 67*: 599.

Simpson, J. R., and Gibson, A. H. (1970). *Soil Biol. Biochem. 2*: 295.

Sinclair, M., and Eaglesham, A. R. J. (1984). *Soil Biol. Biochem. 16*: 247.

Small, J. G. C. (1968). *S. Afr. J. Agric. Sci. 11*: 441.

Smith, S. E., Nicholas, D. J. D., and Smith, F. A. (1979). *Aust. J. Plant Physiol. 6*: 305.

Sprent, J. I. (1973). *New Phytol. 72*: 1005.

Sprent, J. I. (1976). In *Water Deficits and Plant Growth*, Vol. 4, T. T. Kozlowski, ed. Academic Press, Inc., New York, pp. 291-315.

Stein, M., Bromfield, E. S. P., and Dye, M. (1982). *Ann. Appl. Biol. 101*: 261.

Summerfield, R. J., Dart, P. J., Huxley, P. A., Eaglesham, A. R. J., Minchin, F. R., and Day, J. M. (1977a). *Exp. Agric. 13*: 129.

Summerfield, R. J., Minchin, F. R., Roberts, E. H., and Wien, H. C. (1977b). *Plant Sci. Lett. 8*: 355.

Tanner, J. W., and Anderson, I. C. (1963). *Can. J. Plant Sci. 43*: 542.

Thornton, H. G. (1930). *Proc. Roy. Soc. London Ser. B 106*: 110.
Tisdale, S. L., and Nelson, W. L. (1966). *Soil Fertility and Fertilizers*, 2nd ed., Macmillan Publishing Co., Inc., New York.
Trinchant, J. C., and Rigaud, J. (1984). *J. Plant Physiol. 116*: 209.
Turgeon, B. G., and Bauer, W. D. (1982). *Can. J. Bot. 60*: 152.
Turgeon, B. G., and Bauer, W. D. (1983). *Protoplasma 115*: 122.
van Brussel, A. A. N., Tak, T., Wetselaar, A., Pees, E., and Wijffelman, C. A. (1982). *Plant Sci. Lett. 27*: 317.
Vance, C. P., Stade, S., and Maxwell, C. A. (1983). *Plant Physiol. 72*: 469.
Vasse, J. M., and Truchet, G. L. (1984). *Planta 161*: 487.
Vincent, J. M. (1970). *A Manual for the Practical Study of Root-Nodule Bacteria* (IBP Handbook 15). Blackwell Scientific Publications Ltd., Oxford, pp. 164.
Wacek, T. J., and Alm, D. (1978). *Crop Sci. 17*: 514.
Wahua, T. A. T., and Miller, D. A. (1978). *Agron. J. 70*: 387.
Weaver, R. W., and Frederick, L. R. (1972). *Plant Soil 36*: 219.
Wilson, P. W., Fred, E. B., and Salmon, M. R. (1933). *Soil. Sci. 35*: 145.
Worrall, V. S., and Roughley, R. J. (1976). *J. Exp. Bot. 27*: 1233.
Zablotowicz, R. M., Focht, D. D., and Cannell, G. H. (1981). *Agron. J. 73*: 9.
Zobel, R. W., Tredici, P. D., and Torrey, J. G. (1976). *Plant Physiol. 57*: 344.

15

Principles and Practice of Field Designs to Evaluate Symbiotic Nitrogen Fixation

JOHNNY C. WYNNE *North Carolina State University, Raleigh, North Carolina*

FREDERICK A. BLISS and JUAN CARLOS ROSAS* *University of Wisconsin–Madison, Madison, Wisconsin*

*Present affiliation: Escuela Agricola Panamericana, Tegucigalpa, Honduras

I. PURPOSE OF FIELD TRIALS

The capacity of a specific plant-*Rhizobium* association to fix nitrogen under optimal conditions can be defined as "potential nitrogen fixation" as opposed to "actual nitrogen fixation" (Lie, 1981). Actual nitrogen fixation is measured in the field, where factors such as soil moisture, temperature and light may prevent the full expression of the symbiosis. Field testing is an essential step in research designed to assess nitrogen-fixing capacity of a host plant or a *Rhizobium* strain; to determine strain competitiveness or persistence; to determine the need for inoculation; and to select for superior host plants, rhizobial strains, or specific host genotype-rhizobial strain combinations.

A. Considerations in Field Experimentation

Field experiments involving nitrogen fixation generally represent a large investment of time and resources. The failure of a field experiment is, therefore, usually quite costly. Thus field studies should be planned and conducted carefully to reduce the potential for failure.

Several factors should be considered in planning a field study. The inherent variability of the soil, the nutritional level of the soil, the previous and present cropping system, the selection of treatments, the evaluation procedures, and the experimental design can affect the reliability and usefulness of the data. Brockwell et al. (1982) suggest that the characteristics of naturally occurring populations of rhizobia at the field site, an effective inoculation technique with known numbers of viable rhizobia which can be identified when reisolated from the experiment, good-quality seed, and adequate controls are also necessary. The objectives of this chapter are to provide guidelines for planning and conducting field experiments in symbiotic nitrogen fixation. It is impossible to provide all the details necessary to suit all field experiments, but important considerations are discussed. The researcher is provided with references to important literature citations for additional detail.

II. PLANNING FIELD EXPERIMENTS

A. Objectives of the Study

A clear, concise reason for conducting a study is essential to plan a field experiment. The purpose of the study will affect selection of the field site, the treatments employed, the traits evaluated, the experimental design, and the interpretation of the data. For example, the objectives of legume inoculation trials may be to answer the following questions (FAO, 1983):

Is it possible to increase traditional legume crop yields by using inoculation?
What is the effect of inoculation on the yields of introduced crops, and is it possible to grow new legumes effectively by relying solely on inoculation as a source of nitrogen?
What savings can be made by using inoculation rather than applying chemical nitrogenous fertilizer?
What is the effect of inoculation, and what savings can be made in terms of the following crop in the rotational system?
Is one inoculum more efficient than another for a given region?

If the objectives are complex, several experiments may be required to answer all the desired questions. For example, the international legume inoculation trials originally proposed by NifTAL consist of three experiments (Harris, 1979). The best available rhizobial strains for a particular crop species are tested in an initial experiment designated experiment A. Since the strains used for experiment A have not been selected for the local cultivar or for the local agroclimatic conditions, a lack of response to inoculation does not necessarily mean that inoculation is of no value. A second experiment is conducted in pot culture using a local rhizobial strain(s) adapted to the legume(s) being tested under the specific agroclimatic conditions. The third experiment is conducted in the field with a *Rhizobium* strain or mixture of strains selected for the local cultivar under its own growing conditions.

B. Selection of a Field Site

The selection of a field site for symbiotic nitrogen fixation research should be influenced by practical considerations. It may be chosen because it represents an area or possible area for legume crop production which requires investigations, or it may be a site chosen because it meets the requirements of the prescribed experimental conditions. For example, an experimental site for *Rhizobium* strain testing may be selected because it has a low level of soil nitrogen with few or no naturally occurring rhizobia, whereas a site designed to test strain competitiveness and persistence may be chosen because it has an established population of appropriate rhizobia.

Regardless of whether the field site is chosen for optimum experimental conditions or because it is representative of the production area, the field site should be nearly level with a homogeneous, uniform soil type since soil heterogeneity due to differences in soil topography, soil moisture, soil fertility, and previous crop is a source of considerable variability in field experiments (LeClerg et al., 1962).

Selection of Site for Optimum Conditions

The optimum field site will vary with the objectives of the experiment. For an inoculation response test, or one to compare rhizobial strains, or a test to evaluate specific host-strain combinations, an ideal site should have a low level of soil nitrogen, no drainage problems, no previous legume crop, no particular soil salinity or acidity problem, and have no or a limited population of *Rhizobium*, usually less than 1000 per gram of soil (NifTAL) (Harris, 1979; Weaver and Frederick, 1972). Sites for experiments to evaluate strain competitiveness and/or persistence and to identify host plants with superior nitrogen-fixing ability in symbioses with indigenous rhizobia should have similar characteristics except that the site should have an established population of desirable rhizobia.

Selection of Representative Site

If the experiment is designed to demonstrate the effect of symbiotic nitrogen-fixation treatments or to select genotypes for an already established cropping system, the site selected should be as representative of the area or cropping system as possible. Not only the soil characteristics but the growing season, and whether the legume will be sole- or intercropped, should also be considered for experiments in a tropical environment with multiple growing seasons.

C. Analysis of Soil Fertility

Several soil characteristics, including the status of soil nutrients, have been identified that affect survival of rhizobia or efficacy of a resulting symbiosis. Soil pH, nitrogen, and phosphorus often have large effects on growth, survival, and invasion of rhizobia (Munns, 1978; Robson, 1978). At low soil pH, aluminum and manganese are often present at levels that are phytotoxic to certain plants and rhizobial strains. The lack of phosphorus may limit legume growth, particularly in tropical soils. The effect of nitrogen, mineralized in the soil or applied as fertilizer, depends on the level. A low-level application is often recommended to stimulate seedling growth prior to effective nodulation, while higher levels inhibit nodulation and subsequent nitrogen fixation. In addition, calcium, boron, magnesium, sulfur, and molybdenum are required for efficient nitrogen fixation, while copper, zinc, and cobalt may also be involved (Robson, 1978). The extent to which any or all of these nutrients affect the symbiotic relationship is usually site specific and must be considered when evaluating the effectiveness of symbiotic nitrogen fixation. Because the soil nutrient status often limits nitrogen fixation, a soil analysis should be made to determine the fertility requirements for each site. NifTAL (FAO, 1983) recommends that each analysis include the following in priority order:

1. Free nitrogen, available phosphorus, exchangeable potassium, calcium, and magnesium
2. Exchangeable aluminum and manganese if the respective pH is lower than 5.2 and 5.6

3. pH and level of organic matter
4. Available proportions of certain mineral elements if deficiencies are suspected

A technical handbook prepared for FAO (1983) recommends that a more detailed preliminary soil analysis be completed before conducting a field trial on symbiotic nitrogen fixation. The following analyses were suggested: soil particle size (clay + silt); soil acidity in pH (colorimetric and potentiometric method-glass electrode); exchangeable bases (calcium, magnesium, potassium, and sodium); if the pH is lower than 5.2 and 5.6, respectively--exchangeable aluminum and manganese; total phosphorus; available phosphorus; total nitrogen; NH_3 and NO_2; organic carbon; if absence is suspected—cobalt and boron; if absence or excess is suspected—molybdenum.

Although the latter analysis is more complete, it is often difficult to obtain even the simpler analysis recommended by NifTAL. In that case, the treatments should be planned to determine whether fertility or nutrient status of the soil is limiting nitrogen fixation. In most studies of symbiotic nitrogen fixation, the fertility level for the legume crop should be the optimal level for the crop species under test since the plant must live and grow before the symbiotic process can be evaluated.

D. Selection of Treatments

Singularly or in Combination

Experiments can be simple or factorial based on the number of factors or variables to be studied simultaneously. The simple experiment is often used because of its simplicity. For the simple experiment, one variable, such as the effect of rhizobial strain inoculation, is studied separately. However, information from a simple experiment is of limited usefulness because inferences cannot be made about potential interaction between factors such as inoculation with rhizobial strains onto different host cultivars or their response to different fertilizer treatments.

Two or more simple experiments can be combined into a single factorial experiment so that the response of the different factors singularly as well as different interactions can be determined. The factorial experiment has greater efficiency and greater comprehensiveness than simple experiments.

Response to Inoculation Trials

Knowledge of nodulation specificity and effectiveness of symbioses between tropical legumes and *Rhizobium* has increased substantially. The continued study of both familiar and poorly understood genera, in a wide range of soil and climatic conditions, has demonstrated meaningful interaction between host and strain as well as favorable responses to inoculation with effective strains in the field. These studies contradict a popular consensus that many tropical legumes nodulate effectively with indigenous strains and consequently do not respond to inoculation.

Simple inoculation trials which are most often simple experiments, in that only inoculation is considered as the treatment factor, are essential prerequisites for introduction of legumes into new areas and the selection of appropriate strains of *Rhizobium* for production (Nutman, 1976; Date, 1982). Treatments in inoculation trials consist of uninoculated control plants and inoculated and inoculated-plus-nitrogen plants. The uninoculated control plants indicate the presence or absence of indigenous strains and if indigenous strains are present, provide some assessment of their symbiotic effectiveness compared to plants inoculated with selected rhizobial strains. The inoculated treatment provides a comparison of the ability of test strains to colonize the host root environment and to compete for nodule sites. An inoculated-plus-nitrogen treatment should be included to ensure that the legume plant has the potential to grow well once nitrogen is provided and that growth is not limited by some other deficiency.

Results from this set of treatments can be interpreted as follows (Date, 1982):

1. No nodulation, poor plant growth, and severe nitrogen deficiency of uninoculated, inoculated, and inoculated-plus-nitrogen control plants indicate an absence of suitable native rhizobia and factor(s) other than nitrogen are limiting plant growth.
2. Absence of nodules and poor growth of uninoculated and inoculated plants but healthy growth of plants receiving added nitrogen indicates an unsatisfactory inoculum strain or soil conditions inadequate for growth and infection of rhizobia.
3. Absence of nodules and poor growth of uninoculated plants, but effective nodulation and growth of inoculated plants similar to the nitrogen treatment, indicates successful nodulation and N_2 fixation by inoculated strain(s).
4. No nodulation and poor growth of uninoculated plants along with nodulation but poor growth of inoculated plants relative to the nitrogen treatment indicates that the inoculum strain is infective but ineffective for N_2 fixation.
5. Nodulation but poor growth of uninoculated treatment indicates the presence of unsuitable native rhizobia. Inoculated plants effectively nodulated and well grown indicates that inoculum strain(s) is both competitive and effective.
6. Both uninoculated and inoculated plants well nodulated but where growth is poor relative to nitrogen treatment indicates

that (a) native rhizobia are ineffective but compete with introduced strain, or (b) both introduced and native rhizobia are ineffective.

7. All treatments grow well; no response. Nodules on inoculated plant treatment may be native or introduced. Tests to identify nodule occupancy are needed.

Information obtained from rhizobial assessment trials with the three sets of treatments as just described will aid the researcher in determining criteria for assessing suitable strains of *Rhizobium*. For example, when the uninoculated plants have no nodules and poor growth but the inoculated plants are nodulated and have growth similar to the nitrogen treatment, the criterion for strain selection is effectiveness for nitrogen fixation. However, when both uninoculated and inoculated plants are nodulated but growth is poor, selection for competitive ability is essential. When all treatments grow well, inoculation is not essential.

Inoculation response trials can be factorial and include additional treatments. For example, an experiment can consist of three replications of each inoculant treatment with main plots split for phosphorus application.

For perennial species the persistence of a strain is an important criterion for selecting suitable strains. Experiments must be maintained and sampled over several years. For annual legumes plots can be resown each year with uninoculated seed.

Selection of Improved Host Plants

The genotype of the host plant can affect N_2 fixation through direct effects on symbiosis and indirectly through the growth habit and plant vigor. The direct effect of host genotype on symbiosis is generally described in terms of nodulation effectiveness and selectivity of the host for certain rhizobia from a mixed population (Vincent, 1984). The growth habit and plant vigor affects N_2 fixation indirectly through influence of the size of the root system, time to flowering and seed set, and tolerance to edaphic factors limiting plant growth and development. In addition to choosing improved rhizobia, selection of superior host plants is essential to ensure high levels of fixation.

Improved grain legumes, forage, and pasture legumes should be adapted to the production area, contain disease, and pest resistance and produce satisfactory yield of high quality grain, forage, or pasture. Considerable variation for nodulation and N_2 fixation has been observed in most legumes. However, few breeding programs have exploited this variability to develop improved cultivars with increased N_2-fixation potential.

The breeding methods used to develop improved host plants vary depending on the reproductive system of the plant. The crop plants within the Leguminosae include both self- and cross-pollinated species as well as those that are predominantly one or the other (Table 1). For self-pollinated crops, use of breeding methods that allow the development of segregating populations of lines (families) such as the modified pedigree method (MPM) (Brim, 1966) or the inbred backcross line method (IBLM) (Bliss, 1981) offer several advantages. The IBLM has been used successfully for increasing fixation in common bean (McFerson et al., 1981; Bliss et al., 1985). A population of nearly homozygous lines was developed through repeated backcrosses to an adapted parent after an initial cross to a high-fixing donor parent. After the last backcross two to three generations of selfing by single-seed descent produced the population of families. The resulting inbred backcross lines are similar to the recurrent parent for most traits, but differ for the traits of the donor parent. Selection for differences in fixation are made at this time. Because the seeds of each family are not limited, they can be replicated several times at different locations to increase precision and the selected families used for further crossing, additional studies, or released as potential new cultivars. Original parents and adapted local cultivars should be included for comparisons.

Table 1 Genera, Major Crops, and Type of Reproduction of Family Leguminosae

Genus	Crop	Reproduction
Arachis	Peanut	Self-pollinated
Cajanus	Pigeon pea	Usually self-pollinated
Cicer	Chickpea	Self-pollinated
Glycine	Soybean	Self-pollinated
Lens	Lentil	Self-pollinated
Medicago	Alfalfa	Cross-pollinated
Phaseolus	Common bean	Self-pollinated
	Runner bean	Self-pollinated
	Tepary bean	Usually self-pollinated
	Lima bean	Self-pollinated
Pisum	Pea	Self-pollinated
Vicia	Bean	Cross-pollinated
Vigna	Cowpea	Self-pollinated

Breeding procedures for increasing nitrogen fixation of cross-pollinated species such as alfalfa may be more complicated. Phenotypic recurrent selection for increased N_2 fixation in the greenhouse was ineffective when the populations were evaluated in the field (Barnes et al., 1981; Heichel et al., 1981). A breeding procedure where the progenies of individual clones are evaluated in the field will probably be required to increase nitrogen fixation. The progenies could be evaluated similarly to the inbred lines described for self-pollinated species.

Host-Strain Specificity

Field studies designed to estimate host-strain specificity are an extension of an experiment for determining the response to inoculation. Treatments consist of several host genotypes inoculated with different rhizobial strains. An uninoculated check and a nitrogen treatment for each genotype should be included. Generally, the two factors (hosts and rhizobial strains) are evaluated in all combinations. This may increase the number of treatments substantially requiring the experiment to be planted in a split-plot design. Generally, rhizobial strains should be assigned to main plots and host genotypes to subplots because the application of inoculum to whole areas within a block will reduce the risk of contamination.

Other Studies

Field studies to estimate the amount of nitrogen fixed by legumes are often conducted. These studies and the methods useful in determining nitrogen fixation of legumes are described by LaRue and Patterson (1981) and elsewhere in this text.

E. Design of a Field Experiment

Experimental design involves planning a field experiment to permit the collection of unbiased data with maximum accuracy and precision at a reasonable cost. The planning stage includes consideration of soil heterogeneity, nature and importance of experimental error, principles of field experimentation, field plot technique, and appropriate experimental designs (LeClerg, 1966).

The greatest source of experimental error in field experiments is generally due to soil heterogeneity (LeClerg et al., 1962). Since soil heterogeneity has long been recognized as a source of experimental error, experimental designs have been developed to eliminate, or at least minimize, the effect of soil heterogeneity. The general practice for controlling soil heterogeneity in field experiments is to decrease the size of the plot and increase the number of replications. The repetitions of a plot are then scattered over the test area in order to sample different soil conditions. Although efficient experimental designs can remove a portion of the variability due to soil conditions, careful selection of the experimental site is also important in studies of symbiotic nitrogen fixation.

Variability in field experiments comes from two principal sources: (a) inherent variability which exists in the experimental material, and (b) variability that results from lack of uniformity in the physical conduct of an experiment (Steel and Torrie, 1960). The failure of plots treated alike to perform similarly is experimental error (Fisher, 1951). The degree of confidence associated with field data depends on the magnitude of experimental error. Experimental error should be reduced as much as possible by reducing the inherent variability of the experimental material and using careful experimental techniques. A good experimental design simply estimates the magnitude of the remaining uncontrolled experimental error.

Principles of Field Experimentation

Field experiments are designed and performed carefully so that appropriate inferences can be made from the data. To make inferences as precise as possible, replication and randomization of the experiment are practiced. Since the soil of an experimental site is assumed to be variable, treatments must be replicated. Replication provides an estimate of experimental error and reduces the uncertainty of the data. Multiple observations from a plot do not represent true replication but are subsamples and allow estimation of sampling error.

Because soil heterogeneity is universal in field sites, treatments cannot be compared on an identical soil at the same time but must be compared on different plots of an experimental site. This procedure is satisfactory only if treatments are assigned at random to the plots of the site.

Certain restrictions on the random arrangement of treatments in a replicated experiment can be imposed such that some of the total variability not relevant to treatment comparisons may be eliminated. This is called local control and is synonymous with proper experimental design.

Experimental Designs

Each field study presents conditions that must be considered in the choice of an appropriate experimental design. However, the simplest design that fulfills the requirements of the experiment should be used (LeClerg et al., 1962). The randomized complete block design can be utilized for most nitrogen-fixation experiments. A Latin square design can be used to remove variability from soil heterogeneity in two directions. It is very efficient but it is usually not practiced with more than eight treatments (Little and Hills, 1978). A randomized complete block design is usually effective for experiments having up to 15 treatments. For greater numbers of treatments in a single experiment, incomplete block designs should be considered. When it is convenient to arrange the plots of a multiple-factor experiment so that all factors are not distributed over the replication and the factors are not of equal importance, a split-plot design can be used. The split-plot design provides a more precise comparison of treatments assigned to subplots and a less precise comparison of the whole-plot treatments.

Most inoculation trials such as the INLIT trials conducted by NifTAL (Harris, 1979) or the FAO legume inoculation trials (FAO, 1983) specify a randomized complete block design, although Date (1976) suggests that split plots might be useful in inoculation response trials where a trial is extended to include additional treatments, such as the presence or absence of phosphorus.

A statistician should be consulted on the proper experimental design and data analysis if the treatments are numerous or if the experiment is not one of the routine symbiotic nitrogen-fixation field studies.

III. PLANTING A FIELD EXPERIMENT

A. Preparation of Experimental Site

The entire experimental area, including borders, should be treated uniformly. Cultural practices such as plowing, disking, harrowing, and any basic fertilizer treatment used over the entire test should be done before planting using procedures common to the region. Seedbed preparation should be appropriate for planting the legume crop chosen for the experiment.

B. Plot Layout

The specific layout of a trial designed to determine the response to inoculation should vary according to the specific objectives. An example of a standardized experimental method for testing the value of inoculation has been developed by NifTAL (Harris, 1979). NifTAL initiates an experiment with rhizobial strains considered to be the best available for the test legume. Since the rhizobial strains are not selected initially for the local cultivar or for the local agroclimatic conditions, negative results do not necessarily constitute a definitive evaluation. Thus NifTAL recommends a three-phase trial as follows:

1. Experiment A, based on the best available rhizobial strains and local cultivar of the host species.
2. Experiment B, a pot experiment using site soil which compares local strains with strains tested in experiment A to select a strain adapted to the legume being tested under specific agroclimatic conditions.

3. Experiment C, conducted with a *Rhizobium* strain (or mixture) specifically selected for the local cultivar grown under local agroclimatic conditions. The plot layout and arrangement of plots for experiment C are shown in Figure 1.

The experiment is laid out in a randomized complete block design with four replicates. Rows are 7.5 m long with four rows per plot. The width of the plot is from 2 to 4 m, depending on the legume species being evaluated. The plots are designed for early season sampling of nitrogen fixation and a final harvest of yield (Figure 2).

C. Methods of Inoculating Field Studies

Types of Inoculant

Numerous methods of inoculating legumes have been employed depending on the type of inoculant available. Most commercial inoculants are based on solid carriers, mainly soil or peat. The survival of rhizobia has generally been better when solid-based inoculants are applied to the seed than when other forms of inoculant are used. However, several other types of inoculant, such as agar cultures, liquid cultures, or lyophilized cultures, have been used experimentally.

Application of Inoculant

Principles of inoculation for field experiments: Brockwell (1982) lists some general principles of legume inoculation that apply to field experimentation regardless of the method of inoculation. They include:

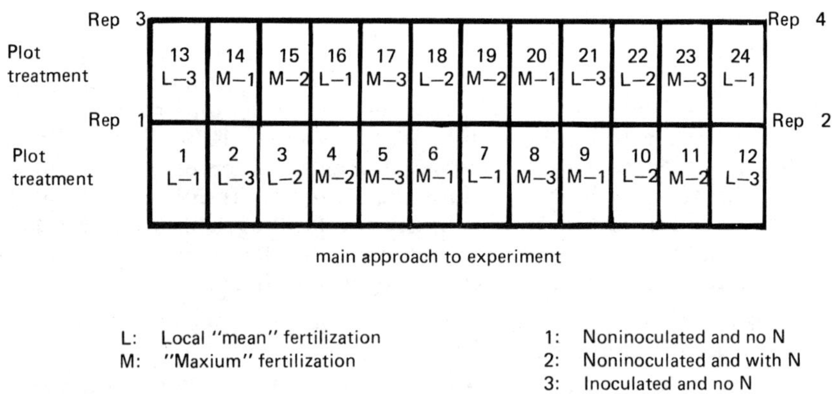

Figure 1 Trial plot layout.

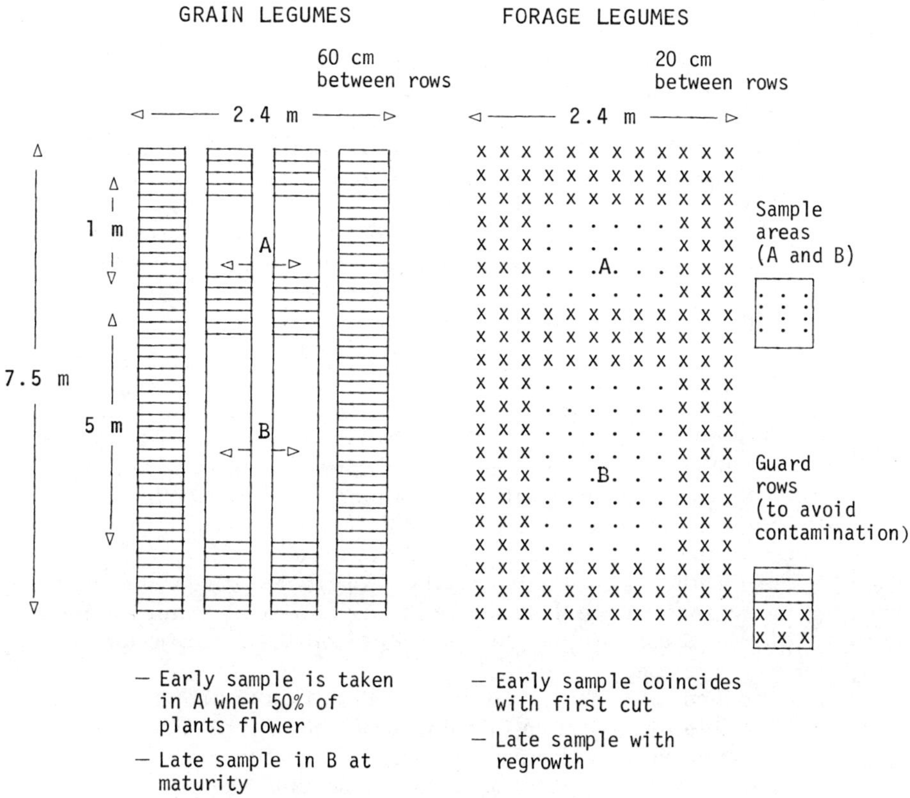

Figure 2 Plot dimensions and sampling area.

1. When uninoculated controls are included, they should be handled before inoculated treatments to reduce the risk of contamination of the controls.
2. The number of inoculant rhizobia should be as high as feasible depending on the objective of the experiment since the presence of large numbers of inoculant rhizobia reduces contamination by other rhizobia.
3. Designing and managing a field experiment with the awareness that rhizobia are mobile organisms and can be moved easily from plot to plot.
4. Rhizobia are incompatible with many pesticides used as seed treatments.

Methods of application

1. *Seed inoculation.* Inoculants can be applied to the seed before planting or they can be applied directly into the seedbed. The most widely used methods of inoculating seeds are slurry inoculation and seed pelleting, while solid carrier-based inoculants and liquid inoculations are usually applied into the seedbed.
   a. *Slurry inoculation.* The rhizobial culture is applied to the seed as a water suspension or the inoculant is mixed with moistened seed. The seed must be dried before planting, thus killing a proportion of the rhizobia. More inoculant can be attached to the seed by using an adhesive such as a 10% solution of sugars or gum arabic in the slurry.
   b. *Seed pelleting.* Legume seeds may be coated with lime to promote nodulation. Advantages include protection of the rhizobia against toxic substances contained in some legume seed coats, unfavorable physical and chemical conditions in the soil, competition from soil microflora, and the effects of acid fertilizers (Brockwell, 1977). Brockwell (1982) described one method for pelleting seeds using standard laboratory equipment that can be used for small lots of seeds. Materials used for seed coating include calcium carbonate, dolomitic limestone, and gypsum.
2. *Seedbed inoculation.* In several situations the application of rhizobia to seed is an inefficient method of inoculation (Burton, 1974; Brockwell et al., 1982). An alternative method of inoculation such as application of the rhizobia to the soil can be used. Inoculant can be applied to the seedbed as a solid inoculant or as a liquid.
   a. *Solid inoculant.* A solid inoculant or soil implant inoculant is produced by coating a solid granulated material with peat inoculant in an adhesive. A suitable adhesive is a 25% aqueous solution of gum arabic without preservatives. The peat inoculant is thoroughly stirred into the adhesive. The resulting suspension is poured onto a granulated material, which can be any inert, particulate free-flowing material, and mixed until the material is evenly coated. For large experiments solid inoculant can be applied through a fertilizer attachment or pesticide applicator of the planter. For smaller experiments it can be applied by hand to the row along with the seed. Solid inoculant is especially suited for experiments on the rate of inoculant or for the inoculation of numerous small samples.
   b. *Liquid inoculation.* Excellent nodulation can be obtained by applying a liquid inoculum into the row beside or beneath the seed. A peat culture of rhizobia from frozen or concen-

trated broth culture is mixed with water and can be applied by a gravity-flow applicator or sprayer. The method works well with small experiments but can create problems with blocked nozzles when used with spray equipment. Brockwell (1982) appraised several inoculation methods suitable for field experimentation (Table 2).

IV. EVALUATION OF TREATMENTS

A. Data Collection

Appropriate measures of the response to inoculation are plant dry matter, total plant nitrogen accumulation, and seed yield for grain legumes and dry matter yield for forage legumes. However, there are many traits associated with nitrogen fixation that can be described during plant growth in the course of an experiment. These include earliness to nodulation, nodule number, nodule mass, nodule color, distribution of the nodule population, longevity of the nodule population, and visual nodulation scores.

It is common for legumes to show a visual response to inoculation that does not necessarily correspond directly to seed yield increase. Thus it is desirable to obtain some measure of the early response to a treatment to distinguish between no response and a potential for response.

B. Measurement of Traits

Nodulation is generally assessed from 6 to 20 weeks after planting, depending on the legume and the seasonal conditions, by sampling a number of plants in each treatment. Nodule number is a valuable criterion in the early stages of plant development, but it is of little use thereafter. At that point the best criterion is nodule dry mass or nodule dry mass per unit dry weight of the whole plant or root system. In experiments where plants with different growth rates are being compared, it is important that measurements be made at similar stages of maturity (e.g., the R3 stage) rather than at the same time after emergence (McFerson et al., 1982). After determining the nodule mass one should determine whether these nodules are effective. As a general rule, effectiveness can be gauged by the degree of pink or red coloration of the nitrogen-fixing bacteroid tissue inside the nodule—white or green being inactive. Another measure of effectiveness is use of acetylene reduction methods (Hardy and Holsten, 1977; Masterson and Murphy, 1976, 1980; Ham, 1977). The acetylene reduction assay has the advantage of sensitivity, speed, and economy (LaRue and Patterson, 1981), although it has been criticized by some as being of questionable value (Minchin et al., 1983; Sinclair et al., 1976). Because

Table 2 Methods and Appraisal of Legume Inoculation

Method	Procedure	Appraisal
Dusting	Dry, peat-based inoculant poured on seed in planter	Quick, simple; ineffective since much of the inoculant is lost before the seed reaches the soil, thus no longer recommended
Slurry	Peat-based inoculant suspended in water and applied to seed, mixed, and allowed to dry before planting	No special equipment required; necessary to dry seed away from direct sunlight; some splitting of seed may occur during handling; slow and labor consuming; reinoculation necessary if not planted within 48 hr; not suitable when fungicides or insecticides applied to seed
Solid inoculant	Polyethylene beads, marble chips, or other freely flowing material coated with peat-based inoculant or peat-based inoculant extruded into freely flowing pellets and applied in the row with seed	Gives excellent nodulation; maintain viability well; suitable for inoculation of pesticide-treated seed
Seed pelleting	Peat-based inoculant suspended in adhesive, applied to seed, mixed together, then finely ground coating material added to sticky seed and mixed until each seed is evenly coated	Protects inoculants against deleterious soil environmental factors; time-consuming preparation; coating material may come off and cause blockage in planters; high inoculant mortality rate observed in many pelleted seed samples
Spray inoculation	Peat-based inoculant suspended in water and sprayed into seedbed alongside seed	Gives excellent nodulation; no storage problems provided that peat inoculant kept cool until suspended in water; suitable for inoculation of pesticide-treated seed; no complicated preparation procedure; easy to apply; blocked delivery nozzles can be a problem
Pre-inoculation	Inoculation of seed with rhizobia prior to sale	Convenient for farmers; mortality rate of inoculant often high, resulting in low levels of inoculation

the plant is usually destroyed during the assay, acetylene reduction is a single-point, indirect estimate of nitrogen-fixation potential. Thus an estimate of fixation by a crop or the identification of superior treatments requires a summation of several assays over a growing season (Arrendell et al., 1985), or other more accurate methods (e.g., ^{15}N isotope application).

The effect of treatments can also be measured by total plant nitrogen or the nitrogen content of any plant part. The final evaluation of a treatment effect is to measure economic yield at harvest (i.e., grain yield or forage yield).

C. Analysis and Interpretation of Data

No study is complete until the data have been summarized, analyzed, and interpreted properly. An analysis of variance or regression analysis providing a test of significance of treatment means is the minimum analysis required. Additional analysis such as covariance or trend analysis may be needed to interpret the results.

REFERENCES

Arrendell, S., Wynne, J. C., Elkan, G. H., and Isleib, T. G. (1985). *Crop Sci. 25*: 865.

Barnes, D. K., Heichel, G. H., Vance, C. P., Viands, D. R., and Hardarson, G. (1981). Successes and problems encountered while breeding for enhanced N_2 fixation in alfalfa. In *Genetic Engineering of Symbiotic Nitrogen Fixation and Conservation of Fixed Nitrogen*, J. M. Lyons, R. C. Valentine, D. A. Phillips, D. W. Rains, and R. C. Huffaker, eds. Plenum Press, New York, pp. 223-248.

Bliss, F. A., Rosas, J. C., St. Clair, D. A., and Kmiecik, K. A. (1985). *HortScience 20*: 544.

Brim, C. A. (1966). *Crop Sci. 6*: 220.

Brockwell, J. (1977). Application of legume seed inoculants. In *A Treatise on Dinitrogen Fixation*, Sec. 4, *Agronomy and Ecology*, R. W. F. Hardy and A. H. Gibson, eds. John Wiley & Sons, Inc., New York, pp. 277-309.

Brockwell, J. (1982). Inoculation methods for field experimenters and farmers. In *Nitrogen Fixation in Legumes*, J. M. Vincent, ed. Academic Press, Inc., New York, pp. 211-221.

Brockwell, J., Diatloff, A., Roughley, R. J., and Date, R. A. (1982). Selection of rhizobia for inoculants. In *Nitrogen Fixation in Legumes*, J. M. Vincent, ed. Academic Press, Inc., New York, pp. 173-191.

Burton, J. C. (1974). Pragmatic aspects of the *Rhizobium*: Leguminous plant association. In *Proceedings of 1st International Symposium on Nitrogen Fixation*, Vol. 2, W. E. Newton and C. J. Nyman, eds. Washington State University Press, Pullman, Wash., pp. 429-446.

Date, R. A. (1976). Principles of *Rhizobium* strain selection, in *Symbiotic Nitrogen Fixation in Plants*, P. S. Nutman, ed. Cambridge University Press, Cambridge, pp. 137-150.

Date, R. A. (1982). Assessment of rhizobial status of the soil. In *Nitrogen Fixation in Legumes*, J. M. Vincent, ed. Academic Press, Inc., New York, pp. 85-94.

FAO (1983). *Technical Handbook on Symbiotic Nitrogen Fixation Legume/Rhizobium*. Food Agricultural Organization of the United Nations, Rome.

Fisher, R. A. (1951). *Design of Experiments*, 6th ed. Hafner Press, New York.

Ham, G. E. (1977). The acetylene-ethylene assay and other measures of nitrogen fixation in field experiments. In *Biological Nitrogen Fixation in Farming Systems of the Tropics*, A. Ayanaba and P. J. Dart, eds. John Wiley & Sons, Inc., New York, pp. 325-334.

Hardy, R. W. F., and Holsten, R. D. (1977). Methods for measurement of dinitrogen fixation. In *A Treatise on Dinitrogen Fixation*, Sec. 4, *Agronomy and Ecology*, R. W. F. Hardy and A. H. Gibson, eds. John Wiley & Sons, Inc., New York, pp. 451-486.

Harris, S. C., ed. (1979). Planning an international network of legume inoculation trials. In *Proceedings of a Workshop Held at Kahului, Maui, Hawaii, Jan. 15-19*. NifTAL Project, Maui, Hawaii.

Heichel, G. H., Vance, C. P., and Barnes, D. K. (1981). Evaluating elite alfalfa lines for N_2-fixation under field conditions. In *Genetic Engineering of Symbiotic Nitrogen Fixation and Conservation of Fixed Nitrogen*, J. M. Lyons, R. C. Valentine, D. A. Phillips, D. W. Rains, and R. C. Huffaker, eds. Plenum Press, New York, pp. 217-232.

LaRue, T. A., and Patterson, T. G. (1981). How much nitrogen do legumes fix? In *Advances in Agronomy*, N. C. Brady, ed. Academic Press, Inc., New York, pp. 15-38.

LeClerg, E. L. (1966). Significance of experimental design in plant breeding. In *Plant Breeding*, K. J. Frey, ed. Iowa State University Press, Ames, Iowa, pp. 243-313.

LeClerg, E. L., Leonard, W. H., and Clark, A. G. (1962). *Field Plot Technique*. Burgess Publishing Company, Minneapolis, Minn.

Lie, T. A. (1981). Environmental physiology of the legume-*Rhizobium* symbiosis. In *Nitrogen Fixation in Plants*, Vol. 1, *Ecology*, W. J. Broughton, ed. Oxford University Press, Oxford, pp. 104-134.

Little, T. M., and Hills, F. J. (1978). *Agricultural Experimentation, Design and Analysis*. John Wiley & Sons, Inc., New York.

Masterson, C. L., and Murphy, P. M. (1976). Application of the acetylene reduction technique to the study of nitrogen fixation by white clover in the field. In *Symbiotic Nitrogen Fixation in Plants*, P. S. Nutman, ed. Cambridge University Press, Cambridge, pp. 299-316.

Masterson, C. L., and Murphy, P. M. (1980). The acetylene reduction technique. In *Recent Advances in Biological Nitrogen Fixation*, N. S. Subba Rao, ed. Holmes and Meier Publishers, Inc., New York, pp. 8-33.

McFerson, J., Bliss, F. A., and Rosas, J. C. (1982). Selection for enhanced nitrogen fixation in common beans, *Phaseolus vulgaris*. In *Biological Nitrogen Fixation Technology for Tropical Agriculture: Papers Presented at a Workshop Held at CIAT, March 9-13, 1982*, P. H. Graham and S. C. Harris, eds. Centro Internacional de Agricultura Tropical, Cali, Colombia, pp. 39-48.

Minchin, F. R., Witty, J. F., Sheehy, J. E., and Muller, M. (1983). *J. Exp. Bot. 34*: 641.

Munns, D. N. (1978). Soil acidity and nodulation. In *Mineral Nutrition of Legumes in Tropical and Subtropical Soils*, C. S. Andrews and E. J. Kamprath, eds. Commonwealth Scientific and Industrial Research Organization, Melbourne, Australia, pp. 247-263.

Nutman, P. S. (1976). IBP field experiments on nitrogen fixation by nodulated legumes. In *Symbiotic Nitrogen Fixation in Plants*, P. S. Nutman, ed. Cambridge University Press, Cambridge, pp. 211-238.

Robson, A. D. (1978). Mineral nutrients limiting nitrogen fixation in legumes. In *Mineral Nutrition of Legumes in Tropical and Subtropical Soils*, C. S. Anderson and E. J. Kamprath, eds. Commonwealth Scientific and Industrial Research Organization, Melbourne, Australia, pp. 277-311.

Sinclair, A. G., Hannagon, R. B., and Risk, W. H. (1976). *N.Z. J. Agric. Res. 19*: 451.

Steel, R. G. D., and Torrie, J. H. (1960). *Principles and Procedures of Statistics*. McGraw-Hill Book Company, New York.

Vincent, J. M. (1984). Potential for enhancing biological nitrogen fixation. In *Crop Breeding: A Contemporary Basis*, P. B. Vose and S. G. Blixt, eds. Pergamon Press Ltd., Oxford, pp. 185-215.

Weaver, R. W., and Frederick, L. R. (1972). *Agron. J. 64*: 597.

16

Production and Quality Control of Inoculants

R. STEWART SMITH *The Nitragin Co., Inc., Milwaukee, Wisconsin*

I. INTRODUCTION

Symbiotic nitrogen fixation provides legumes with a mechanism for obtaining nitrogen from the air. This process utilizes rhizobia bacteria which invade the legume root system and form nodules in which nitrogen fixation occurs. In many situations, establishment of effectively nodulated legumes requires the addition of specific effective rhizobia in the process known as legume inoculation. To effectively establish legume nodulation, the inoculant must have a high-viable population of the proper rhizobia. High-quality inoculants are essential both in soils free of rhizobia for the intended host or in soils that already have an indigenous population. A high population is required to establish sufficient nodulation in the first situation, and a higher population may be required to compete with the indigenous rhizobia.

The production of high-quality inoculant involves the selection of efficient strains followed by process technology steps, including satisfactory carrier selection and preparation, successful culture production, inoculant mixing, curing, and packaging. Quality control standards and procedures ensure the consistent production of high-quality inoculants.

II. INOCULANT PLANT DESIGN

A. General Plant Layout

The following is a general format for the design and layout of an inoculant production plant. The design, layout, and equipment selection must be decisions made for each production facility to meet their specific needs.

A flow-through plant layout is advisable to minimize space, help maintain cleanliness in the laboratory and fermentor room, and allow for minimum ingredient movement and efficient manufacturing. The flow-through design can be either vertical (from top to bottom on different floors) or horizontal (on the same floor). This establishes the laboratory, transfer room, and fermentor room on the top floor, or an isolated section in the one-floor design, with mixing, curing, packaging, and finished goods warehousing flowing in a consecutive manner. It is desirable at the time of the initial plant establishment to allow for the expansion of production capabilities without disrupting the flow-through design.

B. Laboratory

The laboratory should be designed for bacteriological work. There should be adequate facilities and equipment for doing basic and applied research, plus quality control of the cultures and inoculant.

Laboratory bench space, gas outlets, sinks, laboratory equipment, large autoclave, and storage facilities are essential components.

It is necessary to provide an area in which the transferring of cultures between tubes and flasks, and dilution plating, can be done with a minimum chance of contamination. The need for making aseptic bacterial transfers is very important in the maintenance of a culture collection as well as during the transferring for culture production. A laminar-flow sterile air bench in a separate transfer room provides excellent protection for aseptic bacteriological transfers. The transfer room should have ultraviolet lights, gas outlets for laboratory burners, and electrical outlets. There should not be any windows in the transfer room unless they are completely sealed, and there should be only one door from the transfer room into the laboratory.

C. Fermentor Room

The fermentor room should be designed for large-scale culture production and must maintain a high degree of sanitation. It is advisable at the time of initial plant installation to locate equipment so that all plumbing and controls have convenient access, and to allow space for additional equipment at a later time when increased production needs warrant.

The system recommended for rhizobial culture production utilizes batch fermentations with aseptic serial transfer of grown culture. Stainless steel fermentors of increasing size to allow for a 1 to 10% inoculum level should be adequate.

The location of electrical outlets, water lines, air lines, and other utility outlets should be determined before the fermentors are installed. Air conditioning or a large exhaust fan near the ceiling should be installed to control high temperatures generated by steam utilization. Ultraviolet lights may have to be installed if unexpected problems of contamination arise.

D. Steam and Sterile Air Room

A room must be provided close to the fermentor room to house the boiler for steam generation and an air compressor to supply the fermentors with sterile air. Sufficient steam capacity is required to sterilize the fermentors, operate the autoclave, and steam sterilize permanent culture transfer lines.

E. Mixing and Curing Room

Two basic methods of inoculant production currently are utilized. The primary method, which generally employs a processed but nonsterile peat carrier, is more cost-efficient and lends itself to large-scale production. The second system utilizes a presterilized peat carrier (gamma irradiated or autoclaved). The presterilized carrier method produces an inoculant with a higher rhizobial population and provides an opportunity to extend the produced culture by diluting the broth before adding to the peat.

The presterilized carrier system requires the filling and sealing of the bags of dry carrier, followed by sterilization, and then the aseptic culture addition. A clean environment is essential in the room used for the aseptic culture addition in this sterile production system. The remainder of this discussion will describe inoculant production with the nonsterile carrier system.

The location of the room utilized for mixing of the rhizobial culture with the carrier should be adjacent to the fermentor room. However, because of the dust generated in the mixing process, it is essential not to have direct air passage, or doors, connecting the fermentor room with the mixing room. The mixed inoculant may be cured in suitable containers in the mixing room or in an adjacent environmentally controlled curing room. Temperature and humidity in the curing area should be maintained to provide optimum culture growth in the carrier without appreciable moisture loss.

F. Packaging Room

The packaging area should be next to the curing room. Supplies required for packaging inoculant, such as polyethylene bags, cartons, and labeling materials, should be readily accessible to the packaging area. The packaging operation, whether hand filled or automated, is most efficient in an assembly line operation. Sufficient room for filling the bags, heat sealing the bags, and assembling into a carton and case is required.

G. Finished Goods Storage Room

Adjacent to the packaging room should be the finished goods storage room providing temporary storage of the product. In temperate climates ambient temperatures and humidity may be utilized; however, air conditioning is advisable in warm weather, and refrigerated storage is ideal. Inoculants are a perishable product which should not be submitted to conditions that enhance drying, such as increased temperatures and direct sunlight.

A staging area next to the finished goods storage room is required for the assembling, addressing, and shipping of inoculant orders. It should have direct access to the outside of the building, and if large quantities of inoculants are produced and shipped, a truck loading dock is required. A set of scales (the size will depend on the quantities of inoculant shipped) for determining shipping weights will be needed.

III. INOCULANT PLANT OPERATION

A. Fermentation

Equipment

Rhizobia can be grown in equipment of varying size and design depending on the production needs, available resources, and funding. Investigators (Burton, 1981; Roughley, 1970) list the simplest of containers as bottles, round-bottomed flasks, shake flasks, or any of a wide variety of small improvised fermentors. Date (1974) describes a small fermentor which is fabricated from steel drums (10 to 100 liters) fitted with brass inlet, outlet, inoculation, and sampling ports. The whole unit can be placed in an autoclave for sterilization. The size of the small fermentor is often determined by the available autoclaving facilities. Similar small fermentors have been obtained by modifying a portable steam sterilizer (Gulati and Seth, 1973) and a dairy milk can (K. P. Goswami, personal communication, 1985).

Large-scale fermentation equipment should be fabricated from stainless steel as a double-walled pressure boiler with a dimpled jacket for rapid heating and cooling. The fermentor will be stationary and must be supplied with sterile air and pressurized steam.

Fermentors of any capacity must be capable of being sterilized and, with the exception of small units on a shaker, have a supply of sterile air, aseptic air exhaust unit, inoculation port, sampling port, agitation capability (air sparging suffices on smaller units), and a culture discharge device.

Culture Media

Media used in the cultivation of rhizobia must contain all nutritional requirements in a form suitable for rapid cell production. The objective, contrary to many fermentation systems, is for cell production, not for metabolic products.

The rhizobia are placed into two nutritional groups based on growth rate. *Rhizobium meliloti, R. trifolii, R. leguminosarum, R. phaseoli,* and others which are considered to be fast growers produce turbidity in broth, or colonies on solid media, in 3 to 5 days at 28°C. Rhizobia with slow growth, *Bradyrhizobium* (Jordan, 1982), which nodulate soybean, cowpea, peanuts, and other leguminous plants, produce turbidity in broth, or colonies on solid media, in 5 to 7 days at 28°C.

The extensive literature on energy metabolism is reviewed by Elkan (1984). Most fast growers utilize disaccharides (sucrose being the most common source) as growth substrates, whereas slow-growers cannot grow well on these substrates. Mannitol is the traditional carbon and energy source used for small-volume cultivation of all *Rhizobium*, although some slow growers prefer a pentose, such as arabinose. The triose, glycerol, has been shown (Arias and Martinez-Drets, 1976;

Elkan and Kwik, 1968) suitable for *Bradyrhizobium japonicum* and is an inexpensive source for large-scale production. Iswaran and Jauhri (1969) reported malt extract as another carbon source for large-scale production, and Nandi and Sinha (1974) found molasses to also be an efficient, inexpensive carbon source.

Yeast extract is almost universally used as a source of organic nitrogen and growth factors. The source and concentration of yeast extract should be evaluated since higher concentrations (0.35% or greater) have been shown to cause cell distortion (Skinner et al., 1977), reduced cell numbers (Date, 1972), and loss of symbiotic effectiveness (Strijdom and Allen, 1966). A suitable yeast extract is also a good source and quantity of the necessary growth factors, such as biotin, niacin, pyridoxine, riboflavin, and thiamine.

In addition to the many inorganic salts supplied by the yeast extract, most media for rhizobia supply a phosphate buffer, magnesium sulfate, sodium chloride, and usually calcium. Commonly used media for rhizobia production are listed in Table 1.

Table 1 Media Composition for Rhizobia Production

1. Norris and Date (1976); YEM		2. Burton (1967)	
Mannitol	10.0 g	Sucrose	10.0 g
Yeast extract (Difco)	1.0 g	Mannitol	2.0 g
K_2HPO_4	0.5 g	K_2HPO_4	0.5 g
$MgSO_4 \cdot 7H_2O$	0.8 g	$MgSO_4 \cdot 7H_2O$	0.2 g
NaCl	0.2 g	NaCl	0.1 g
$FeCl_3 \cdot 6H_2O$	0.01 g	$CaCO_3$	0.25 g
Distilled water	1000 ml	Yeast autolysate	1.0 g
		Distilled water	1000 ml
3. Fred et al. (1932)		**4. Arias and Martinez-Drets (1976)**	
Sucrose	10.0 g	Glycerol	10.0 g
Yeast extract	3.0 g	Yeast extract	3.0 g
K_2HPO_4	0.75 g	KH_2PO_4	1.0 g
$MgSO_4 \cdot 7H_2O$	0.4 g	K_2HPO_4	1.0 g
$CaCO_3$	0.4 g	$MgSO_4$	0.18 g
Distilled water	1000 ml	NaCl	0.2 g
		$CaCl_2$	0.13 g
		$FeCl_3 \cdot 6H_2O$	0.004 g
		NH_4Cl	0.5 g
		Distilled water	1000 ml

Culture Production

Culturing systems: The batch system, which utilizes consecutive transfers of grown culture from smaller vessels to larger fermentors, is recommended for rhizobia culture production. The recommended inoculum level is between 1 and 10%; however, a higher level of inoculum will decrease the time in a given fermentor and is an advantage with the slow-growing rhizobia because it decreases the opportunity for contamination.

The semicontinuous or continuous fermentation system is not recommended with rhizobia because of the increased chance of contamination with increased fermentation time, and the need to mix the broth with the carrier in a batch operation is best facilitated with a batch fermentation system.

Starters: The following reports provide evidence for genetic instability in *R. trifolii* (Vincent, 1944; Labandera and Vincent, 1975; Zurkowski, 1982), *B. japonicum* (Kuykendall and Elkan, 1976; Meyer and Pueppke, 1980; Upchurch and Elkan, 1977), *R. meliloti* (Scherrer and Denarie, 1971), and cowpea rhizobia (Herridge and Roughley, 1975). Spontaneous mutations occur with repeated transfers and growth, and have been shown to alter the colony morphology, nodulation and fixation capabilities, carbohydrate utilization, salt tolerance, antibiotic sensitivity patterns, lectin binding, and cell plasmid contents. It is therefore essential to keep commercial strains in a long-term storage system with the objective of maintaining genetic stability. The number of transfers during culture production should be kept to a minimum.

Culture production begins by transferring the selected strains from permanent storage to either a small broth volume either in test tubes or flasks, or streaked on agar slopes within a 500-ml sidearm Roux flask. Liquid broth starters are transferred aseptically to a small fermentor after adequate growth is obtained. The agar flask system provides the advantages of being able to examine the streaks visually for cleanliness and sufficient growth, and the grown flasks can be inventoried under refrigeration for a short period until further culture production is required. When the small fermentor is inoculated from a Roux flask, the agar growth is washed with a sterile 1% Bacto peptone solution by gently swirling the flask, taking care not to wet the cotton plugs. Aseptic transfer from the Roux flask to the fermentor is accomplished by gravity feeding through presterilized rubber tubing.

Growth conditions and operations of the small fermentor should be approximately the same as in the succeeding larger fermentors, which are used if larger quantities of culture are required for inoculant production. The medium is formulated and sterilized in the fermentor. After the medium is cooled to 30°C the fermentor is aseptically inoculated with an active culture.

Temperature: Exact temperatures are not required for rhizobia production. Date and Roughley (1977) recommended fermentation temperatures between 26 and 28°C as suitable, whereas Burton (1967) suggests using 30 to 32°C, except for strains able to nodulate *Medicago* species, in which a 35°C temperature is preferred. With small fermentors without temperature control capabilities the ambient room temperature should be controlled to maintain adequate temperature for the fermentor. Larger fermentors provide temperature control capabilities through internal coils or outer jackets.

Aeration: Aeration is required for satisfactory growth of rhizobia in liquid culture. A flask with large surface area-to-volume ratio is adequate for rotary shaker production; however, forced sterile air is necessary in stationary fermentors. Small fermentors can be supplied with various types of air systems, including aquarium pumps, aspirators, compressed air in tanks, or air compressors. Larger fermentors require air blowers or air compressors. Filtering devices packaged with cotton or commercial air filters are required to provide sterile air to the fermentor. The air supply is positioned through the top of the fermentor with a fine pore-size sparger dispersing the air throughout the broth. Mechanical agitation is frequently provided on larger fermentors to facilitate aeration, and media circulation during the sterilization stage, cooling stage, and may be used during culture growth. An antifoam (white mineral oil or commercial antifoam) in the medium is recommended if excessive foaming occurs during the fermentation processes. Air exhaust from the fermentor is usually passed to the environment through a gooseneck within a solution containing a bactericide.

Airflow rates have been reported from as little as 0.2 to 60 liters of air per hour per liter of medium (Date, 1974). Roughley (1970) reported 5 liters of air per hour per liter as satisfactory.

Cell population: Under good fermentation conditions a cell population of 1 to 5×10^9 per milliliter can be obtained. The grown culture to be used for inoculant production should be mixed with the carrier during the active late log to early stationary growth phase. Most inoculants are produced with culture directly from the fermentor. However, when a presterilized peat is used the culture can be diluted with sterile water (Somasegaran, 1985) to contain a final count of 1×10^6 cells/ml, then aseptically incorporated into the peat. After 4 weeks of incubation, the final rhizobia population in the inoculant produced with diluted culture has been reported as being the same as peat inoculant produced with straight culture (Roughley, 1970; Somasegaran and Halliday, 1982). Inoculant production with diluted culture effectively increases the amount of finished inoculant produced from a given fermentation capacity. However, the potential production cost savings with this system will depend on the method of carrier sterilization and the charges for the labor-intensive culture addition operation. This system is not conducive to large inoculant production systems.

B. Inoculant Carrier

Carrier Function and Characteristics

Rhizobia are usually incorporated into a solid carrier which serves several functions during the period from manufacturing through product use. The carrier becomes the major portion of the product and provides a convenient base to facilitate product packaging and final product application and use. In moist formulations the carrier also absorbs and retains the moisture, which is essential to the survival of rhizobia, and provides a surface for growth of the bacteria.

Characteristics of an excellent solid carrier would include: high water-holding capacity; high water retention; no heat of wetting; nearly sterile; uniform chemically; uniform physically; nontoxic, biodegradable, and nonpolluting; nearly neutral pH or easily adjusted; supports rhizobial growth; amenable to nutrient supplements; rapid release of rhizobia in the soil; manageable in the mixing, curing, and packaging operations; suitable for all rhizobia; available in powder and granular forms; adequate supply; and reasonable cost.

Evaluated Carriers

Peat has been the standard carrier around the world; however, peat is not available in many countries, especially in the tropics. When available, the specific peat source must be evaluated to ensure suitability as an inoculant carrier.

Many solid materials have been evaluated as inoculant carriers. These are grouped in the following general categories: (a) soils—peat and peat plus additives, coal and coal with additives (Crawford and Berryhill, 1983; Dube et al., 1975) and inorganic soils; (b) plant materials—compost made from bagasse (Philpotts, 1976), sawdust, rice husk (Khatri et al., 1973), plant compost (Iswaran et al., 1972), corncob (McLeod and Roughley, 1961), farmyard manure (Chakarapani and Tilak, 1974), both soybean meal and coir dust (Iswaran, 1972), and cellulose (Pugashetti et al., 1971); and (c) inert materials—vermiculite (Sparrow and Ham, 1983), perlite, ground rock phosphate, calcium sulfate, and polyacrylamide gel (Dommergues et al., 1979). Besides culture addition to solid carriers, inoculants have been formed by first lyphilizing culture (Scott and Bumgarner, 1965) or oil drying the bacteria cells (Johnston, 1962) before incorporating into a solid carrier. Kremer et al. (1982) proposed incorporating these two processes by first lyophilizing the bacteria cells before combining the dried cells in an oil carrier. This process had the proposed advantage of combining the lyophilized cells with fertilizer without detrimental effects to rhizobia survival. Other inoculants have been produced as bottled culture held at room or refrigerated temperature, and frozen concentrate cells.

Most of the mentioned carriers are either naturally abundant resources or available waste materials. Little research has been conducted with the objective of synthesizing a carrier with superior characteristics.

It is essential in all carrier evaluations to determine the ability of rhizobia to survive over time. Growth in the carrier is beneficial, but the level of the long-term viable population is the most important factor.

Sterilized Carriers

Inoculants produced with sterilized peat carriers support a higher population of rhizobia and maintain better longevity. In addition, presterilized carriers contain few, if any, contaminants and provide the option for extending culture production by diluting the broth without sacrificing the final quality of the inoculant (Somasegaran and Halliday, 1982). Disadvantages with a sterilized carrier include a higher cost of production, more labor intensive, a sterilizing unit of sufficient capacity to meet production needs, and aseptic precautions during culture addition to the presterilized packaged carrier. These may be acceptable and present no limitation to small-scale production but become a costly problem to very large manufacturers.

Various sterilizing methods have been employed. Gamma irradiation (5 mrad) has generally be superior to autoclaving (Roughley and Vincent, 1967) or considered to be more uniform in final product quality than autoclaving (Strijdom and Van Rensburg, 1981). This level of irradiation may not produce a sterile product but ensures very low numbers of surviving contaminants. Fumigation with ethylene oxide or methyl bromide (Deschodt and Strijdom, 1974) provided poorer rhizobia survival than autoclaving, although the harmful effect was not due to residual fumigant. Microwave radiation has been found to be an effective method for reducing soil borne pathogens in mineral soil (Ferriss, 1984) but has not been fully investigated with organic soil (peat). Compared with autoclaving or methyl bromide-chloropicrin fumigation, the microwave treatment released less nutrient into the soil solution and resulted in less recolonization of soil by fungi, but it has less effect on soil prokaryotes.

Carrier Form

The form of the produced carrier depends on the intended method of application. The majority of inoculants have been in a powder form for seed application before planting. The particle size must pass through at least a 0.25-mm sieve (Burton, 1967), some require 50% to pass through a 0.075-mm sieve (Strijdom and Deschodt, 1976). Better seed adhesion is obtained with a finer particle size.

A granular form of inoculant allows application directly into the furrow with the seed. A particle size between 1.18 and 0.35 mm provides a granule that absorbs the culture rapidly and after being cured flows uniformly through a granular applicator. A granular inoculant contains approximately the same population of rhizobia per gram as the powdered inoculant. When applied at a rate between 5 and 30 kg/ha, depending on the row width, the granular form will provide significantly more rhizobia than a seed-applied application, which supplies only 210 g of inoculant on the quantity of seed planted in 1 ha. Granular in-furrow application also avoids direct contact of the inoculant with chemically treated seed (fungicide, insecticide, or molybdenum) and avoids disrupting a delicate seed coat such as *Arachis hypogaea*.

Other formulations for in-furrow application include broth cultures or frozen concentrates used as aqueous suspensions, or powdered peat-based inoculant suspended in water and dripped into the furrow (Gault et al., 1982).

The form of inoculant to be produced should depend on the general method of planting, equipment used for planting, and acceptable cost. The granular form of inoculant will be more costly but under many conditions, especially those with environmental stresses immediately after planting, will be superior in providing adequate nodulation.

C. Product Mixing and Curing

Mixing

Legume inoculants are formulated by adding the rhizobia broth to the processed nonsterile carrier in either a batch or continuous mixing system. Generally, this blending operation is in a ribbon or paddle-type batch mixer or a concrete mixer. Many peats are acid and must be neutralized at the time of culture addition by supplying sufficient $CaCO_3$ to adjust the inoculant to a pH of 6.8 to 7.2. The proportions of broth and peat to be mixed, commonly 1:1.5 to 1:2, are generally determined by the nature and moisture-holding capacity of the peat.

Moisture

The moisture content of the peat is generally increased from approximately 7 to 10% to between 32 and 40% moisture status (wet weight basis). Moisture content can be expressed as a percent (Pw) on either a wet weight or dry weight basis. Inoculants have generally been reported on a wet weight basis, which is easy and rapid to determine. When reporting inoculant moisture in the literature, it is important to specify the method of moisture determination.

Because peats from different sources vary widely in their ability to absorb moisture and in their ideal percent moisture for rhizobial growth and survival, Date and Roughley (1977) proposed that inoculant moistures be expressed on the basis of water potential (pF). Fawcett and Collis-George (1967) described a simple filter-paper method for determining water potential in soils.

In the preparation of inoculants it is essential that the appropriate moisture content for a specific carrier be determined. The ideal moisture range for the inoculant must consider the bacterial growth and maximum population attainable after mixing, and the bacterial survival and anticipated moisture loss from the package over the period of the product shelf life.

Curing

After mixing, the inoculant should be spread in a shallow tub or tray for a curing period of 48 to 72 hr at 22 to 24°C (Burton, 1967). The procedure in Australia (Thompson, 1980) is to cover the inoculant with plastic to prevent desiccation. During this period the heat of wetting generated by moistening the dry peat carrier is dissipated, moisture is uniformly absorbed, and the rhizobia begin a growth cycle in the inoculant. Heat of wetting occurs with culture addition to the dry carrier, because during the carrier drying process, heavily bound water in the peat has been removed by high-energy inputs. There is a positive relationship between temperature of drying and heat of wetting (Roughley and Vincent, 1967; Hoseh, 1937).

Following the curing period the inoculant should be passed through a coarse screen or hammer mill to disintegrate any aggregates. This can be performed as a beginning step in the packaging process. If the inoculant is to be stored in bulk before packaging, it should be removed from the curing tub and placed in a container providing moisture retention.

The operation of inoculant production with a sterile carrier is totally different. First the processed dry peat powder is packaged and sterilized by 5-Mrad gamma irradiation (Date and Roughley, 1977; Somasegaran, 1985) or autoclaved for 60 min (Somasegaran, 1985). In either case the rhizobia broth is introduced aseptically with a hollow needle directly thorugh the wall of the polyethylene bag. The surface of the bag is sterilized around the point of injection and the small hole is immediately covered with a self-adhesive label. Immediately after sealing, the inoculum and carrier are mixed by shaking or kneading between the fingers, and then incubated.

D. Inoculant Packaging

Packaging Equipment

The task of filling the inoculant bag with the cured inoculant can be labor intensive with hand filling or a semiautomatic weight dumping machine. Both processes require personnel filling a preformed bag, checking for an appropriate weight, placing pinholes in the bag for aeration, and then sealing the top of each bag. This complete process can be automated with a bag form, fill, and seal machine. A continuous roll of preprinted film is automatically formed into a bag, volumetrically filled, pinholed, and heat sealed with great speed. However, large volumes must be produced to warrant investment in an automatic system.

Containers

In the early years of the inoculant industry, various containers were used to package the product. These include glass bottles, metal-foil-lined cardboard cans, metal cans, and rigid plastic containers. Flexible low-density polyethylene film (0.038 to 0.076 mm) is now widely used throughout the world. Polyethylene provides high moisture retention, sufficiently high gas exchange, and is heat sealable.

Aeration

It is generally agreed (Burton, 1967; Roughley, 1968) that aeration through the package is desirable for optimum growth and survival. This is achieved with the low-density polyethylene film plus pinholes in the bag. Pinholing allows explusion of excess packaged air from the bag for more condensed packaging in the carton. This also prevents both ballooning of the bag or obtaining a partial vacuum due to insufficient gas exchange either way through polyethylene film. Rapid gas exchange may be caused by temperature changes and microbial activity. When sterilized carriers are used, pinholing is not recommended, due to the chance for contamination.

Package Labeling

The package should provide the proper product identification, specifying the legumes on which it may be used, detailed application instructions for the user, an expiration date, quantity of seed or area to be treated, and the manufacturing company.

Inoculants are generally manufactured to be used during the next planting season. Most manufacturers recommend using the product up to a maximum of 12 months from production. To ensure that product is not carried over to a second planting season, an expiration date is placed on each package.

Also included on the inoculant package should be a lot number identifying the batch of inoculant produced from each fermentor run. Lot numbers are important for quality control and allow tracing the manufacturing history of any bag of material.

E. Storage and Distribution

The quality of inoculant provided to the user is as dependent on factors and conditions that occur after packaging as it is to those that occur during manufacturing. This includes the storage conditions at the site of manufacturing, during transportation, and in the distribution network. It is common in the United States that the inoculant travel from the site of manufacturing to a regional distribution warehouse, then to a distributor warehouse, is transferred to a dealer location before being purchased by a farmer, and is stored on the farm before being used. Each segment of the distribution chain must be aware of proper inoculant storage conditions.

Peat-based inoculants are generally cured at 20 to 27°C for 4 to 5 weeks to provide the maximum population (Burton, 1967). This usually extends past the packaging phase, in which case the initial storage temperature should allow the completion of curing. Survival of the rhizobia on the seed is also improved by at least 4 weeks of curing (Materon and Weaver, 1985). Increasing the time of curing reduces the rate of rhizobial population decline on seed in a curvilinear manner (Figure 1).

After proper curing, inoculants are best stored at 4°C or lower, which results in large numbers of viable rhizobia maintained for 12 months. Viability may be lost in a few weeks at temperatures of 35°C or higher. Temperature effects are frequently confounded with the effects of moisture loss. Van Schreven (1970) reported increased survival and reduced desiccation when stored at 2°C when compared with a storage temperature between 10 and 28°C (Figure 2). Smith et al.

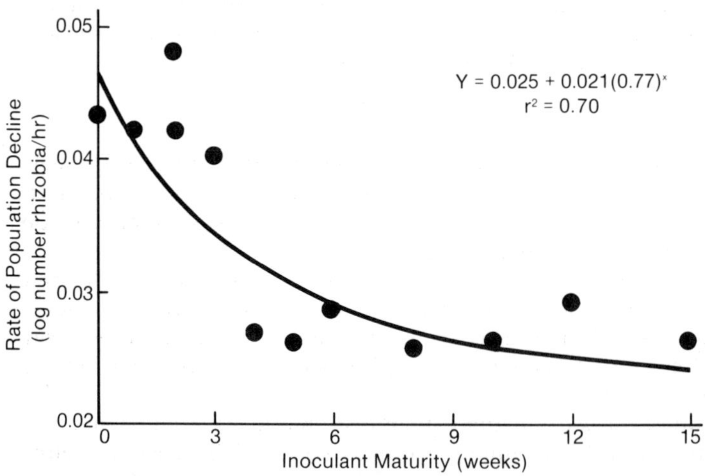

Figure 1 Curvilinear relationships between the rate of rhizobial population decline on clover seed with the maturity of peat inoculants. (From Materon and Weaver, 1985.)

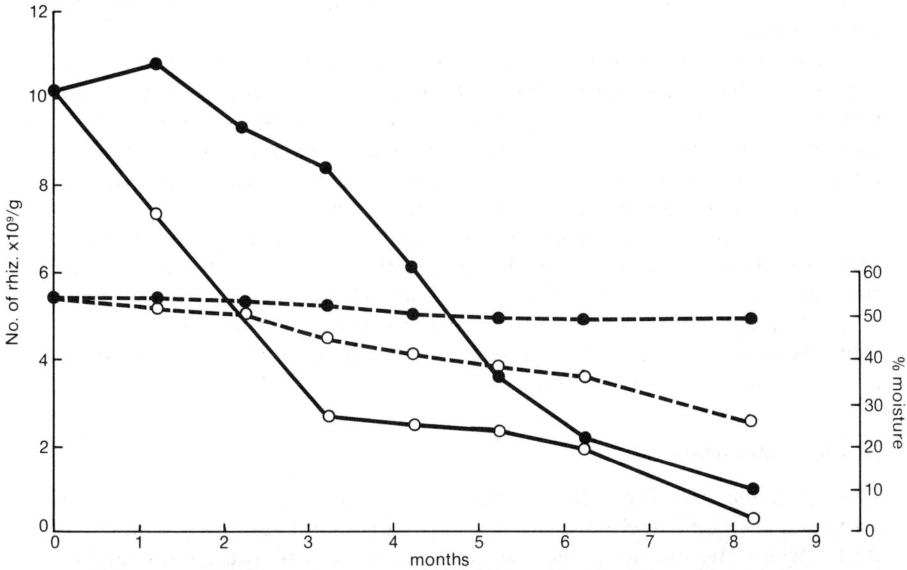

Figure 2 Effect of storage temperature on the survival of a mixture of *Rhizobium meliloti* strains (A145, A148, and A151) in soil-peat culture and on the moisture content of the cultures. Closed circles, cultures kept at 2°C; open circles, cultures kept in the loft of the laboratory at 10 to 28°C. (From Van Schreven, 1970.)

(1983) evaluated the effects of number of days in shipment, maximum exposed temperature, and final inoculant moisture on the population of rhizobia in an international inoculant shipping evaluation. Number of days in shipment and maximum exposed temperature were inversely related to the final rhizobia population, while inoculant moisture at the completion of the trial was directly related and exerted the largest influence on the inoculant quality. Date (1974) also demonstrated the decline in numbers with a decrease in inoculant moisture content.

IV. QUALITY CONTROL

A. Organization

An effective quality control program is essential for the consistent manufacturing of highly effective inoculants. The product quality is the responsibility of the manufacturer, with legislative control agencies only establishing minimum standards they deem necessary for the inoculant product. Without official regulatory agencies and their testing program, the standards set by the manufacturer may be very low or inconsistent.

The manufacturing unit should have quality control personnel to implement the program who are technically capable and without production responsibility. A quality control manual should be prepared which assigns responsibilities, specifies standards, defines testing methods, illustrates the necessary forms, and provides the company's policy and commitment to the quality control program.

An effective quality control program begins with specified standards for all aspects of production and reliable methods for evaluating the product or component against the specification. Record keeping of quality control tests are essential for the production season, and it is recommended that at least a summary from the production season be kept for long-term reference.

B. Raw Materials

The specifications for raw materials, both produced by the inoculant manufacturer or purchased from a supplier, must be checked and verified. When the carrier, such as peat, is produced the final carrier product should be examined for proper moisture content and particle size analysis. Because of the inherent variability with peat, it is advisable to assign lot numbers to a specified batch quantity and to maintain the quality control records for each lot number. Packaging materials such as preprinted polyethylene bags, multiwalled paper-plastic bags for granular inoculants, and cases should be inspected for proper printing, thickness, and color.

The stock culture collection of strains used in inoculant production is an important raw ingredient and must be incorporated into the quality control program. Strains used for inoculant production should be examined for purity and checked with the host in a greenhouse trial to varify nodulation and fixation efficiency. These steps should be taken before placing strains into a long-term culture storage system, such as ultracold freezers (-70°C or below) or freeze-drying, to maintain cell viability and genetic stability. More commonly used systems for maintaining stock cultures are agar slopes and porcelain beads. Both systems are reliable for maintaining cell vaibility for shorter durations but require more frequent transfer and regrowth. By these processes the chances for contamination or genetic change are increased. The porcelain bead method is recommended for operations with limited resources.

It is desirable to remove strains from the permanent storage system, again verify purity and fixation, before mass culturing for inoculant production. There should be the fewest possible transfers in this process, with the retrieval of another source from the permanent collection if a concern develops with any working strain.

C. Inoculant Production

During each phase of inoculant production, the person performing the task is the first-line quality control inspector. They must know the procedures, specifications, and standards for the production at their station and should notify a foreman or quality control inspector if the product falls below the standards. Full-time quality control personnel must be an integral part of production and have authorization to detain or reject inferior material.

Culture Production

Broth samples should be aseptically drawn from the fermentor each day and inspected for appropriate growth and the lack of contamination. A microscopic inspection allows rapid examination for cell morphology and contamination. Several microscopic procedures are available, providing various features. A Gram stain allows differentiation of gram-negative rhizobia from gram-positive contaminants, plus an examination for appropriate cell size and morphology. A wet mount under phase contract provides a cell density estimate, examination for viable cells, evaluation of cell size and morphology, and an inspection for contamination. The Petroff-Hausser chamber allows an estimate of the total cell count and also examination for contaminants which are morphologically distinct from rhizobia.

A pH measurement that varies from the expected indicates a potential contamination problem. Fast-growing rhizobia produce acid so that the broth should be between pH 6.2 and 6.8, while the pH of slow-growing rhizobia, which produce alkaline, should be between 6.8 and 7.5.

Growth curves may be established for each strain, which then provides the opportunity to examine the optical density of the broth and to assess normal growth. Deviation above the standard growth curve indicates possible contamination, while deviation below suggests a poor inoculum or abnormal growth.

Confirmation of both the total viable rhizobia population and possible contamination is best obtained by plating both low and high dilutions on yeast extract-mannitol agar. Low dilutions allow examination for faster-growing contaminants at low concentrations, while the higher dilution (10^{-7} to 10^{-8}) generally is suitable for counting rhizobia or distinguishing contaminants at higher concentrations which have a different rate of growth or colony morphology. Microscopic examination of individual colonies may help identify contaminants. Confirmation of effective rhizobia can only be obtained by employing a plant growout test. Both the plating procedure and greenhouse growout test require substantial time and thus have limited value in making decisions concerning the utilization of broth from a fermentor.

The final rhizobia broth to be used for inoculant production should be free of contamination and contain at least 1×10^9 colony-forming units (CFU)/ml.

Inoculant Mixing

During the mixing operation, moisture determinations allow adjustments in culture addition to maintain the product moisture within specifications. At this stage, it is also important to label the product properly as it enters the curing phase. Lot numbers assigned to each final fermentor should code for the product type (soybean, alfalfa, clover, etc.) and indicate in consecutive order each batch for that type.

Packaging

When the inoculant is being packaged it is important to check to see that the proper material is entering the packaging system. Varification of the proper bags to be filled, lot number, expiration date, and weight of the filled bag are standard quality control checks. A sample bag of finished product from each lot should be saved for detailed laboratory examination.

Finished Product

Sample packages from the packaging line may need to be incubated at 28°C until approximately 3 to 4 weeks after mixing before performing the quality control tests. This allows the completion of the curing process and the opportunity for growth in the product. Inspection and evaluation of each lot of finished product should involve as a minimum the determination of the moisture content and the population of rhizobia.

Moisture values should be within specified standards, which will depend on the type and source of the carrier. Lots outside the acceptable moisture range should be corrected or rejected.

The estimation of viable rhizobia may be obtained by either a plate count, or plant growout most probably number (MPN) estimation. The plate count is suitable for products made of a sterile carrier and high-quality products made with a nonsterile carrier. Acceptable plate counts with a nonsterile carrier are possible when the rhizobial population is very high and the contaminant population does not interfere with the rhizobia colony growth and identification. When the plate count is not possible, the reliable MPN method is recommended. This requires a longer testing period but also allows verification of both infectivity and nitrogen-fixing capability. Serological identity of specific strains may also be obtained from the nodules of plants in the MPN test.

D. Standards

Legume inoculants should contain sufficient viable rhizobia so that the intended host will be satisfactorily nodulated in a *Rhizobium*-free soil and effectively compete with indigenous rhizobia in soils where the crop has previously been grown. These are idealistic standards and involve many soil, environmental, and biological variables. Insufficient knowledge and understanding exists concerning the responses of the microsymbiont and host under these many conditions. Therefore, artificial standards for acceptable levels of viable rhizobia in inoculants have been established in many countries.

In the United States the standards are established by the manufacturer and range from 10^7 to 10^8 rhizobia/g. A few states established testing programs which required samples to provide nodulation in minimal greenhouse tests. The function was to protect the farmer from very poor quality inoculants, but did not establish high standards or monitor inoculants uniformly across the legume production region.

In Australia, where sterilized peat is the carrier, the standard for all inoculants (escept *Lotononis*, 5×10^8 per gram) is 10^9 per gram. Such peats are also required to be free of contaminants at the 10^{-6} dilution (Thompson, 1980). Similar standards have been set in New Zealand (Anon., 1979). South Africa set a standard of 10^8 per gram (Van Rensburg and Strijdom, 1974). Canadian standards (Anon., 1980) specify that the inoculant must contain, per gram of product, sufficient viable cells of the nodule inducing species to provide, when used according to directions, at least the following number of viable cells per seed: 1000 per seed for small seeded forage legumes, 10,000 per seed for sanfoin, and 100,000 per seed for large-seeded legumes. This is evaluated with packaged inoculants by applying the manufacturer's recommended rate and determining the viable cells per seed with the MPN method.

REFERENCES

Anon. (1979). New Zealand Ministry of Agriculture and Fisheries, Seed Testing Station, Palmerston North, New Zealand.

Anon. (1980). Canada Department of Agriculture, Ottawa, Canada.

Arias, A., and Martinez-Drets, G. (1976). *Can. J. Microbiol. 22*: 150.

Burton, J. C. (1967). In *Microbial Technology*, H. J. Peppler, ed. Van Nostrand Reinhold Company, Inc., New York.

Burton, J. C. (1981). *Trop. Agric. Trinidad 58*: 291.

Chakarapani, B., and Tilak, B. V. B. R. (1974). *Sci. Cult. 40*: 433.

Crawford, S. L., and Berryhill, D. L. (1983). *Appl. Environ. Microbiol. 45*: 703.

Date, R. A. (1972). *J. Appl. Bacteriol.* *35*: 379.
Date, R. A. (1974). *Proc. Indian Nat. Sci. Acad.* *40*: 667.
Date, R. A., and Roughley, R. J. (1977). In *A Treatise on Dinitrogen Fixation*, Sec. 4, *Agronomy and Ecology*, R. W. F. Hardy and A. H. Gibson, eds. John Wiley & Sons, Inc., New York.
Deschodt, C. C., and Strijdom, B. W. (1974). *Phytophylactica 6*: 229.
Dommergues, Y. R., Diem, H. G., and Divies, C. (1979). *Appl. Environ. Microbiol.* *37*: 779.
Dube, J. N., Namdeo, S. L., and Johar, M. S. (1975). *Curr. Sci.* *44*: 434.
Elkan, G. H. (1984). In *Biological Nitrogen Fixation—Ecology, Technology, and Physiology*, M. Alexander, ed. Plenum Press, New York.
Elkan, G. H., and Kwik, I. (1968). *J. Appl. Bacteriol.* *31*: 399.
Fawcett, R. G., and Collis-George, N. (1967). *Aust. J. Exp. Agric. Anim. Husb.* *7*: 162.
Ferriss, R. S. (1984). *Phytopathology 74*: 121.
Fred, E. B., Baldwin, I. L., and McCoy, E. (1932). *Univ. Wis. Stud. Sci.* *5*: 41.
Gault, R. R., Chase, D. L., and Brockwell, J. (1982). *Aust. J. Exp. Agric. Anim. Husb.* *22*: 299.
Gulati, S. L., and Seth, V. K. (1973). *Curr. Sci.* *42*: 394.
Herridge, D. F., and Roughley, R. J. (1975). *J. Appl. Bacteriol.* *38*: 19.
Hoseh, M. (1937). *Soil Sci.* *43*: 257.
Iswaran, V. (1972). *Madras Agric. J.* *59*: 52.
Iswaran, V., and Jauhri, K. S. (1969). *Curr. Sci.* *38*: 67.
Iswaran, V., Sen, A., and Apte, R. (1972). *Curr. Sci.* *41*: 299.
Johnston, W. R. (1962). U.S. Patent 3,034,968.
Jordan, D. C. (1982). *Int. J. Syst. Bacteriol.* *32*: 136.
Khatri, A. H., Choksey, M., and D'Silva, E. (1973). *Sci. Cult.* *39*: 194.
Kremer, R. J., Polo, J., and Peterson, H. L. (1982). *Soil Sci. Soc. Am. J.* *46*: 439.
Kuykendall, L. D., and Elkan, G. H. (1976). *Appl. Environ. Microbiol.* *32*: 511.
Labandera, C. A., and Vincent, J. M. (1975). *J. Appl. Bacteriol.* *39*: 209.
Materon, L. A., and Weaver, R. W. (1985). *Appl. Environ. Microbiol.* *49*: 465.
McLeod, R. W., and Roughley, R. J. (1961). *Aust. J. Exp. Agric. Anim. Husb.* *1*: 29.
Meyer, M. C., and Pueppke, S. G. (1980). *Can. J. Microbiol.* *26*: 606.

Nandi, P., and Sinha, N. (1974). *Proc. Indian Nat. Sci. Acad. 40*: 479.

Norris, D. O., and Date, R. A. (1976). In *Tropical Pasture Research, Principles and Methods*, N. H. Shaw and W. W. Bryan, eds., Bulletin 51. Commonwealth Agriculture Bureau, Canberra, Australia.

Philpotts, H. (1976). *J. Appl. Bacteriol. 41*: 277.

Pugashetti, B. K., Gopulgouda, H. S., and Patil, R. B. (1971). *Curr. Sci. 40*: 494.

Roughley, R. J. (1968). *J. Appl. Bacteriol. 31*: 259.

Roughley, R. J. (1970). *Plant Soil 32*: 675.

Roughley, R. J., and Vincent, J. M. (1967). *J. Appl. Bacteriol. 30*: 362.

Scherrer, A., and Denaire, J. (1971). *Plant Soil, Spec. Vol.*: 39.

Scott, J. R., and Bumgarner, H. R. (1965). U.S. Patent 3,168,796.

Skinner, F. A., Roughley, R. J., and Chandler, M. R. (1977). *J. Appl. Bacteriol. 43*: 287.

Smith, R. S., Judy, W. H., and Stearn, W. C. (1983). *INTSOY Series Number 23*, International Agriculture Publications, University of Illinois, Champaign, Ill.

Somasegaran, P. (1985). *Appl. Environ. Microbiol. 50*: 398.

Somasegaran, P., and Halliday, J. (1982). *Appl. Environ. Microbiol. 44*: 330.

Sparrow, S. D., and Ham, G. E. (1983). *Agron. J. 75*: 181.

Strijdom, B. W., and Allen, O. N. (1966). *Can. J. Microbiol. 12*: 275.

Strijdom, B. W., and Deschodt, C. C. (1976). In *Symbiotic Nitrogen Fixation in Plants*, P. S. Nutman, ed. Cambridge University Press, Cambridge.

Strijdom, B. W., and Van Rensburg, H. J. (1981). *Appl. Environ. Microbiol. 41*: 1344.

Thompson, J. A. (1980). In *Methods for Evaluating Biological Nitrogen Fixation*, F. J. Bergersen, ed. John Wiley & Sons, Inc., New York.

Upchurch, R. G., and Elkan, G. H. (1977). *Can. J. Microbiol. 23*: 1118.

Van Rensburg, H. J., and Strijdom, B. W. (1974). *Phytophylactica 6*: 307.

Van Schreven, D. A. (1970). *Plant Soil 32*: 113.

Vincent, J. M. (1944). *Nature 153*: 496.

Zurkowski, W. (1982). *J. Bacteriol. 150*: 999.

17

The Role of Culture Collections in Biological Nitrogen Fixation

HAROLD H. KEYSER *Agricultural Research Service, U.S. Department of Agriculture, Beltsville, Maryland*

I. INTRODUCTION

Historically, collections of microorganisms were established for use in taxonomy (Clark and Loegering, 1967). The usefulness of such collections now extends into many areas of microbiology, so that collections can be considered a living library of reference strains. They provide biological standards for research in genetics, biochemistry, serology, and symbiotic interactions. As pointed out by Dye (1979), our present knowledge of the *Rhizobiaceae* and the symbiotic association of rhizobia with legumes has depended on the availability from collections of well-characterized strains.

In addition to serving the needs of the research community, culture collections support the requirements of teaching and industry. Several *Rhizobium* inoculant manufacturers produce material containing cultures obtained from public collections.

The maintenance of microbial germplasm in collections serves two purposes: the collection and identification of biological strains and types, and the retention of genetic homogeneity or stability (Rogosa, 1981). Of equal importance to the addition of unique germplasm is the evaluation of cultures currently in the collection. Together these activities largely determine the value of a collection to its users.

In this chapter we address the role of culture collections in supporting research and application of biological nitrogen fixation. An excellent review exists on the functions and maintenance of a *Rhizobium* collection (Dye, 1979), and several chapters in this book deal with aspects directly related to activities of a collection (isolation, cultivation, maintenance, and symbiotic evaluation). Therefore, in this chapter we emphasize other issues related to the role and value of culture collections. Due to the experience of the author, the discussion is based on *Rhizobium*, although the principles discussed are relevant to other symbiotic and associative nitrogen-fixing microbes.

II. AVAILABILITY OF CULTURES, INFORMATION, AND SERVICES

For the research worker in the area of biological nitrogen fixation, the availability of cultures and related information is often critical. This is especially so for the novice in this area. Several collections of *Rhizobium* exist throughout the world, and the breadth of germplasm represented is impressive, as is the accumulated information on the cultures. Information about collections, their contents, and services has been relatively scant in the past. However, in the last 15 years there has been a significant increase worldwide in research and application of biological nitrogen fixation. With this have come more collections and information about them. The larger, service-oriented collections represent a substantial input of money and effort, usually supported by a large national or international program.

A. Types of Collections

Collections of *Rhizobium* can be placed in three arbitrary groups: specialized research collections, national reference collections, and major germplasm collections. Specialized research collections are synonymous with personal working collections. Such collections may exist only as long as a particular project, while some may become a career-long hobby. Certainly, the value of such collections cannot be underestimated.

However, knowledge of their existence may be limited until the research worker publishes results citing the cultures. At this point the cultures become available to others through the accepted practice of providing material for verification of published findings. Such collections may represent a narrow range of isolates, or they may be quite comprehensive, like the collection of the late O. N. Allen.

In many countries there are national agricultural institutes which house reference collections of *Rhizobium*. They usually contain cultures for the legumes of national importance, although other reference cultures are often represented. In some countries, such national collections may be a source of cultures for commercial inoculum production. For example, the Australian Inoculants Research and Control Service in Gosford maintains a collection of *Rhizobium* representative of the important host legumes of Australia. They evaluate and select strains for use in commercial inoculant production, and monitor quality of the product.

Certain national collections have become regional collections (serving the needs of several countries) by joining the Microbiological Resource Center (MIRCEN) network. The MIRCENs were established in 1975 through the combined support of the United Nations Educational, Scientific and Cultural Organization, the International Cell Research Organization, and the United Nations Environment Program. The first two *Rhizobium* MIRCENs were established at Porto Alegre, Brazil, and Nairobi, Kenya. Part of their mission is to collect, maintain, evaluate, and distribute for use cultures of *Rhizobium* for important legumes in their region of the world. Recently, three more *Rhizobium* MIRCENs have been established in Paia, Hawaii, Beltsville, Maryland, and Dakar, Senegal.

Major germplasm collections of *Rhizobium* have a large comprehensive collection of isolates for many legume hosts or an extensive collection of isolates for fewer (even one) host. Examples of the former are the collections at NifTAL (Paia, Hawaii), USDA (Beltsville, Maryland), Rothamsted (Harpenden, England), and the two CSIRO collections (Canberra and Brisbane, Australia). Major collections which specialize in cultures for only a few hosts would be those at the international research centers; CIAT in Cali, Colombia, has collections of rhizobia for bean and tropical forages; IITA in Ibadan, Nigeria, for cowpea; and ICRISAT in Hyderābād, India for peanut, chickpea, and pigeon pea.

B. Information on Collections

The biotechnological importance of culture collections can be exploited only if information on the properties of the holdings is known, properly documented, and easily accessible (Kirsop, 1983). Unfortunately, it is often difficult to find such information. A recent survey by the

American Society for Microbiology (Rogosa, 1981) found that only 8% of the microbial collections in the United States and 14% of the foreign collections had published catalogs available for general distribution. The situation is similar for collections of *Rhizobium*. Date (personal communication, 1981) noted that of approximately 125 collections in the world, as few as six had published catalogs. Since then, the author is aware of five more catalogs that have been published. Certainly, at a minimum, the larger collections (MIRCENs, international research centers, and recognized major resource banks) should provide such catalogs for the community of users.

The best compilation on existing *Rhizobium* collections is the World Catalogue of *Rhizobium* Collections, edited by Skerman (1983), available from the World Data Center for Microorganisms, University of Queensland, St. Lucia, Queensland 4067, Australia. The catalog lists 69 rhizobial culture collections in 38 countries. Many excellent research and national collections are listed, and all of the major resource collections, with the exception of ICRISAT, are listed. The best characterized strains in each collection are listed, and more important, it provides the interested researcher with names and addresses for gathering additional information. The World Data Center periodically updates its information and makes it available upon request.

C. Other Services

Some of the larger collections provide services in addition to providing cultures, including advisory, training, taxonomy and identification, and duplicate storage.

The five *Rhizobium* MIRCENs (and probably other collections as well) provide training to qualified persons in the area of practical methodology useful in *Rhizobium*-legume research and application. This training includes activities associated with collection maintenance; authentication, purification, symbiotic evaluation, preservation, and inoculant preparation and use. The specialized training programs are critical for the enhancement of agriculture through symbiotic nitrogen fixation, especially in developing countries.

Some of the large, well-supported collections also serve as a repository for smaller but valuable collections. For example, the retirement of a curator of a research collection can leave the collection in jeopardy if the institution in charge decides not to continue support. The NifTAL collection acquired the research collection of O. N. Allen, and the USDA collection acquired the research collection of J. C. Burton. These collections are fairly large and probably not all the strains are valuable. However, some of the strains are unique and the ease of permanent preservation (by lyophilization or freezing) allows the collections to continue in safety so that further evaluation can proceed on a routine basis. Researchers must be encouraged to deposit unique or otherwise valuable cultures, and curators of major collections must seek strains that will enhance the value of a collection. On the other hand, discretion (a reasonable policy for culling) is required, as accessioning is expensive and time consuming.

Valuable collections should be duplicated in a separate repository to prevent loss due to a local hazard such as fire. This can easily be facilitated by having another collection hold a set of lyophilized cultures.

The enthusiastic nonmicrobiologist working with rhizobia may wish to have a particular strain verified as a particular species type or strain. Collections specializing in certain species may provide this service through serology, phage susceptibility, antibiotic typing, or other techniques.

Finally, collections act as advisory and information centers. They can recommend a particular strain for a specific purpose. Through experience, collection staff gain expertise in taxonomy and identification, preservation, and biochemical-physiological characteristics of specific cultures. Exchange of information about cultures by users back to the collection also enhances the data base of the collection. The value of such information from the user cannot be underestimated, as the abundant literature on *Rhizobium* precludes curators from collecting all the published research on their cultures.

III. ACCESSIONING OF CULTURES

Accessioning is more than the simple acquisition of viable and pure cultures. It includes characterization and evaluation of the usefulness of a culture, preservation, and documentation of all data gathered for a culture (Rogosa, 1981).

A. Acquisition

Most *Rhizobium* culture collections have some accessions that are not sufficiently evaluated. Therefore, certain criteria should be established regarding the acquisition of even more germplasm. Certainly, acquisition for a specific research endeavor is the most common criterion for the individual researcher. For large culture collections acquisition should also enhance the collection in a specific manner. Criteria suggested for selecting strains to be acquired include important local strains; strains with properties important for agriculture; strains with unusual properties; strains about which there is much published work; strains from unusual host plants, rare species, and type strains; and strains representing ranges of host-bacteria interactions (Skerman, 1983; Rogosa, 1981). Also, strains from geographic centers of diversity of their homologous hosts can be considered germplasm worthy of acquisition (Lie, 1978; Keyser et al., 1981). Table 1 summarizes examples from the literature of valuable germplasm in each of these categories. Strains of *Rhizobium* can be acquired as pure cultures, from nodules, or from soil.

Cultures received from an individual or collection are presumed to be pure and authentic; however, it is worth the effort to verify this before proceeding with detailed research. Inspection of growth from a streak plate on a routine medium and a nodulation test on an appropriate host would be sufficient for most purposes. However, a strain passing these tests might still be a mixed culture with regard to serogroup, phage susceptibility, or other characteristic that is not obvious upon visual inspection. The best assurance of a culture being pure is a rigorous examination by the researcher with as many diagnostic tests available as possible.

The isolation of rhizobia from nodules is discussed in detail in Chapter 1. However, two points about isolation of rhizobia should be emphasized. First, isolation from nodules from a particular soil environment may be enhanced if the medium is modified to resemble more closely the soil conditions. For example, some strains from very acid soils grew only on acidified medium (Date and Halliday, 1979). Second, Date et al. (1979) found that the predictability of the suitability of a given *Rhizobium* for a given host genotype is increased when information concerning the soil from which both were collected is considered. They found that *Stylosanthes* accessions from alkaline soils formed effective symbioses only with rhizobia from similar conditions. Therefore, careful consideration of specific objectives prior to isolation from nodules should increase the chances of obtaining the desired cultures.

Soil samples can be collected and later used to isolate rhizobia through recovery from nodules of a given host. This may be necessary if a collection trip is made when plants have senesced or nodule-bearing roots are too deep to recover. Most rhizobia survive very well in soil, of course, and even small samples may have a plethora of strains and species which survive for long periods if protected from heat and desiccation. In the author's laboratory, 5-year-old soil samples from China still contain several different species of *Rhizobium* and *Bradyrhizobium*. To import soil from abroad invariably requires a government certificate. In the United States this is issued for certain purposes, with the requirement that the soil, media and plants grown in it be destroyed after testing to prevent escape of disease organisms.

B. Characterization

Culture collections are of value primarily because they can provide to users germplasm that is well characterized. Collections of *Rhizobium* have traditionally emphasized characterization of strains for important

Table 1 Suggested Criteria for Acquisition of *Rhizobium* Germplasm and Relevant Examples

Criterion	Example	References
Important local strains	Indigenous ineffective isolates of *R. leguminosarum* biovar. *trifolii*	Jones et al. (1978)
Strains with properties important for agriculture	Indigenous effective and competitive isolates of *R. leguminosarum* biovar. *trifolii*	Jones et al. (1978)
Strains with unusual properties	Acidophilic isolates of *B.* sp. (*Stylosanthes*)	Date and Halliday (1979)
Strains frequently referenced in publications	*B.* sp. strain 32H1	Gibson et al. (1976)
Strains from unusual host plants	Isolates from nodules of the nonlegume *Parasponia* sp.; stem nodules of *Sesbania rostrata*	Trinick (1973) Dreyfus and Dommergues (1981)
Rare species	Strains of *R. fredii* from Asia	Keyser et al. (1982), Dowdle and Bohlool (1985)
Type strains	Taxonomic type strains of *Rhizobium* and *Bradyrhizobium* species	Jordan (1984)
Strains representing ranges of host-bacteria interactions	Strains of *B. japonicum* and *R. fredii*; strains of *B.* sp. (*Stylosanthes*)	Keyser and Cregan (1984), Date et al. (1979)
Strains from geographic centers of diversity	Indigenous isolates of *R. leguminosarum* biovar *viceae* from the Middle East	Lie (1978)

symbiotic traits, with the goal of increasing legume production through an optimal nitrogen-fixing symbiosis. While research on rhizobia has expanded into many basic disciplines that involve many types of characterization, service collections must necessarily limit characterization to those traits most useful to a majority of its users. Since this book addresses practical methodology in symbiotic nitrogen fixation, the following discussion is limited to the practical characterization of cultures commonly carried out in collections.

Evaluation of cultures begins with pure culture studies in the laboratory. Culture purity and authenticity must be verified initially, followed by characterization for optimal growth conditions (e.g., preferred medium and temperature). Even in a pure culture one may encounter variation in colony types which may have significance regarding stability and symbiotic effectiveness. Herridge and Roughley (1975) examined 17 separate cultures of the Australian commercial inoculant strain *Bradyrhizobium* sp. CB756 for colony characteristics and symbiotic effectiveness. Twelve of the cultures contained two distinct colony types, and one culture yielded three colony types. All substrains were verified serologically. Examination of single-colony isolates from one culture indicated that colony characteristics were unstable and did not breed true. Also, there was a relationship between symbiotic effectiveness with *Macrotyloma axilare* and the colony type of the original isolate. The three most ineffective substrains were from large, gummy colonies, and the most effective from pinpoint, dry colonies. Such variation in an otherwise pure culture cautions against routine use of single colonies from a strain for general purposes. Similar colony variation in *B. japonicum* USDA 110 was found by Kuykendall and Elkan (1976), although the colony variants were stable and differed in growth on mannitol and arabinose. The USDA 110 variants were found by use of a surfactant that separated the clones.

Subsequent to pure culture evaluation, strains are evaluated for their nitrogen-fixing ability on a particular legume host. Commonly, a strain's symbiotic effectiveness is compared to that of the most effective reference strain. A discussion of symbiotic effectiveness is provided in Chapter 14.

Strains showing superior symbiotic effectiveness under controlled conditions can be characterized further for other desirable qualities. These qualities, as summarized by Brockwell et al. (1982), include:

1. Competitive ability with other strains for nodulation of the host
2. Nitrogen-fixing ability over a range of environmental conditions
3. Nodulation and nitrogen-fixing abilities in the presence of soil nitrogen
4. Ability to multiply in broth and survive in inoculant carriers
5. Ability to survive when incorporated in seed pellets

6. Persistence in the soil
7. Ability to migrate from the initial site of inoculation
8. Ability to colonize soil away from the influence of host roots
9. Ability to survive adverse physical conditions such as desiccation, heat, or freezing
10. Strain stability during storage and growth

The ability of a strain to survive and persist in the soil away from the host, termed "saprophytic competence" (Chatel et al., 1968), includes factors 6 through 9 above. The importance of some of these factors under certain conditions in Australia is well documented (Chatel and Parker, 1973a,b; Chatel and Greenwood, 1973). In many instances the competitiveness of an effective strain may be its most important quality (Ham, 1980; Trinick, 1982). The value of strain evaluation for competitiveness has been demonstrated by Jones et al. (1978). In field trials against native ineffective rhizobia, isolates of competitive *R. leguminosarum* biovar *trifolii* used as inocula accounted for the successful establishment of subclover.

C. Preservation

One of the prime responsibilities of a microbial culture collection is the preservation of strains in a viable and genetically stable state. Preservation techniques retard or stop metabolic activity by (a) withholding essential nutrients, water, and oxygen; (b) significantly reducing ambient temperature; or (c) combinations of these (Rogosa, 1981). In such states of preservation microorganisms are less likely to undergo genetic change than if they are maintained by successive transfer.

In the early days of culture collections, before freezing and freeze-drying techniques were employed, strains were maintained by means of test tube culture. Later, other methods were introduced, such as layering under mineral oil or air drying. With all these methods there was a chance of contamination and mutation. Today, the best means of preserving cultures are freeze-drying and freezing at very low temperatures, which are effective in providing long-term viability and genetic stability (Clark and Loegering, 1967). These and other preservation techniques used with *Rhizobium* are discussed here.

Comparison of Preservation Methods

Each laboratory should maintain a "mother culture" collection to serve as the ultimate source and reference (Date, 1982). This collection should be preserved by the best method available, which will vary according to the resources of each laboratory.

For routine daily use in research, most workers keep a working collection of cultures in agar tubes, maintained in a refrigerator (5°C).

The traditional agar slope culture is convenient and most rhizobia survive well for 6 months to 2 years (Vincent, 1970). The experience of the author is that the fast growers *R. fredii* and *R. phaseoli* require more frequent subculture than other fast and slow growers. Still, as a method of permanent preservation, the agar culture is quite unsatisfactory, because of the potential for the accumulation of mutants within a culture, including symbiotic mutants.

A variation of the agar tube culture is that of an agar slope or deep which is layered with paraffin oil. The oil prevents dehydration of agar and restricts oxygen. Although this method is satisfactory for relatively short storage periods, Means and Erdman (1963) found that after 7 to 8 years without subculturing, about 70% of their cultures were not viable. The paraffin-covered cultures in this test were stored at room temperature, and survival may have been better at 2 to 5°C (in a refrigerator) as recommended by Vincent (1970).

Norris (1963) introduced a method for preserving rhizobia by drying liquid culture-impregnated porcelain beads by exposure to silica gel in a small bottle. This method is relatively simple and inexpensive. Norris found that all cultures were viable after 2 years of storage and no change in colony morphology or symbiotic characteristics were found.

Jensen (1961) reported that *R. meliloti* remained viable after 30 to 45 years at room temperature in sterile soil, although other strains of *Rhizobium* perished within 4 to 10 years. However, given the textural and chemical variability of different soils, this would be a method of preservation impossible to standardize. Vincent (1970) suggests that a dried (but not less than 10% mositure), sterilized, neutral fertile loam should be chosen in addition to storage at low temperature (presumably at 2 to 5°C).

Rhizobia can be frozen at low temperatures. At about -20°C (obtained in a deep freezer) Vincent (1958) found survival of *R. trifolii* to be superior to storage at 2°C. In the *Rhizobium* collection at USDA Beltsville, cultures are frozen and stored at -50 to -80°C in mechanical freezers. Cultures sampled after 4 years of storage were viable and retained their symbiotic properties. Mechanical freezers are expensive though quite dependable. Cultures can also be stored in liquid nitrogen (-196°C) or liquid nitrogen vapor (-150 to 180°C). At such temperatures cultures can be kept indefinitely, and this is considered the best method of perservation (Rogosa, 1981). However, use of liquid nitrogen storage requires a well-equipped laboratory and a regular supply of liquid nitrogen.

Rhizobia can also be preserved very well by freeze-drying (lyophilization). Cultures are frozen in ampoules and then dried by vacuum. This is the most widely used method among culture collections and is considered second only to liquid nitrogen storage. In addition, a freeze-drying assembly is not too expensive. The labor input is high initially due to the limited number of ampoules that can be attached to a manifold at one time. Once the cultures are freeze-dried and ampoules flame sealed, they can be stored at room temperature, although cold temperature storage is preferable.

Dye (1979) has described in detail the vacuum drying of rhizobia from the liquid state (without prefreezing). A liquid culture is absorbed on cellulose fibers and then vacuum dried in ampoules. This method is considered as good as traditional freeze-drying for rhizobia and is used to preserve the germplasm in the Rothamsted Collection of *Rhizobium*. The assembly is relatively simple, consisting of a vacuum pump, desiccant chamber, ampoule manifolds, isolation valves, and a vacuum gauge. This assembly would be less expensive than a purchased freeze-drier. Cultures dried and stored for 13 years were all viable, and no symbiotic changes were noticed (Dye, 1979).

Table 2 provides a comparison of preservation methods for *Rhizobium* based on cost, longevity, and genetic stability.

Description of Preservation Methods

The recommended preservation methods for use with *Rhizobium* are described by Vincent (1970) and Dye (1979).

Freezing: Fresh cells are harvested from an agar slope or flat with 20% sterile glycerol (or other suitable cryoprotectant). The volume of the cryoprotectant should approximately equal the volume of the cells to be harvested. Alternatively, a liquid culture can be centrifuged

Table 2 Comparison of Methods of Preservation of Rhizobium

Method	Cost: Material	Cost: Labor	Longevity (yr)	Genetic stability
Agar transfer	Low	High	$\frac{1}{2}$-2	Poor
Oil storage	Low	Medium	2-20	Poor
Sterile soil	Low	Medium	5-45	Moderate
Porcelain beads	Low	Low	>3	Good
Freezing				
-20°C	Medium	Low	4-5	Good
-80°C	High	Low	>4	Good
-196°C	High	Low	Infinite	Good
Freeze-drying	High	Initial high	4-40	Good
Vacuum-drying	Medium	Initial high	>13	Good

and the cell pellet taken up with the cryoprotectant. The dense suspension is placed in small, sterile glass or plastic vials with a ½- to 2-ml capacity. The labeled vials are then capped and placed in racks in a deep freezer (-20°C), mechanical freezer (-60 to -80°C), or liquid nitrogen refrigerator. For revival, frozen material is thawed rapidly in a 35°C water bath until all ice is dissipated. The sample is then transferred to a liquid broth medium for incubation.

Freezing of cultures is discussed in detail by Hatt in the "ATCC Laboratory Manual on Preservation: Freezing and Freeze-Drying," available from American Type Culture Collection, 12301 Parklawn Drive, Rockville, Maryland 20852 (price U.S. $4.00).

Freeze-drying: The cultures of *Rhizobium* are grown and harvested as described above in the freezing method. The cryoprotectants used to harvest cells are solutions of 10% sucrose and 5% peptone or 20% skim milk. Aliquots of 0.1 to 0.5 ml are dispensed into sterile glass ampoules and plugged with cotton. Ampoules may be labeled on the outside with permanent ink or inside on a strip of filter paper. Initial freezing of the suspension can be accomplished by evaporative freezing (application of vacuum in connection with a condensor or desiccant), an alcohol-dry ice solution, or by placing vials in a mechanical freezer (-60°C) for 1 hr. Ampoules are exposed to vacuum for 16 to 24 hr; the drying is complete when the final moisture content is 1%. Ampoules are then flame sealed under vacuum with the use of a twin-jet burner. Alternatively, flat-bottomed ampoules fitted with a split-ring rubber stopper can be used in a chamber that permits self-sealing by pressing in the stopper from a flat tray above. The stopper is then capped manually with an aluminum crimped seal. Although ampoules can be stored at room temperature, it is preferable to use the lowest temperature storage available. Revival of a freeze-dried culture is made by aseptically opening the ampoule and dissolving the pellet with liquid medium. This can then be transferred to a larger-volume medium for incubation.

Vacuum-drying: This is the method described in detail by Dye (1979). Cultures are harvested from an agar slope with a solution of 10% sucrose and 5% peptone, pH 6.8. A 0.1-ml aliquot is inoculated onto absorbent cotton within the ampoules, and a nonabsorbent cotton ball is pushed into the ampoule just above the label. The ampoule is then constricted using an "ampoule constrictor" and attached to a manifold. Vacuum is applied for 16 to 22 hr, with the P_2O_5 desiccant changed after 4 to 5 hr. Storage and revival are as described above for freeze-dried cultures.

D. Documentation

Detailed records should be kept on all strains in a collection. These should contain as much information as possible about original isolation

Table 3 Record of Strain Origin and Characterization Used by World Federation of Culture Collections

NOMENCLATURE DATA

Genus RHIZOBIUM	Species, subspecies, infrasubspecies	Type B
=100 Author(s) of the name (*with date(s)*) (*e.g. Rhizobium phaseoli* Dangeard 1926)		
=110 World Directory Collection Number	Collection Acronym	Collection Accession Number

HISTORY

=120 Received as (Genus, species, subspecies, infrasubspecies)		=125 Date received / / Day Month Year
=130 Received from (*Name and address of person or collection*)	As	=140 Strain designation
=150 Who received it from (*Name and address of person or collection*)	As	=160 Strain designation
=170 Who received it from (*Name and address of person or collection*)	As	=180 Strain designation
=185 Any previous history (include accession number)		
=190 Also held in following permanent collections (*Include acronym and accession number, e.g. ATCC 12345*)		

ORIGIN

=200 Isolated from

=201 Genus ________ =202 Species ________ =203 Cultivar ________

=210 Part of plant (*Tick applicable block*)

☐ Root nodule ☐ Other (specify) ....

=220 Original collection site =221 ________ Country

=222 ________ State/Province =223 ________ Nearest town / =224 ________ Distance and direction from town

=225 ________ Altitude =226 ____ ° ' " Latitude (indicate North or South) =227 ____ ° ' " Longitude (indicate East or West)

Collection site soil properties

=350 Soil family ________	=358 Phosphorous (ppm) ________
=352 pH (water) ________	=360 P-extraction solution used ________
=354 pH (N KCl) ________	=362 Salinity (m moles/cm) ________
=356 pH (.01 M $CaCl_2$) ________	=364 Base saturation (%) ________

=230 Isolated by (*Name and address*)	As	= 235 Strain designation	=240 Date / / Day Month Year
=250 Identified by			=260 Date / / Day Month Year

Reproduced by permission of V.B.D. Skerman.

Table 3 *(Continued)*

MAINTENANCE WFCC Form SCC-4(R)

=270 Method of preservation *(Please tick whichever is applicable)* and =280 Temperature of storage

Original source material		°C	Agar culture		°C
Lyophilised culture		°C	Liquid nitrogen		°C
In soil		°C	Under oil		°C
Desiccated on beads		°C	Other *(Specify)* ______		°C

=290 Preservation suspending medium

=300 Growth medium *(Name and reference, or give details on separate page)*	=310 Temperature

=320 Growth conditions *(e.g. aerobic, anaerobic, special gas mixture, light)*

=322 Incubation time *(Tick applicable block)* Slow (> 5 days) ☐ Fast (< 5 days) ☐	=327 Period between subcultures

SPECIAL FEATURES AND USAGE *(e.g. for citric acid production, Type strain etc.)*

=370 Other legumes nodulated	=373 Effectiveness *(E, PE, I)*	=376 Suitability as inoculant *(Excellent, good, fair, poor)*

Tolerance to stress: Indicate either (a) *range* or (b) *high, medium, low* tolerance

=380 As free-living organisms:

pH______, Al^{+++}______, Mn^{++}______, NaCl ______, Temp.______ °C

=390 As symbionts in host nodules:

pH______, Al^{+++}______, Mn^{++}______, NaCl ______, Temp.______ °C

=330 Other special features *(CHO metabolism, etc.)*	=335 Reaction in standard YMA ☐ Acidic ☐ Basic

REFERENCES *(Journal, volume, pages, year)*

=340

If space is inadequate to reply to any question, please submit additional information on separate pages identified by appropriate reference number *(e.g. =310)*.

E = Effective PE = Partially effective I = Ineffective

details, cultural and symbiotic characterization, preservation history, and references of publications using a given strain. Dye (1979) describes the documentation used at Rothamsted, which includes for each strain a master record sheet, a customer card, a stock card, a work card, a drying card, and plant test record. The recording of all work done on a strain is more easily facilitated with the use of desktop microcomputers. Table 3 is a strain data record used for *Rhizobium* at the World Data Center for microorganisms.

It is important to retain the original collection number of a strain. The individual researcher has little reason to renumber a strain received from another lab. The large collections commonly assign their own accession numbers to strains received from others because it is easier to catalog and code for computer records. The best practice for all workers is to use the original number in reports and publications, and any renumbered strain should always be identified as synonymous to the original strain (Vincent, 1970). Otherwise, the potential for confusion is significant. This is best exemplified by the work of Kingsley and Bohlool (1983), who found that six cultures they received of chickpea rhizobia were in fact the same culture (USDA 3HOal = ATCC11444 = Rothamsted 3827 = ICRISAT 3827 = CC 1189 = TAL 385 = TAL 619).

REFERENCES

Brockwell, J., Diatloff, A., Roughley, R. J., and Date, R. A. (1982). In *Nitrogen Fixation in Legumes*, J. M. Vincent, ed. Academic Press, Inc., New York, p. 173.

Chatel, D. L., and Greenwood, R. M. (1973). *Soil Biol. Biochem.* 5: 433.

Chatel, D. L., and Parker, C. A. (1973a). *Soil Biol. Biochem.* 5: 415.

Chatel, D. L., and Parker, C. A. (1973b). *Soil Biol. Biochem.* 5: 425.

Chatel, D. L., Greenwood, R. M., and Parker, C. A. (1968). *Transactions of the Ninth International Congress of Soil Science, Adelaide*, Vol. 2, p. 65.

Clark, W. A., and Loegering, W. Q. (1967). *Ann. Rev. Phytopathol.* 5: 319.

Date, R. A. (1982). In *Nitrogen Fixation in Legumes*, J. M. Vincent, ed. Academic Press, Inc., New York, p. 25.

Date, R. A., and Halliday, J. (1979). *Nature London 277*: 62.

Date, R. A., Burt, R. L., and Williams, W. T. (1979). *Agro-Ecosystems 5*: 57.

Dowdle, S. F., and Bohlool, B. B. (1985). *Appl. Environ. Microbiol.* *50*: 1171.

Dreyfus, B. L., and Dommergues, Y. R. (1981). *FEMS Microbiol. Lett. 10*: 313.

Dye, M. (1979). In *Recent Advances in Biological Nitrogen Fixation*, N. S. Subba Rao, ed. Oxford & IBH Publishing Co., New Delhi, p. 435.

Gibson, A. H., Scowcroft, W. R., Child, J. J., and Pagan, J. D. (1976). *Arch. Microbiol. 108*: 45.

Ham, G. E. (1980). In *Nitrogen Fixation*, Vol. 2, W. E. Newton and W. H. Orme-Johnson, eds. University Park Press, Baltimore, p. 131.

Herridge, D. F., and Roughley, R. J. (1975). *J. Appl. Bacteriol. 38*: 19.

Jensen, H. L. (1961). *Nature London 192*: 682.

Jones, M. B., Burton, J. C., and Vaughn, C. E. (1978). *Agron. J. 70*: 1081.

Jordan, D. C. (1984). In *Bergey's Manual of Systematic Bacteriology*, Vol. 1, N. R. Krieg and J. G. Holt, eds. Williams & Wilkins, Baltimore, p. 234.

Keyser, H. H., and Cregan, P. B. (1984). *Crop Sci. 24*: 1059.

Keyser, H. H., Bohlool, B. B., Hu, T. S., and Weber, D. F. (1982). *Science Washington D.C. 215*: 1631.

Kingsley, M. T., and Bohlool, B. B. (1983). *Can. J. Microbiol. 29*: 518.

Kirsop, B. (1983). *Trends Biotechnol. 1*: 4.

Kuykendall, L. D., and Elkan, G. H. (1976). *Appl. Environ. Microbiol. 32*: 511.

Lie, T. A. (1978). *Ann. Appl. Biol. 88*: 445.

Means, U. M., and Erdman, L. W. (1963). *Soil Sci. Soc. Amer. Proc. 27*: 305.

Norris, D. O. (1963). *Emp. J. Exp. Agric. 31*: 255.

Rogosa, M. (1981). *Microbiol Collecitons of Major Importance to Agriculture*. American Phytopathological Society, St. Paul, Minn.

Skerman, V. B. D. (1983). *World Catalogue of Rhizobium Collections*. World Data Center for Microorganisms, University of Queensland, St. Lucia, Queensland, Australia.

Trinick, M. J. (1973). *Nature London 244*: 459.

Trinick, M. J. (1982). In *Nitrogen Fixation in Legumes*, J. M. Vincent, ed. Academic Press, Inc., New York, p. 229.

Vincent, J. M. (1958). In *Nutrition of the Legume*, G. H. Hallsworth, ed. Butterworth Scientific Publications, London, pp. 108-123.

Vincent, J. M. (1970). *A Manual for the Practical Study of the Root-Nodule Bacteria* (IBP Handbook 15). Blackwell Scientific Publications Ltd., Oxford.
Oxford.

Index

A